Fundamental and Applied Aspects of Bacterial Spores

Academic Press Rapid Manuscript Reproduction

Based on an International Symposium held under the auspices of the
Federation of European Microbiological Societies and
the U.K. Society of Applied Bacteriology
in Cambridge, England, 13–17 September 1982

Fundamental and Applied Aspects of Bacterial Spores

ADYA

Edited by

G. J. Dring

Unilever Research Laboratory
Colworth House
Sharnbrook
Bedford, England

D. J. Ellar

Department of Biochemistry
University of Cambridge
Cambridge, England

G. W. Gould

Unilever Research Laboratory
Colworth House
Sharnbrook
Bedford, England

1985

ACADEMIC PRESS
(Harcourt Brace Jovanovich, Publishers)
London Orlando San Diego New York
Toronto Montreal Sydney Tokyo

ACADEMIC PRESS INC. (LONDON) LTD.
24–28 Oval Road
LONDON NW1 7DX

United States Edition published by
ACADEMIC PRESS, INC.
Orlando, Florida 32887

BRITISH LIBRARY CATALOGUING IN PUBLICATION DATA
Fundamental and applied aspects of bacterial spores.
—(FEMS symposium)
1. Wind-pressure
I. Dring, G.J. II. Gould, G.W. III. Ellar, D.J.
IV Series
589.9'046 TA654.5

ISBN 0-12-222080-3

LIBRARY OF CONGRESS CATALOGING-IN-PUBLICATION DATA
Main entry under title:
Fundamental and applied aspects of bacterial spores.
Based on an International Symposium on "Fundamental and Applied Aspects of Bacterial Spores," held Sept. 13-17, 1982 at the University of Cambridge, England; arranged and sponsored by the Federation of European Microbiological Societies with the U.K. Society of Applied Bacteriology.
Includes index.
1. Spores (Bacteria)—Congresses. 2. Bacteria, Sporeforming—Congresses. I. Dring, G. J. II. Gould, G. W. (Grahame Warwick) III. Ellar, D. J. IV. International Symposium on "Fundamental and Applied Aspects of Bacterial Spores" (1982 : University of Cambridge) V. Federation of European Microbiological Societies. VI. Society for Applied Bacteriology.
[DNLM: 1. Spores, Bacterial—congresses.
W1 F21 / QW 51 F981 1982]
QR79.F86 1985 589.9'0165 85-47844
ISBN 0-12-222080-3 (alk. paper)

PRINTED IN THE UNITED STATES OF AMERICA

85 86 87 88 9 8 7 6 5 4 3 2 1

Contents

Spore Resistance and Dormancy

Germination and Outgrowth

Applied Aspects of Spores

Contributors

Numbers in parentheses indicate the pages on which the authors' contributions begin.

S. J. Alcock (261), *The Campden Food Preservation Research Association, Chipping Campden, Gloucestershire GL55 6LD, England*

É. Andrássy (397), *Central Food Research Institute, Budapest, Hungary*

A. F. Antippa (309), *Université du Québec à Trois-Rivières, Trois-Rivières, Québec, Canada G9A 5H7*

G. R. Bender (227), *Departments of Microbiology and Electrical Engineering, The University of Rochester, Rochester, New York 14642, USA*

H. Boschwitz (139), *Department of Biological Chemistry, Hebrew University of Jerusalem, Givat-Ram, Jerusalem, Israel*

A. F. Braña (197), *Departamento de Microbiología, Universidad de Oviedo, Oviedo, Spain*

H. Brantner (463), *Department of Microbiology, Hygiene Institute, The University of Graz, Graz, Austria*

K. L. Brown (261, 275), *The Campden Food Preservation Research Association, Chipping Campden, Gloucestershire GL55 6LD, England*

M. R. W. Brown (317), *Department of Pharmacy, The University of Aston in Birmingham, Birmingham, England*

H. Bücker (241), *DFVLR, FF-ME, Abt. Biophysik, D-5000 Cologne 90, Federal Republic of Germany*

E. L. Carstensen (227), *Departments of Microbiology and Electrical Engineering, The University of Rochester, Rochester, New York 14642, USA*

H.-Y. Cheung (317), *Laboratory of Molecular Biology, National Institute of Neurological and Communicative Disorders and Stroke, National Institutes of Health, Bethesda, Maryland 20205, USA*

S. Z. Child (227), *Departments of Microbiology and Electrical Engineering, The University of Rochester, Rochester, New York 14642, USA*

M. J. Cloutier (187), *Department of Biology, Syracuse University, Syracuse, New York, 13210, USA*

J. L. Cordier (475), *Institute of Microbiology, Swiss Federal Institute of Technology, 8092-Zurich, Switzerland*

T. Deak (455), *University of Horticulture, Budapest, Hungary*

D. H. Dean (61), *Departments of Microbiology, Genetics and Program in Molecular, Cellular and Developmental Biology, The Ohio State University, Columbus, Ohio 43210, USA*

K. R. Dhariwal (177), *Laboratory of Molecular Biology, National Institute of Neurological and Communicative Disorders and Stroke, National Institutes of Health, Bethesda, Maryland 20205, USA*

H. A. Douthit (297), *Division of Natural Science, The University of Michigan, Ann Arbor, Michigan 48109, USA*

H.-R. Ebersold (475), *Institute of Microbiology, Swiss Federal Institute of Technology, 8092-Zurich, Switzerland*

I. Fabri (455), *Centre for Food Control for the Ministry of Agriculture, H-1355 Budapest, Hungary*

J. Farkas (397), *International Facility for Food Irradiation Technology, Wageningen, The Netherlands*

E. Ferrari (117), *Istituto di Genetica, Universita di Pavia, Pavia, Italy*

H. M. Fischer (475), *Institute of Microbiology, Swiss Federal Institute of Technology, 8092-Zurich, Switzerland*

M. Frangou-Lazuridis (329), *Department of Biological Chemistry, University of Ioannina, Ioannina, Greece*

F. Franks (3), *Department of Botany, University of Cambridge, Cambridge CN3 3EA, England*

E. Freese (177), *Laboratory of Molecular Biology, National Institute of Neurological and Communicative Disorders and Stroke, National Institutes of Health, Bethesda, Maryland 20205, USA*

A. Galizzi (117), *Istituto di Genetica, Universita di Pavia, Pavia, Italy*

M. Gianni (117), *Istituto di Genetica, Universita di Pavia, Pavia, Italy*

D. Goldman (177), *Laboratory of Clinical Science, National Institute of Mental Health, Bethesda, Maryland 20205*

G. W. Gould (371), *Unilever Research Laboratory, Colworth House, Sharnbrook, Bedford MK44 1LQ, England*

J. A. Guijarro (341), *Departamento de Microbiología, Universidad de Oviedo, Oviedo, Spain*

W. A. Hamilton (35), *Department of Microbiology, Marischal College, The University, Aberdeen AB9 1AS, Scotland*

J. H. Hanlin (187), *Department of Biology, Syracuse University, Syracuse, New York 13210, USA*

C. Hardisson (197, 341), *Departamento de Microbiología, Universidad de Oviedo, Oviedo, Spain*

L. Hirschbein (77), *Institut de Microbiologie associé au C.N.R.S., Université de Paris XI, F–91405, Orsay, France*

G. Horneck (241), *DFVLR, FF-ME, Abt. Biophysik, D-5000 Cologne 90, Federal Republic of Germany*

E. J. Hsu (485), *Institute of Botany, Academia Sinica, Taiwan*

K. Jayaraman (145), *Department of Molecular Biology, School of Biological Sciences, Madurai Kamaraj University, Madurai 625021, India*

H. F. Jenkinson (129), *Microbiology Unit, Department of Biochemistry, University of Oxford, Oxford OX1 3QH, England*

P. Kerjan (145), *Laboratoire d'Enzymologie du C.N.R.S., 91190 Gif-sur-Yvette, France*

A. Keynan (139), *Department of Biological Chemistry, Hebrew University of Jerusalem, Givat-Ram, Jerusalem, Israel*

U. Kirschner (157), *Institut für Mikrobiologie, Molekulare Genetik, Universität Frankfurt, Frankfurt, Federal Republic of Germany*

S. L. Landuyt (485), *Department of Biology, University of Missouri, Kansas City, Missouri 64110, USA*

I. Lazaridis (329), *Department of General Biology, University of Ioannina, Ioannina, Greece*

G. M. Lefebvre (309), *Université du Québec a Trois-Rivières, Trois-Rivières, Québec, Canada G9A 5H7*

B. A. Levine (47), *Department of Inorganic Chemistry, The University, Oxford OX1 3QR, England*

P. López (251), *Departamento de Higiene y Microbiología de los Alimentos, Facultad de Veterinaria, Universidad Complutense, Madrid–3, Spain*

F.-K. Lücke (409), *Federal Centre for Meat Research, D-8650 Kulmbach, Federal Republic of Germany*

P. Lüthy (475), *Institute of Microbiology, Swiss Federal Institute of Technology, 8092-Zurich, Switzerland*

J. Mandelstam (129), *Microbiology Unit, Department of Biochemistry, University of Oxford, Oxford OX1 3QH, England*

M. B. Manzanal (197), *Departamento de Microbiología, Universidad de Oviedo, Oviedo, Spain*

R. E. Marquis (227), *Departments of Microbiology and Electrical Engineering, The University of Rochester, Rochester, New York 14642, USA*

A. Moir (89), *Microbiology Department, University of Sheffield, Sheffield, England*

R. Morse (101), *Genetics Department, University of Birmingham, Birmingham B15 2TT, England*

S. Murthy (145), *Department of Molecular Biology, School of Biological Sciences, Madurai Kamaraj University, Madurai 625021, India*

V. Nagel (455), *Centre for Food Control for the Ministry of Agriculture, H-1355 Budapest, Hungary*

S. Nandi (329), *Department of Molecular Biology, Madurai Kamaraj University, Madurai 625021, India*

J. S. Novak (187), *Department of Biology, Syracuse University, Syracuse, New York, 13210, USA*

J. A. Ordóñez (251), *Departamento de Higiene y Microbiología de los Alimentos, Facultad de Veterinaria, Universidad Complutense, Madrid–3, Spain*

P. Palacios (251), *Departamento de Higiene y Microbiología de los Alimentos, Facultad de Veterinaria, Universidad Complutense, Madrid–3, Spain*

R. A. Preston (297), *Division of Natural Science, The University of Michigan, Ann Arbor, Michigan 48109, USA*

Henriette M. C. Put (421), *Conway Laboratories, Thomassen and Drijver-Verblifa N.V., 7400 GB Deventer, The Netherlands*

G. Reitz (241), *DFVLR, FF-ME, Abt. Biophysik, D-5000 Cologne 90, Federal Republic of Germany*

H. J. Rhaese (157), *Institut für Mikrobiologie, Molekulare Genetik, Universität Frankfurt, Frankfurt, Federal Republic of Germany*

H. J. Rogers (21), *Division of Microbiology, National Institute for Medical Research, London NW7 1AA, England*

J. A. Salas (341), *Departamento de Microbiología, Universidad de Oviedo, Oviedo, Spain*

B. Sanz (251), *Departamento de Higiene y Microbiología de los Alimentos, Facultad de Veterinaria, Universidad Complutense, Madrid–3, Spain*

P. Schaeffer (77), *Institut de Microbiologie associé au C.N.R.S., Université de Paris XI, F–91405, Orsay, France*

F. Scoffone (117), *Istituto di Genetica Biochimica ed Evoluzionistica del C.N.R., Pavia, Italy*

B. Seddon (329), *Department of Developmental Biology, University of Aberdeen, Aberdeen AB9 2UE, Scotland*

P. Setlow (285), *Department of Biochemisty, University of Connecticut Health Center, Farmington, Connecticut 06032, USA*

R. A. Slepecky (187), *Department of Biology, Syracuse University, Syracuse, New York 13210, USA*

D. A. Smith (89, 101), *Genetics Department, University of Birmingham, Birmingham B15 2TT, England*

J. Szabad (455), *Paprika Manufacturing Company, Szeged, Hungary*

J. Szulmajster (145), *Laboratoire d'Enzymologie du C.N.R.S., 91190 Gif-sur-Yvette, France*

V. Tabajdi-Pinter (455), *Centre for Food Control for the Ministry of Agriculture, H-1355 Budapest, Hungary*

R. Vetter (157), *Institut für Mikrobiologie, Molekulare Genetik, Universität Frankfurt, Frankfurt, Federal Republic of Germany*

L. Vitković (177), *Laboratory of Molecular Biology, National Institute of Neurological and Communicative Disorders and Stroke, National Institutes of Health, Bethesda, Maryland 20205, USA*

W. M. Waites (383), *Agricultural Research Council Food Research Institute, Norwich NR4 7UA, England*

P. D. Walker (353), *Bacteriology Research and Development, Wellcome Research Laboratories, Beckenham, Kent, England*

A. D. Warth (209), *C.S.I.R.O., Division of Food Research, North Ryde, New South Wales 2113, Australia*

M. D. Yudkin (139), *Microbiology Unit, Department of Biochemistry, Oxford OX1 3QU, England*

Zs. Zalavari (455), *Centre for Food Control of the Ministry of Agriculture, H-1355 Budapest, Hungary*

Preface

The Federation of European Microbiological Societies with the U.K. Society of Applied Bacteriology arranged and sponsored an International Symposium on Fundamental and Applied Aspects of Bacterial Spores which was held at the University of Cambridge, England, 13–17 September 1982. The symposium contained 'Keynote' addresses by invited speakers who reviewed broad areas of specific new developments in spore research, as well as a wide variety of contributed papers and posters. The symposium also included a number of addresses on topics deliberately chosen to be outside the usual scope of spore-related conferences and yet of such general scientific intent and importance as to stimulate some 'lateral thinking'. These papers are collected at the beginning of this volume as 'Spore-related Topics'.

Most of the invited contributions and a broad selection of the contributed papers have been collected in this volume and represent an impressive up-to-date assembly of current spore-related research. The papers, as the symposium title suggests, cover new fundamental advances in genetics, sporulation, resistance and germination of spore formers. They also cover recent and important practical aspects of spore research relevant to the food and pharmaceutical industries, agriculture, health and medicine.

We would like to thank FEMS and SAB for their encouragement and support, and the contributors for their cooperation in the publication of this volume.

During the preparation of this book, we learned of the untimely death of one of the contributors, our dear friend Hans Rhaese. We mourn his passing and hope that his family will find comfort in knowing how much he was loved and respected by his many friends and colleagues.

G.J.D.
D.J.E.
G.W.G.

Spore-related Topics

DEHYDRATION AND THE FUNCTION OF WATER IN THE STABILITY OF BIOLOGICAL MACROMOLECULES

F. FRANKS

Department of Botany, University of Cambridge, Cambridge, CN3 3EA, UK

SUMMARY

Water plays a crucial part in the maintenance of macromolecular native states and in self-assembly processes. Hydration interactions are susceptible to perturbation by temperature, pressure and solutes. A critical degree of hydration is required for the proper functioning of macromolecules. Dehydration, whether caused by drought, freezing or salinity, produces symptoms of injury, but many organisms are able to survive dehydration by the synthesis of solutes which can act as water replacers.

INTRODUCTION

At the outset it will be taken for granted that water is one of the major regulators of life processes. Its involvement can be considered at different levels (1):

1) It is a universal reactant in the biochemical processes that govern synthesis and metabolism. Thus, water participates in all reactions involving hydrolysis, condensation, oxidation and reduction. The mechanisms by which water fulfills its role are not always well understood. For instance, mystery still surrounds the chemistry of the oxidation of water to molecular oxygen during photosynthesis.

2) Water acts as cement in maintaining stable native states of proteins; it may also be implicated in proton transfer processes during enzyme catalysed reactions. At the supramolecular level, hydration and dehydration control the assembly and dissociation of multi-subunit structures, such as complex

enzymes, microtubules, flagella and viruses.

3) Water acts as solvent, lubricant, plasticizer and transport medium, distributing nutrients and removing waste products.

4) Many forms of life exist, and can only exist, in an aqueous environment, and even those organisms that have adapted to life on dry land have had to do so at a considerable cost in energy, having had to develop complex control systems to maintain their water balance.

If evidence is needed for the sensitive way in which water and life are associated, then it is provided by the effects produced by exchanging D for H. D_2O is one of the most potent poisons for all higher forms of life. Yet, from a physical point of view, such an isotopic exchange is trivial (2). It seems, therefore, that all life processes are sensitively attuned to the characteristics of the hydrogen bond, as it exists in H_2O. Since the intermolecular geometry and hydrogen bond energetics existing in water are themselves easily perturbed, it can be inferred that hydration effects involved in life processes are also quite sensitive to such perturbations.

WATER STRUCTURE

This is a term which has been subjected to a great deal of abuse and misuse. It should always be borne in mind that water is a typical liquid as regards its dynamic properties, i.e. any distinct structure has a life time of very short duration (typically 1-10 ps). Currently the favourite structural model for liquid water is a random, three-dimensional network of hydrogen bonds, with a preference for tetrahedral, or approximately tetrahedral coordination of oxygen atoms (3), (4).

Such a time-averaged structure in which each water molecule has only four nearest neighbours is characterized by a large free volume and is therefore quite susceptible to the effects of pressure. The nine stable polymorphs of ice bear witness to this pressure sensitivity (5). The thermodynamic properties of water suggest that the liquid bears a marked resemblance to "ordinary" ice, that is hexagonal ice, with the fundamental difference that in the liquid such molecular configurations decay and reform extremely rapidly. In assessing the function of water in biological processes it is important to remember the combination of time-averaged structure and extreme mobility of protons, hydroxyl ions and H_2O molecules.

STRUCTURE PERTURBATION

Since the water structure is dominated by the spatial and

orientational properties of the hydrogen bond, and since hydrogen bonds are weak interactions, water structure is readily perturbed by changes in pressure and temperature and by the presence of solutes. Three distinct types of solute effects can be identified: those due to ions, "hydrophobic" molecules and molecules that contain several functional groups capable of participating in hydrogen bonding (6).

There is a vast literature on the subject of ion hydration (7), but the development of neutron diffraction has made possible the detailed probing of the nature of the ion hydration sphere (8). Figure 1a shows the disposition of water molecules surrounding a cation. This appears to be the same for

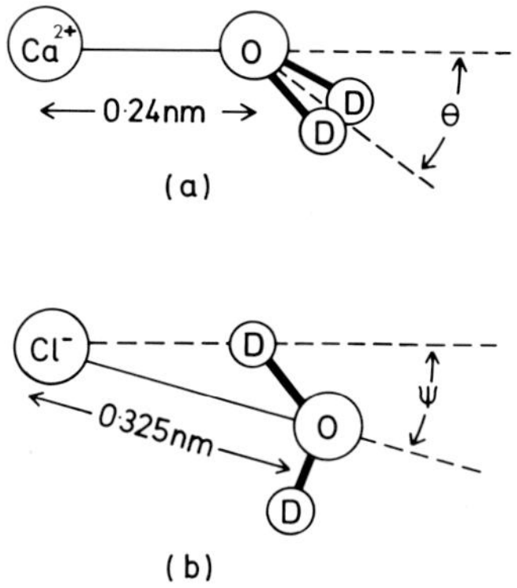

FIG. 1 *Primary hydration shell geometry of Ca^{2+} (a) and Cl^- (b) as obtained from neutron diffraction (6). The angles of tilt θ and ψ depend on the solution concentration and tend to zero at infinite dilution. The hydration shell of both ions consists of six water molecules.*

all cations studied so far: Ni^{2+}, Li^+, Na^+, K^+ and Ca^{2+}, although the ion-oxygen distances and the tilt angles change with the nature of the ion and the concentration. The Cl^- hydration shell geometry is shown in Fig. 1b. Surprisingly, the number of water molecules in the primary hydration shells of all monoatomic ions so far investigated is the same and numbers six. Here again it must be emphasized that the water molecules in the hydration sphere are subject to exchange processes. However, water exchange rates can vary by several orders of magnitude, depending on the ion. Since the electric field generated by an ion decays according to a 1/r dependence, where r is the distance from the ion, ion hydration is a long range effect and extends far beyond the primary hydration sphere. At present not much is known about the configurations and dynamics of water molecules beyond the primary hydration shell, but the effects produced by ions in the surrounding aqueous medium are manifest. They are reflected, for instance, in the viscosity of aqueous solutions. Thus, the effect of Na^+ is to raise the viscosity of water, whereas K^+ produces a lowering of the viscosity (9). The effect of hydration is also implicit in the concentration dependence of activity coefficients of electrolytes (7). The perturbations of water produced by apolar molecules or apolar residues in apparently polar molecules differ radically from those due to ions. So-called hydrophobic effects arise from a reorientation of water molecules in response to the apolar group which cannot interact with water by hydrogen bonding. The new water structure so produced resembles the unperturbed water structure in that each oxygen atom is still hydrogen bonded to four other oxygen atoms, but as a result of a slight distortion of the sp^3 hybridization of the oxygen orbitals, the distribution of water molecules is changed slightly, so that large cavities are created, capable of accommodating the apolar residue (10,11). A typical polyhedral structure is shown in Fig. 2. The net effect of such a redistribution of hydrogen bonds is a reduction in the configurational degrees of freedom of water which manifests itself as a negative entropy of mixing. In practice it is found that the solution properties of hydrophobic solutes are dominated by the magnitude of this entropy, even where water and the solute mix with the evolution of heat (10). Lack of solubility is thus seen to arise from structural effects, rather than from a lack of interactions.

There is at present much discussion about the range of the hydrophobic interaction, with the weight of evidence suggesting that it decays exponentially with distance, with a characteristic decay length of 1 nm. This means that 50% of the

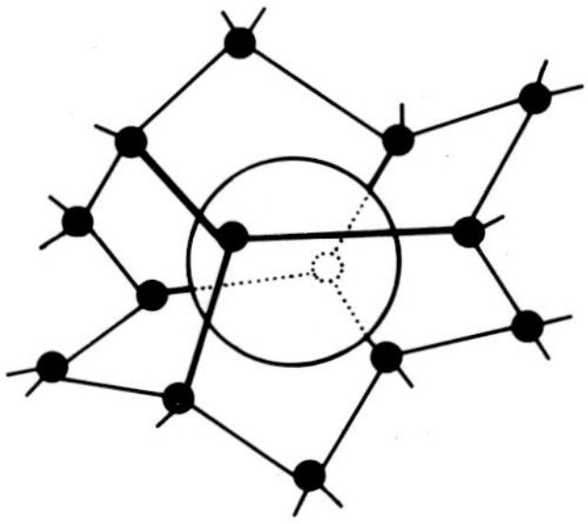

FIG. 2 *Typical hydrogen bonded polyhedral water cage induced by an inert solute (hydrophobic hydration), as identified by a Molecular Dynamics simulation. In the particular cage shown which is one of many different possibilities, the solute is surrounded by 14 water molecules; see (11).*

interaction takes place at distances greater than 0.7 nm, a long range effect indeed (12).

The redistribution of water molecules in response to the presence of an apolar residue is termed hydrophobic hydration. When two (or more) such residues, each with its associated hydration shell, interact, there is a rearrangement of water molecules which is reflected in the thermodynamic functions of mixing or dilution. The effect is known as the hydrophobic interaction and it shows itself as an apparent attraction between apolar molecules or residues. Indeed, it has been so described in standard biochemical texts (13). Such an interpretation is of course erroneous, because the apparent attraction is really the resultant of two repulsions. This distinction is not trivial: whereas it is possible to model the hydrophobic interaction by a potential function, say of a van der Waals type, the weakness of such procedures is illustrated when the temperature dependence of the interaction is considered. Being of an entropic origin, the hydrophobic interaction becomes stronger with rising temperature, and this is a feature that the normal van der Waals type potential cannot reproduce (14).

In the limit, where the concentration of apolar groups becomes high, there is not enough water left to provide the complex hydration cages. The result is a massive aggregation of apolar residues and the "relaxation" of hydrophobic hydration shells to unperturbed water. Examples of such aggregation processes are the phospholipid bilayers that form the basis for biological membranes, surfactant micelles and the folded globular proteins, the interiors of which are largely composed of hydrophobic amino acid residues. The configurational basis of the hydrophobic effect then leads to the apparently anomalous result that such aggregation processes are accompanied by *positive* entropy changes. Finally, the third type of hydration effect involves hydrogen bonding between solutes and water. Since the hydrogen bond is so orientation specific, differences in hydration energies and geometries depend to a large extent on the detailed stereochemistry of the solute molecule. For the series of aldohexapyranose it has been established that equatorially placed -OH groups facilitate interactions with water (15). This is believed to be due to the compatibility (spatial and orientational) of equatorial -OH groups on organic compounds with the -OH distributors in liquid water (16). Such minor configurational changes as the juxtaposition of equatorial and axial -OH groups can lead to significant differences in proton exchange rates between polyhydroxy compounds and water (17). Hydration effects also play an important part in determining the solution configurations of oligosaccharides and polysaccharides, and these, in turn, determine properties such as viscosity, viscoelasticity and propensity for gel formation (18).

WATER AND PROTEIN STABILITY

The native states of proteins and other biological macromolecules possess a stability that is highly marginal (19). For instance, the free energy of the native state relative to the unfolded (denatured) state of small globular proteins is characteristically of the order of -50 to -100 kJ/mol, equivalent to the free energy of three or four hydrogen bonds. Yet, the folded structures contain many hydrogen bonds. The free energy of stabilization is therefore the sum of several contributions, some of which provide stability and others act as destabilizing factors. Another relevant point is that for proteins to function optimally, they require varying amounts of water. The exact amount is difficult to assess, but studies on lysozyme under conditions of controlled water contents have shown that for the enzyme to show any catalytic activity at all, requires 20% of water (20). It is also well known that protein crystals, such as are grown for X-ray diffraction studies, contain up to 60% of water. It follows, therefore, that hydration interactions contribute to the stability of the functional state and to biological performance. In order to ascertain how the solvent performs this function, it is first of all necessary to determine the locations of water molecules within the folded protein structure. Over recent years very high resolution X-ray diffraction and neutron diffraction methods have been increasingly applied to the mapping of water molecules in protein crystals (21). Five types of sites have been identified: 1) as metal ligands, where water often fills one of the coordination sites in the active site of an enzyme and is then replaced by the substrate during the catalytic process. An example is provided by carboxypeptidase in which the Zn atom is surrounded by three histidine residues and one water molecule. 2) Water molecules bridge main chain peptide groups in distant parts of the polypeptide chain. For instance, in papain the NH group of asn-46 is linked to the CO group of glu-35 by a water molecule (22). Such peptide bridges can involve one, two, or even three water molecules. 3) Water molecules are found where charged residues exist in close proximity in the interior of the structure. Thus, a water molecule is located between lys-17, glu-35 and lys-174 in the papain molecule, with the fourth hydrogen bond made with another water molecule. This type of arrangement is referred to as a water mediated salt bridge (22). 4) Water molecules can link main chain peptide groups with polar side chains buried in the protein interior.

TABLE I *Estimated balance sheet for the stability of the native state of a small globular protein (approx. 100 residues). Units: kJ/mol.*

CONTRIBUTION	STABILIZING	DESTABILIZING
Hydrogen effects	600 ?	
Hydrogen bonding		
intrapeptide and hydration		
distortion from ideal geometry		400 - 800
Configurational entropy		300 - 800
"Relaxation" of water	500	
Van der Waals interactions	?	?
Ionic residues in interior		20 ?
Experimental	50 - 100	

Residues which participate in this type of hydration interaction include trp, cys, tyr, ser and thr. 5) Water bridges commonly occur between peptide chains in oligometric structures. The most extensively studied example of such a specific hydration structure is collagen, where regular arrays of water molecules are implicated in the stabilization of the three stranded collagen fibril (22).

It is much more difficult to identify water molecule locations near, or at the periphery of the bipolymer molecule, because of the short life times of such hydration structures and the rather ill defined locations of specific molecules. Nevertheless, dynamic measurements, based mainly on n.m.r. relaxation of oxygen-17 indicate a very distinct perturbation of water molecules in the proximity of proteins, giving rise to reduced diffusion rates and anisotropic rotation (23). The question remains: what is the range of such perturbations and what is the decay behaviour of such interactions. From the above discussion it becomes obvious that the assignment of a so-called hydration shell to a protein, as though it were an ion, is naive and counterproductive.

It is now possible to attempt an estimate of the contribution made by hydration effects to the configurational stability of a small protein of known crystal structure. The balance sheet drawn up in Table I attempts such an estimate. The only quantity that can be measured unambiguously is the free energy of unfolding (or folding), and even this measurement is subject to more uncertainties than a cursory reading of the biochemical literature might suggest (24). Reference to the estimakes in Table I shows that the resultant free energy, though small, is made up of several very large contributions and that the largest degree of uncertainty is associated with the contribution from hydrophobic interactions. As mentioned above, this arises mainly from our ignorance of the range of such effects.

HYDRATION AND BIOLOGICAL VIABILITY

For optimum *in vivo* functioning, proteins require a highly aqueous medium, the properties of which are controlled by pH and various solutes. Because the stabilities of proteins are so marginal, the temperature, pH and concentration tolerances for proper functioning are quite narrow. In this context it is important to note that the well studied thermal transitions (denaturation) of proteins have their counterparts at low temperatures, although the phenomenon of cold inactivation and its impact on life at low temperatures has not yet been subjected to much systematic investigation (25). The point must be made that cold, as a physiological stress condition is widespread, whereas heat denaturation is more of a laboratory curiosity, since living organisms are not normally subjected to thermal environments in which their proteins would denature.

The involvement of hydration in cold inactivation has not yet been clearly defined. That hydration does play a role is beyond doubt, because the physicochemical properties of water itself are so sensitive to temperature, especially at low temperatures and in the undercooled state (26). This is of great importance in survival at subzero temperatures, since many organisms avoid freezing as part of their seasonal acclimation to subzero temperatures.

THE PHYSICAL CHEMISTRY OF DEHYDRATION

Three types of dehydration must be considered: the removal of water by evaporation of by freezing, and the competition for the available water by added solutes (chemical dehydration). Evaporation can be isothermal or combined with a rise in temperature, but the two effects must be clearly distinguished. The same is also true for effects due to freezing in particular and those due primarily to low temperature as such (25). Nearly all chemical processes in homogeneous media are affected by temperature, and for aqueous systems such effects very often become very marked at low temperatures. This is particularly true for the degree of ionization of water, for the viscosity and for the self-diffusion coefficient (25). Freezing, on the other hand, occurs with the separation of pure water as a new, solid phase. As the

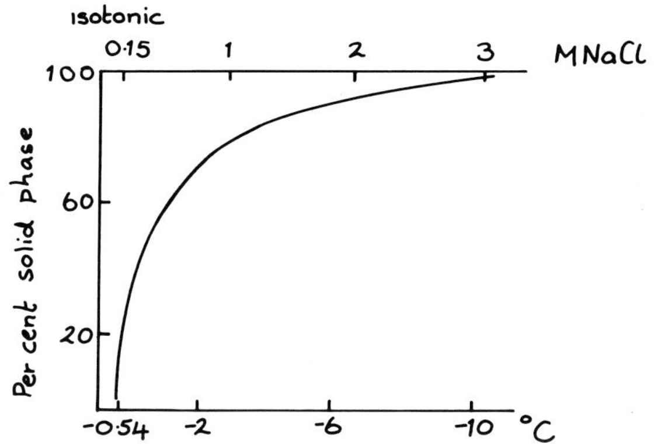

FIG. 3 *The effect of freezing on the concentration of the residual liquid phase in an initially dilute (0.15 M) solution of NaCl.*

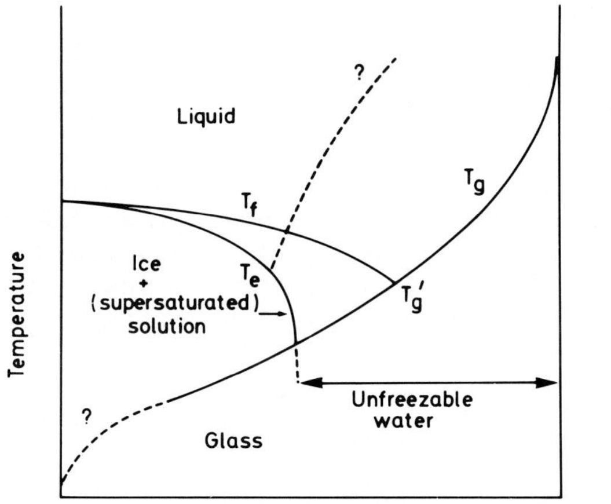

FIG. 4 *The occurrence of unfreezable water either because the liquidus curve (T_m) becomes vertical or because it intersects the glass transition curve (T_g). The point of intersection (T_g') is invariant at constant pressure and yields an estimate of the amount of unfreezable water. For a detailed discussion, see (28).*

temperature of an aqueous solution is lowered, so the solute concentration increases, while ice separates out. This is shown in Fig. 3 for a 0.15 molal solution of NaCl (isotonic with erythrocytes) (27). The freezing point of such a solution is -0.54°C. When the temperature has fallen to -6°C, about 80% of the water has frozen, and the NaCl concentration in the residual liquid phase has increased by more than tenfold. In well behaved systems the lowest temperature at which liquid can persist is the eutectic temperature, but in systems of biochemical significance eutectic separation of water from other (crystalline) components is rarely observed. Instead, such systems exhibit supersaturation, with phase diagrams of the type shown in Fig. 4. An important principle is illustrated by the two liquidus curves shown, that of so-called unfreezable water. Where the liquidus curve becomes vertical, any further reduction in the temperature will not increase the amount of ice in the system. Even where the liquidus curve is not vertical, it will meet the glass transition curve which sets the limit to crystallization (at least within any reasonable time). Here again the system will contain a certain proportion of unfreezable water, although mechanically the mixture will exhibit the properties of a solid.

Unfreezable water, sometimes referred to (erroneously) as *bound* water, is thus seen to be a manifestation of metastability, either due to supersaturation or to rapidly increasing viscosity with decreasing temperature (glass formation) (28). The phenomenon is of considerable biological and technological importance; it is at the basis of the prevention of complete dehydration of aqueous systems. The combination of dehydration and low temperature (freezing) affects the kinetics of reactions in a complex manner (29). Thus, the dependence of reaction rate on temperature alone is given by the Arrhenius relationship, with ln k decreasing as a function of 1/T. Increasing concentration, on the other hand, produces an increase in k. The onset of freezing therefore produces a discontinuity in the k(T) relationship. In some cases the rate actually rises steeply as freeze concentration first sets in. It then passes through a maximum and eventually declines when the solution has reached its limiting concentration (29). Such rate anomalies are quite pronounced for dilute solutions, because the freeze concentration factor is then much larger than it is for initially more concentrated solutions. Partial dehydration as it occurs at fairly high subzero temperatures can thus give rise to considerably enhanced reaction rates.

The pronounced curvature of the liquidus curves, shown diagrammatically in Fig. 4 is indicative of solute-solute interactions. In some cases such interactions can be virtually irreversible, because of the high activation barriers which make the reverse reaction improbable. Such irreversible aggregation effects have been observed with globular proteins in the partly or wholly denatured state. In such cases the unfolding becomes irreversible, because refolding is interfered with by chance interactions between certain amino acid residues on different polypeptide chains. Actually, such unfolding, followed by controlled aggregation and insolubilization can be utilized in the fabrication of food products from plant proteins (30). Indirect dehydration of one solution component by the addition of other soluble substances (e.g. salts) is conventionally expressed by a lowering of the water activity a_w. Thus, a_w is regarded as a measure of biological viability, product quality and, generally, the thermodynamic state of macromolecules in solution. There is, however, good reason to doubt whether any of these attributes are really governed by a_w (31). By definition, a_w depends only on the *number*, but not on the kind of dissolved particles (molecules or ions). In practice, however, specific effects are the rule rather than the exception; witness, for instance, the well known lytropic ion series (32,33). Thermodynamic arguments cannot account for the fact that for two solutions of the same activity, one (e.g. a sulphate) will salt-out while the other one (e.g. a nitrate) will salt-in a protein. While salting-out can be accounted for (although not very convincingly) on the basis of electrostatics, this is not so for salting-in. It is even harder to explain the "salting out" of carbohydrates by certain polyols, both species being nonionic.

It appears then, that polymer dehydration, through the action of added solutes, is a much more subtle process than can be accounted for by the water activity. Detailed studies of ternary systems water/protein/additive have shown that some ions are preferentially sorbed by the polymer, at the expense of water; they include the typically salting-in ions. The salting-out ions are preferentially excluded from the polymer hydration shell, and such an effect can be regarded as protection of the hydrated polymer (34). When the third component is a polyhydroxy compound, interesting effects can occur: even where a dehydration effect is observed, the protein is still protected against unfolding and aggregation. Such additives can therefore enable a protein (and other biological structures) to withstand the dehydrating effects of heat, freezing or high salt concentrations (2).

BIOLOGICAL CONSEQUENCES OF DEHYDRATION

As a first order approximation it can be stated that living organisms maintain a water balance that promotes their optimum functioning. It follows that any change in the physical and/or chemical environment that interferes with this water balance must be detrimental. Of these, seasonal changes in temperature

very large increases in the concentrations of certain solutes over short periods of time. If it were simply a matter of equalizing the osmotic pressure set up by the applied stress, there could be no question of solute specificity. It has also been established that the same compatible solutes are able to protect *in vitro* enzymes against inactivation by heat, freezing or high salt concentrations, i.e. under conditions where no transmembrane osmotic stress exists (43,44). This protection is therefore achieved by a further *increase* in the concentration of soluble species in the solution (salt + compatible solute). A closer study of the types of molecules that are implicated in osmoregulation suggests that the phenomenon is more likely related to subtle reorganizations of macromolecular hydration shells by compatible solutes.

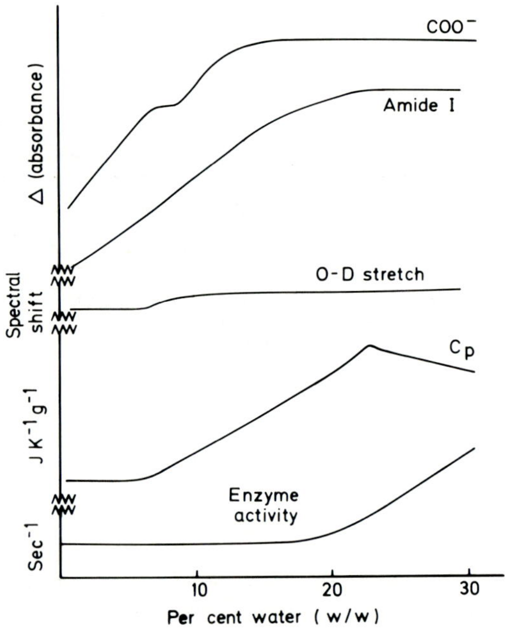

FIG. 6 *Effects of hydration on physicochemical and enzymatic properties of lysozyme: absorbance increases in the infrared carboxylate and amide I bands, frequency shift of the maximum in the symmetric O-D stretch band of the adsorbed water (total shift is from 2550 to 2580 cm^{-1}); apparent specific heat of lysozyme + water (total rise is from 1.25 to 1.55 $J\ K^{-1}\ g^{-1}$); enzyme activity (total increase shown is from zero to $5 \times 10^{-6}\ s^{-1}$). The dilute solution value is $40 \times 10^{-6}\ s^{-1}$. For details, see (20).*

EXTREME DEHYDRATION

Reference has already been made to recent structural studies of the role of water in the stabilization of globular and oligomeric proteins at the single water molecule level. Parallel investigations have attempted to define a minimum water content for biological activity. It is possible to prepare enzyme-substrate complexes at low temperatures and to subject them to isothermal dehydration. Water can then be readmitted at a controlled rate and the physicochemical and biochemical changes studied as a function of water content. This type of experiment has been performed on lysozyme, and the results are summarized in Fig. 6 (20). Most spectroscopic monitors exhibit gradual changes with increasing hydration, levelling off at a water content near 20% w/w. This is also the water content at which lysozyme first begins to exhibit any activity. The subsequent increase in activity with increasing hydration is complex; it reaches its dilute solution value at a water content of 33% w/w.

The very fact that some organisms are able to survive for long periods in an almost completely dehydrated state has stimulated research into the mechanisms of protection and survival of intact organisms; it is believed that such studies will also provide an insight into the role of water in maintaining the integrity of fully hydrated cells. In parallel with whole organism studies there is an increasing research activity in the general area of *in vitro* dehydration of labile structures and of the means of protecting them from the ravages of drying. Thus, the insight first gained by the environmental biologist is likely to be translated into considerable biochemical and technological benefits. It is also possible that the newer biological techniques will make possible the inducement of resistance to a degree of dehydration in tender organisms.

ACKNOWLEDGEMENTS

It is a pleasure to thank colleagues and former colleagues for many illuminating discussions: Graham Gould, David Reid, Helen Skaer, Patrick Echlin and Sheila Mathias. Our work on the biophysics of low temperature resistance is supported by the Leverhulme Trust, The Royal Society and The Agricultural Research Council.

REFERENCES

1. Franks, F. (1981) in *Handbook of Water Purification* (Lorch, W., ed.) pp.1-72, McGraw-Hill, London.

2. Franks, F. (1981) in *Water Biophysics* (Franks, F. and Mathias, S.F., eds.), pp.279-294, John Wiley & Sons, Chichester.
3. Rahman, A. and Stillinger, F.H. (1971) *J. Chem. Phys.* **55**, 3336-3359.
4. Rahman, A. and Stillinger, F.H. (1973) *J. Amer. Chem. Soc.* **95**, 7943-7948.
5. Whalley, E., Heath, J.B.R. and Davidson, D.W. (1968) *J. Chem. Phys.* **48**, 2362-2370.
6. Franks, F. in *Biochemical Thermodynamics* (Jones, M.N., ed.), pp.15-74, Elsevier, Amsterdam.
7. Conway, B.E. (1981) *Ionic Hydration in Chemistry and Biophysics*, pp.409-435, Elsevier, Amsterdam.
8. Enderby, J.E. and Neilson, G.W. (1979) in *Water - A Comprehensive Treatise* (Franks, F., ed.), Vol.6, pp.1-46, Plenum Press, New York.
9. Desnoyers, J.E. and Perron, G. (1972) *J. Solution Chem.* **1**, 199-212.
10. Franks, F. (1975) in *Water - A Comprehensive Treatise* (Franks, F., ed.), Vol.4, pp.1-94, Plenum Press, New York.
11. Geiger, A., Rahman, A. and Stillinger, F.H. (1979) *J. Chem, Phys.* **70**, 263-276.
12. Israelachvili, J. and Pashley, R.G. (1982) private communication.
13. Lehninger, A.L. (1975) *Biochemistry* (2nd ed.), p.43, Worth Publishers, New York.
14. Chan, D.Y.C., Mitchell, D.J., Ninham, B.W. and Pailthorpe, B.A. (1979) in *Water - A Comprehensive Treatise* (Franks, F., ed.), Vol.6, pp.239-277, Plenum Press, New York.
15. Franks, F. (1979) in *Polysaccharides in Foods* (Blanshard, J.M.V. and Mitchell, J.R., eds.), pp.33-50, Butterworths, London.
16. Suggett, A. (1976) *J. Solution Chem.* **5**, 33-46.
17. Bociek, S. and Franks, F. (1979) *J. Chem. Soc. Faraday Trans. I* **75**,262-270.
18. Morris, E.R. (1979) in *Polysaccharides in Foods* (Blanshard, J.M.V. and Mitchell, J.R., eds.), pp.14-31.
19. Pain, R.H. (1979) in *Characterization of Protein Conformation and Function* (Franks, F., ed.), pp.19-36, Symposium Press, London.
20. Rupley, J.A., Yang, P.H. and Tollin, G. (1980) in *Water in Polymers*, A.C.S. Symp. Ser. 127, pp.111-132.
21. Finney, J.L. (1979) in *Water - A Comprehensive Treatise* (Franks, F., ed.), Vol.6, pp.47-122, Plenum Press, New York.
22. Berendsen, H.J.C. (1975) in *Water - A Comprehensive Treatise* (Franks, F., ed.), Vol.5, pp.293-330, Plenum Press, New York.

23. Halle, B., Andersson, T., Forsen, S. and Lindman, B. (1981) *J. Amer. Chem. Soc.* **103**, 500-508.
24. Pfeil, W. and Privalov, P. (1979) in *Biochemical Thermodynamics* (Jones, M.N., ed.), pp.75-115.
25. Franks, F. (1981) in *Effects of Low Temperatures on Biological Membranes* (Morris, G.J. and Clarke, A., eds.), pp.3-20, Academic Press, London.
26. Angell, C.A. (1982) in *Water - A Comprehensive Treatise* (Franks, F., ed.), Vol.7, pp.1-82, Plenum Press, New York.
27. Farrant, J. (1977) *Phil. Trans. Roy. Soc.* **B278**, 191-202.
28. Franks, F. (1982) in *Water - A Comprehensive Treatise* (Franks, F., ed.), Vol.7, pp.215-338, Plenum Press, New York.
29. Fennema, O. (1975) in *Water Relations of Foods* (Duckworth, R.B., ed.), pp.539-556, Academic Press, London.
30. Kinsella, J.E. (1976) *Crit. Rev. Food Sci. Technol.* **7**, 219-280.
31. Franks, F. (1982) *Cereal Foods World*, pp.403-407.
32. Hofmeister, F. (1888) *Arch. Exp. Pathol. Pharmakol.* **23**, 247-260.
33. von Hippel, P.W. and Wong, K.-Y. (1964) *Science* **145**, 577-580.
34. Timasheff, S.N. (1982) in *Water Biophysics* (Franks, F. and Mathias, S.F., eds.), pp.70-72.
35. Levitt, J. (1980) *Response of Plants to Environmental Stresses*, Academic Press, New York.
36. Lyons, J.M. and Raison, J.K. (1970) *Plant Physiol.* **60**, 470-474.
37. Feeney, R.E. (1974) *Amer. Sci.* **62**, 712-719.
38. DeVries, A.L. (1982) in *Water Biophysics* (Franks, F. and Mathias, S.F., eds.), pp.306-308.
39. Crowe, J.H. and Madin, K.A. (1975) *J. Exp. Zool.* **193**, 323-334.
40. Crowe, J.H. and Crowe, L.M. (1982) *Cryobiology* **19**, 317-328.
41. Wyn-Jones, R.G., Storey, R., Leigh, R.A. *et al.* (1977) in *Regulation of Cell Membrane Activities in Plants* (Marre, E. and Ciferri, O., eds.), pp.121-136, North Holland, Amsterdam.
42. Gould, G.W. and Measures, J.C. (1977) *Phil. Trans. Roy. Soc.* **B278**, 151-164.
43. Paleg, L.G., Douglas, T.J., van Daal, A. and Keech, D.B. (1981) *Austral. J. Plant. Physiol.* **B**, 107-114.
44. Pollard, A. and Wyn-Jones, R.G. (1979) *Planta* **144**, 291-298.

PEPTIDOGLYCANS: STRUCTURE, CONFORMATION AND FUNCTION IN WALLS AND SPORE CORTEX

H.J. ROGERS

Division of Microbiology,
National Institute for Medical Research,
The Ridgeway, Mill Hill, London NW7 1AA

The purpose of this paper is to draw attention to some of the less well understood aspects of the structure and conformation of the peptidoglycans in the walls of vegetative spore forming bacteria and in the cortex of spores. In many cases, further understanding is necessary before the behaviour of the walls and cortex can be predicated. Some of the biological observations described here would seem to give point to the necessity for a renewed chemical and physical attack on the peptidoglycans.

PEPTIDOGLYCANS OF THE WALLS OF VEGETATIVE ORGANISMS

Primary structures: The general structures of peptidoglycans as glycan chains made of 1 - > 4 β linked alternating N-acetyl glucosaminyl and N-acetylmuramyl residues that are cross-linked by short peptide chains is well known (1,2,3,4). Many of the aerobic spore forming bacilli as well as Gram-negative bacteria have peptidoglycans called either type I (1,3) or type $A1_{\gamma}$ (2). A diagrammatic representation of these is shown in Fig. 1, in which the residue A_1 is L-alanyl, A_2 is iso-D-glutamyl, A_3 is meso-diaminopimelityl and A_4 is D-alanyl. In a few such as *B. sphaericus*, the vegetative cell wall contains a L-lysyl residue at A_3 and a bridge amino acid (see Fig. 2) in this example the D-isoasparaginyl residue is present (5). In *B. megaterium*, it is the DD isomer of diaminopimelic acid rather than the *meso* form at A_3. The structure of the peptidoglycan of most of the anaerobic spore forming *Clostridia* is also likely to be identical to that of the

FUNDAMENTAL AND APPLIED ASPECTS
OF BACTERIAL SPORES

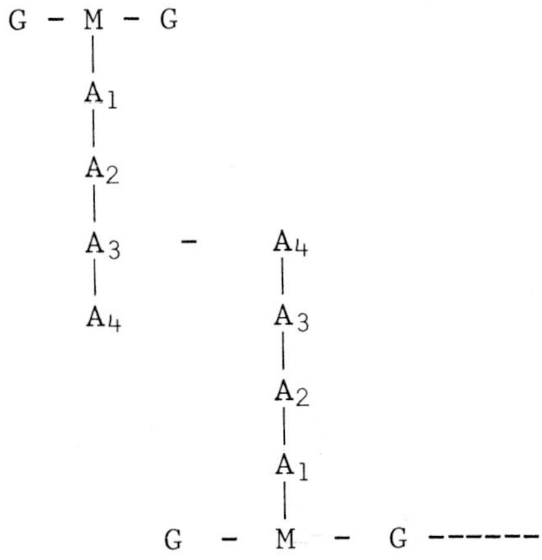

FIG. 1 *Structure of type I peptidoglycans where A_1 - A_4 are amino acid residues. Commonly, A_1 is L-alanyl, A_2 is D-isoglutamyl, A_3 is often diaminopimelityl although a variety of other diamino amino acid residues have been encountered in different species. A_4 is universally D-alanyl. M = N-acetylmuramyl residues. G = N-acetylglucosaminyl residues.*

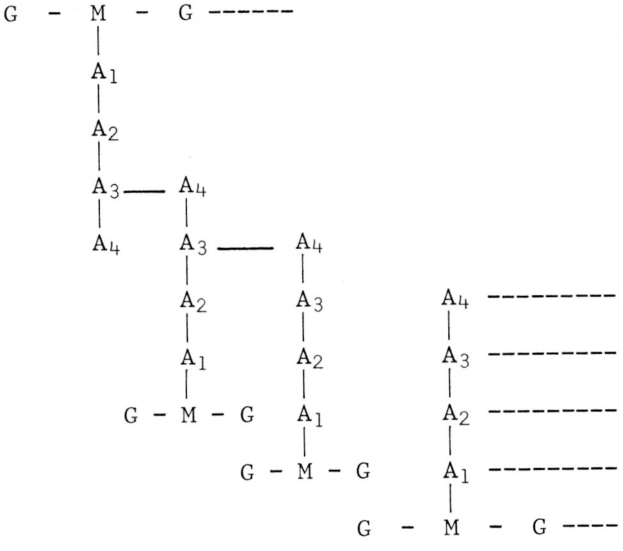

FIG. 2 *Possible multiple cross-linking in peptidoglycans. Abbreviations as in Fig. 1.*

majority of the aerobes (2) although again a few departures occur such as in *Cl. perfringens* where R_3 is diaminopimelic acid in the LL- rather than the *meso* configuration and a bridge is present between the peptide chains of a glycyl residue (6). Other characteristics of primary structure, however, are also likely to be of equal importance for the function of peptidoglycans such as the length of the glycan chains, the proportion of cross-linked peptide chains (as shown in Figs. 1 and 2) and any further substitutions such as amides on otherwise free carboxyl groups.

Early work (summarised in 7) suggested that the average length of the glycan chains was rather short, being 10-15 disaccharides long. However, subsequent work (8) in which autolysins were rapidly inactivated by the application of anionic detergents showed that they were likely to be much longer. In this latter work the reducing end-groups of N-acetylmuramyl and N-acetylglucosaminyl residues were both estimated. Since in the one example so far studied, glycan chains were found to be synthesised by addition at the reducing N-acetyl muramyl terminus, (9) measurement of the proportion of available reducing N-acetylmuramyl residues should indicate the length to which the chains are biosynthesised. That is, providing the growing end groups in wall preparations are not substituted. The results obtained from these measurements were confirmed by determination of the amount of non-reducing N-acetylglucosaminyl residue. The proportion of free N-acetylglucosaminyl reducing groups, on the other hand, indicates the extent to which the biosynthesised chains have been hydrolysed by autolytic *endo*-N-acetylglucosaminidases. The average "biosynthesised" chain was found to vary from 96 for a wild type *Bacillus subtilis* strain to 140 disaccharide residues for an autolytic deficient strain of *Bacillus licheniformis*. In both bacilli the glycan lengths obtained from measurement of the free N-acetylglucosamine groups were about half these values indicating a limited nicking of the strands by the autolytic enzyme. Only in *Staphylococcus aureus* were very short chains of only nine disaccharides found by this latter criterion despite the presence of "biosynthesised" chain lengths of 176. These measurements represent the average of a heterogeneous mixture of glycans with different chain lengths. Fractionation by column chromatography of the glycans from *B. subtilis* walls showed the presence of chains with lengths from 600 to 26 disaccharides. More surprisingly, each of these fractions appeared to have been hydrolysed to about the same extent by the endo-β-glucasaminidase to give lengths of between a half and a third of the original length (unpublished work by J.B. Ward and S.M. Fox, quoted in 10). More work is needed in this area to extend the observations to other species and to ensure that all the glycan chains originally present in the walls have free reducing N-acetylmuramyl groups and that none remain still linked to the isoprenoid intermediate to which they are presumably attached during biosynthesis. Also, perhaps still more precautions could be taken to make quite certain that no enzymic hydrolysis of the chains occurs during their preparation beyond that already present in the walls of the multiplying organisms.

The average degree of cross-linking (usually expressed as the cross-linking index) between peptide chains in peptidoglycan from the species of bacilli that have been examined is 40-50 per cent, this being measured as the proportion of the amino groups of A_3 residues (see Fig. 1) cross-linked to D-alanyl residues compared with the total number. If the peptidoglycan were wholly constructed according to Fig. 1, the index would be 50 per cent. In many bacilli, however, some A_3 residues also occur in uncross-linked tripeptides which lowers the cross-linking index to about 40 per cent. In these the terminal D-alanyl residues present in the original pentapeptide intermediate have been removed, presumably by the action of D-D-carboxypeptidases. In other species, such as *Bacillus cereus*, the uncross-linked peptides are in the form of tetrapeptides (11). Although such knowledge about the *average* degree of peptide cross-linkage is rather complete and clearly of interest, it leaves many important questions unanswered. Among these are:-

The distribution of cross-links among glycan strands. Are, for example, some glycan strands joined together by many cross-bridges whilst others are relatively free? Is the degree of cross-linkage through the thickness of the wall constant or is newly deposited peptidoglycan on the inner face of the wall more, or less, cross-linked than the old outside material? Is the peptidoglycan in the septa and poles of bacteria the same as that in the cylindrical part of the cell? These questions are of great importance in attempting to deduce many functional aspects of the peptidoglycans. Few questions seem to have been raised even as to the functional importance of the average degree of cross-linking of peptidoglycan. In view of the lethal effects of the β-lactam antibiotics, which inhibit the transpeptidases forming cross-links, most would assume that all the cross-links found in a wall are necessary for bacterial growth and survival. Much recent work however has raised the possibility of more than one transpeptidase in a given organism and it has been shown that some bacteria will grow despite greatly reduced linkage of their peptidoglycans. For example, a strain of *Staphylococcus aureus* that was very highly resistant to methicillin would grow at the same exponential rate in a rich medium with or without a high concentration of the β-lactam. In its presence, the peptidoglycan was only about 50-60 per cent cross-linked whereas in its absence it was, as is usual in this species, more than 90 per cent complete (12). This result may suggest that the multiple cross-linking of the type considered below is not entirely necessary to the staphylococcus. No work, however, has yet been done to show whether growth under more adverse conditions is affected, nor to see whether the physical properties of such partially corss-linked walls are different.

Multiple cross-linking (see Fig. 2) can occur because in peptidoglycans, of the types already illustrated, one residue of A_3 has an $-NH_2$ not satisfied and therefore available to accept in transpeptidation a D-alanyl carboxyl group from another peptide chain. By this means, it should be theoretically possible to build up a continuous chain of such cross-linkages giving an index of 100 per cent. Indeed, such a figure is nearly achieved by the staphylococcus. The possibility of multiple cross-linkages complicates interpretation of average values expressed by the index. For example, a value of 50 per cent may mean either a peptidoglycan constructed wholly as in Fig. 1, or that there is a mixture of multiply linked and totally uncross-linked peptides. Such work as has been done to elucidate this problem by isolation of the peptides, is summarised elsewhere (4). Again, it is clear that the presence or absence of multiply cross-linked peptidoglycan could have profound effects upon some wall functions, although it may not always be necessary for growth of staphylococci in rich media.

A common, secondary modification to the primary structures of peptidoglycan outlined so far, is the amidation of free carboxyl groups. The presence of amide groups has commonly been recognised by the liberation of stoichiometric amounts of ammonia by acid hydrolysis of peptidoglycan fragments. The commonest molar proportions seem to be one mole of ammonia per dimer or monomer. Although higher amounts have been found in products from some organisms such as *B. stearothermophilus* (12a). Evidence is often less than satisfactory, however, as to whether it has been substituted on to the free α-carboxyl group of D-glutamyl residues or on to one of the free carboxyls of the diaminopimelic acid molecules (11,13). Consensus suggests that it is commonly the latter which is substituted although in a strain of *B. licheniformis* there was evidence for the former (14) and also for *Cl. perfringens*. More modern methods should be used to investigate this problem, a solution to which may be important in understanding functional aspects of peptidoglycans, particularly in their relation to inorganic cations.

Other modifications of peptidoglycans, such as O-acetylation of the glycan chain or substitution of N-glycolyl for N-acetyl on the amino sugars have not been reported for spore forming bacteria. A modification that has so far been reported only for *B. cereus* (15), but which may be more widespread, is the total loss of substituents on some of the amino groups of the amino sugars in the glycan chains to leave glucosaminyl residues with free $-NH_2$ groups. Apart from the modified charge on such walls, there is the important fact that the loss of acetyl groups makes the peptidoglycans resistant to lysozyme. Four strains of *B. cereus* (15) differed in that their peptidoglycans contained from 27 per cent to 74 per cent un-acetylated glucosamine. The muramyl $-NH_2$ groups are always substituted. The peptidoglycans of *Cl. perfringens* also has un-acetylated glucosamine in its glycan chains (Ward, unpublished work). A potent enzyme de-acetylating the N-acetylglucosamine of peptidoglycan is present (16) in the cytoplasm of *B. cereus*.

SPORE CORTEX PEPTIDOGLYCAN

More drastic modifications of peptidoglycan occur in the spore cortical peptidoglycan (17,18). Up to 60 per cent of the N-acetyl muramyl residues can be lactonised to N-acetylmuramyl lactone, leaving residues with no free -COOH groups, only about 20 per cent of cross bridges of the type shown in Fig. 1 occur whilst 18-20 per cent of the N-acetyl muramyl groups have only L-alanine attached to their COOH groups. Spores from fourteen species were found to contain muramyl lactone. It is thought that the peptide modifications take place after the formation of the polymerised material rather than by the use of separate precursors. Estimation of the average lengths of the glycan chains in the spore cortex differs from 80-100 saccharide units in *Bacillus* (17) to 300 in *B. megaterium* (19). The latter was estimated from the free N-acetyl muramyl groups and therefore is likely to represent the biosynthesised chain length of the glycans in the spore peptidoglycan.

THE CONFORMATION OF PEPTIDOGLYCANS

Probably, because of the unique three-dimensional organisation of peptidoglycans and the presence in the primary structure of both glycans and peptides, the secondary structure has to date resisted attack. A number of speculative models were suggested a decade ago based on arranging the glycan strands by analogy with individual chains in the crystalline regions of α-chitin (20,21,22). The presence of lactyl groups on the three positions of the N-acetylmuramyl residues, however, prevents the formation of alternate 3 - > 5 intramolecular hydrogen bonds present in α-chitin and is likely to allow a degree of rotation of the 1 - > 4 β-glycosidic bond between N-acetylmuramyl and N-acetylglucosaminyl residues in the glycan chain. Tipper (20) suggested a further hydrogen bond between the lactyl carboxyl group and 06 of the contiguous N-acetylglucosamine. One consequence of attempting to adhere to the α-chitin conformation for the glycan chains with a two-fold screw axis (i.e. each sugar residue is rotated 180° in relation to the previous one) is that all the lactyl side chains of the N-acetylmuramyl residues protrude from one side of the chains, therefore demanding a parallel or anti-parallel glycan chain opposite to it for peptide cross-linkage to occur. If the glycoside bonds are to an extent free to rotate, these groups and their attached peptides may be arranged at various angles to the direction of the main axis of the chain, thus allowing greater freedom to the orientation of surrounding chains. The peptides themselves cannot be arranged as an α-helix, since the side chains of both the iso-glutamyl and the diaminopimelityl residues are included into the peptide chain. Based on approximate free-energy considerations, a β-pleated sheet (21) or a 2_7-helix (23) have been suggested. The latter came as a result of the failure to detect — by infra-red spectroscopy — the hydrogen bonding which would correspond to β-pleated sheets. Since the glycan chains are relatively inflexible it would seem possible that the peptide may undergo conformational changes with expansion or contraction of the walls. So far, however, no evidence is available to decide the peptide conformation. Energy calculations (25,26) suggest that neither the β-pleated sheet, nor the 2_7 helix is acceptable.

Despite early support from infra-red spectra for a chitin-like structure for the glycans in peptidoglycan preparations (23,24), the consensus of evidence is now against such a conformation (25,26,28). To summarise the evidence very briefly: The X-ray reflections collected from α-chitin indicate a periodicity along the glycan of 1.028, (27a) whereas for the peptidoglycans, the figure is 0.94 to 0.98, according to the moisture content of the preparation (25,26,27). However, the sharpness of the bands and the amount of information obtained from chitin is very much greater than from any of the wall or peptidoglycan preparations, introducing a degree of uncertainty into all the X-ray measurements. However, although Burge *et al.* (25,26) attribute their X-ray pictures to reflections from the glycan chains — Labischinski *et al.* (27) attribute their 0.7 and 0.94 periodicities to the peptides. The second line of evidence against a chitin conformation for the glycan strand is from the application of modern refined infra-red spectroscopy (28). The picture obtained from chitin was clearly different from those for preparations of peptidoglycan from several species which in very general terms all showed similar features. An important point made by Naumann *et al.* (28) is that only limited parts of polymers, even such as chitin and cellulose are crystalline and outside these regions, it is likely that the individual macro-molecules no longer adhere to the strict two-fold screw symmetry: An interesting model building exercise has been undertaken (25, 26) in which the amino sugars were rotated around the 0 of the 1 -> 4 β-glycoside bonds between N-acetylglucosamyl and N-acetylmuramyl residues, giving the range of conformation angles for the linkage allowable from free energy calculations. Combining this approach with the X-ray diffraction pictures and the use of optical diffraction masks, a model was built. This probably represents the most sophisticated and determined attempt to build a model for peptidoglycan consistent with the very limited data available.

Density determinations clearly lie at the nub of deciding the degree of "crystallinity" present in materials. If macromolecules are arranged in packed, hydrogen-bonded, close fitting orderly arrays as in "crystalline" materials such as areas of α-chitin and cellulose, the volume occupied by a unit mass in necessarily smaller than for the same number of molecules randomly arranged in an amorphous state. Unfortunately, few authors have used the same methods either for measuring densities or for preparation of the cell walls or peptidoclycans from bacteria. Comparing (Table 1), however, the results for dry walls and dry peptidoglycans of *Staphylococcus aureus* for example, it would appear that wall densities may be much lower than those of peptidoglycans. This may mean that the attached anionic polymers in the walls lead to more expanded structures that have much lower densities. Little trust, for generalisations, however can be put into such a limited available comparison. The highest value of 1.46 for peptidoglycan is the same as that for "crystalline" α-chitin whilst the lower values would correspond to highly hydrated amorphous materials. It is not reassuring to note that different authors studying the biophysics of peptidoglycans have used methods of preparation that take very varied degrees of precaution against the action of autolytic enzymes. In some instances these are quite inadequate. Work on the lentghs of glycan chains has shown how much and how rapidly damage can be done to peptidoglycan. Such hydrolysis of bonds may have profound effects on the conformation of the polymer.

The one aspect of conformation that authors appear to agree upon, is that the glycan chains of the peptidoglycans run in the surface of the wall and not radially to the axis of

TABLE 1 *Densities of wet and dry cell wall and peptidoglycan preparations*

MICRO-ORGANISM	DENSITY	METHOD	REFERENCE
Cell walls	g/cm^3		
Micrococcus luteus	Dry 1.14±.07 Wet 1.06±.05	Air displacement	29 30
Staphylococcus aureus	Dry 1.19±.05 Wet 1.07±.02	Air displacement	30
Bacillus megaterium " "	Wet 1.04 Wet 1.10	Black and Gerhardt (1962)	30+31 30+32
Peptidoglycans			
Spirillium serpens	Wet 1.46-1.47	CsCl isopycnic centrifugation	23
Lactobacillus	Wet >1.44	Sucrose-^2H_2O isopycnic centrifugation	23
Escherichia coli Bacillus subtilis 168 Staphylococcus aureus	Dry 1.32, 1.34 1.33±.02.	Ethanol-CCl flotation	28

the cell. This, of course, is consistent with the lengths of the glycan chains which are greater than the thickness of the walls.

SOME RELEVANT BIOLOGICAL CONSIDERATIONS

Damage by autolytic enzymes might be particularly severe if the three-dimensionally organised peptidoglycan as it exists in the living cell has a conformation considerably constrained away from that which would have a minimum free energy such as is necessarily assumed in model building. The presence of such strain in a helically arranged fibrous peptidoglycan might, for example, be used to explain the axial rotation of rod shaped cells, so elegantly demonstrated by studies (33,34, 35) of the rotation of helical forms of filaments of *Bacillus subtilis* formed from individual bacteria that fail to separate because of their deficienty in autolytic enzymes (36). These bacteria are not flagellated and are therefore non-motile in the usual sense (37). If action of the small amounts of residual autolytic endo N-acetyl β-glucosaminidase and L-alanyl N-acetyl-muramidase remaining in the strains hydrolyses bonds and releases energy from a constrained spirally arranged peptidoglycan, this could possibly be the engine that drives the axial rotation of the bacteria as well as possibly the expansion and contraction of the cortex of spores. Other evidence (38) has clearly shown that during reversion from wall-less protoplasts to normal looking bacilli, long fibres of material chemically and immunologically indistinguishable from the bacillary wall are formed. At first, these fibres are arranged radially to the cell but are found later in the plane of the wall. The impression is that such fibres might always be present but are buried in the fully mature wall possibly by other more amorphous material. Other electron microscopic studies (39,40) have also provided rather strong evidence for the fibrous arrangement of wall polymers in fully mature walls of bacilli and in the peptidoglycan layer of *Escherichia coli*.

THE CORTEX OF SPORES IN RELATION TO THE VEGETATIVE WALL

The acknowledged differences in primary structure between cortical and vegetative peptidoglycan have been mentioned as have the shortcomings in our knowledge of peptidoglycan even in vegetative walls. Too little work has been done on cortical peptidoglycan and indicates how little we know. Consequently, its properties, for example, to expand or contract, or to bind cations by its presumed free fixed anionic charges, should be deduced only with the greatest of caution. The binding properties of whole walls of bacilli and many other Gram-positive micro-organisms cannot easily be used as a model for the behaviour of the cortical material. About half the weight of the wall preparations from bacilli consists of anionic polymers such as the teichoic and teichuronic acids, whereas there is no evidence that any of these polymers exist in cortical material which may consist only of peptidoglycan. Much earlier work had made apparent that in vegetative wall preparations, such anionic polymers have prime importance in the fixing of anions. Their role has been very carefully examined recently for walls of *B. licheniformis* isolated under such conditions that autolytic enzymes were inactivated in the bacteria even before they were broken to prepare the walls (41). As can be seen from Table 2, the binding capacity on a

TABLE II *The fixing of the cations by the polymers in the walls of* Bacillus licheniformis *when walls are immersed in 5 mM solutions of salts (from Beveridge* et al. *1982)*

CATION	TEICHOIC ACID µMOLE/MOL.P.	TEICHURONIC ACID µMOLES/ MOLE URONIC ACID	PEPTIDOGLYCAN µMOLE/ MOL A_2 pm.
Na^+	.913	.123	.089
K^+	.251	.801	0
Mg^{2+}	.230	.418	.027
Ca^{2+}	.264	.624	.107
Mu^{2+}	.685	.178	.004

molar basis of both the teichoic acid and the teichuronic acid are very much greater than the peptidoglycan. Indeed, the poor ability of the isolated peptidoglycan to fix cations despite the presumed presence of free negative charges should be noted. Isolated walls and peptidoglycan of *B. subtilis* on the other hand, were shown earlier (42,43) to bind very much larger amounts particularly on monovalent cations and Mg^{2+}. Indeed, the existence of nucleation sites was proposed because the amounts bound were far in excess of those to be expected from the fixed negative charges. Another striking difference between the behaviour of the isolated walls of the two bacilli is revealed by removal of the anionic polymers. When the teichoic and teichuronic acid are removed from walls of *B. licheniformis* their thickness is reduced by about 50 per cent (44,41) whereas *B. subtilis* walls showed no significant change in thickness when teichoic acids were removed (43). The peptidoglycans which presumably remain as the only known materials in both organisms after the removal of the anionic polymers have been thought to date, to have identical primary structures. Their behaviour towards ions and towards the removal of the anionic polymers on the other hand, is apparently different. Such surprising results should indicate great caution in making any assumptions about the behaviour of the fixed negative charges in cortical peptidoglycan which has considerable and obvious differences from the vegetative wall polymers in the known aspects of its primary structure.

A further point that may have relevance to the role of anionic polymers in walls of growing organisms, comes from a study of a phosphoglucomutase negative mutant of *B. licheniformis* (45). When the organism was grown in a chemostat under conditions of phosphate limitation it was unable to replace its wall teichoic acid by teichuronic acid as could the wild type. Instead the wall consisted almost exclusively of peptidoglycan. Such organisms instead of being rod-shaped were spherical. An interesting comparison is the conditional *rod*A mutant of *B. subtilis* which, when grown at its restrictive temperature, is round and its walls too are almost devoid of teichoic acids (46,47). Thus it could be that the shape of spores has some relationship to the absence of teichoic acid in the cortex. The physical properties of the walls of organisms growing with little or no anionic polymers linked to their peptidoglycans have not been explored.

REFERENCES

1. Ghuysen, J.-M. (1968) *Bacteriol. Rev.* **32**, 425-464
2. Schleifer, K.H. and Kandler, O. (1972) *Bacteriol. Rev.* **36**, 407-477.
3. Rogers, H.J. (1974) *Ann. N.Y. Acad. Sci.* **235**, 29-51.
4. Rogers, H.J., Perkins, H.R. and Ward, J.B. (1981) in *Microbial Cell Walls and Membrane*, Chapman and Hall.
5. Hungerer, K.D. and Tipper, D.J. (1969) *Biochemistry* **8**, 3577-3587.
6. Leyh-Bouille, M., Bonaly, R., Ghuysen, J.-M., Tinelli, R. and Tipper, D.J. (1970) *Biochemistry* **9**, 2944-2951.
7. Rogers, H.J. (1970) *Bacteriol. Rev.* **34**, 194-214.
8. Ward, J.B. (1973) *Biochem. J.* **133**, 393-398.
9. Ward, J.B. and Perkins, H.R. (1973) *Biochem. J.* **135**, 728.
10. Rogers, H.J. (1981) in *β-lactam Antibiotics* Shockman, D.G., eds.), Academic Press, New York.
11. Hughes, R.C. (1971) *Biochem J.* **121**, 791-802.
12. Wyke, A.W., Ward, J.B., Orr, D.C. and Hayes, M.V. (1982) *Soc. Gen. Micro* **8**, M12.

12a Grant, W.D. and Wicken, A.J. (1970) *Biochem. J.* **118**, 59.

13. Warth, A.D. and Strominger, J.L. (1971) *Biochemistry* **10**, 4349-4358.

14. Mirelman, D. and Sharon, N. (1968) *J. Biol. Chem.* **243**, 2279-2287.
15. Araki, Y., Nakatani, T., Nakayana, K. and Ito, E. (1972) *J. Biol. Chem.* **247**, 6312-6322.
16. Araki, Y., Ota, S., Araki, A. and Ito, E. (1980) *J. Biochem. (Tokyo)* **88(2)**, 469-479.
17. Warth, A.D. (1978) *Advances in Microb. Physiol.* **17**, 1-15.
18. Warth, A.D. and Strominger, J.L. (1969) *Proc. Natl. Acad. Sci., USA* **64**, 528.
19. Johnstone, K. and Ellar, D.J. (1982) *Biochemica et Biophys. Acta* **714**, 185-191.
20. Tipper, D.J. (1970) *Int. J. Syst. Bacteriol.* **20**, 361-377.
21. Keleman, M.V. and Rogers, H.J. (1971) *Proc. Natl. Acad. Sci., USA* **68**, 992-996.
22. Oldmixon, E.H., Glauser, S. and Higgins, M.L. (1974) *Biopolymers* **13**, 2037-2060.
23. Formanek, H., Formanek, S. and Wawra, H. (1974) *Eur. J. Biochim.* **46**, 279-294.
24. Formanek, H., Schleifer, K.H., Seidl, H.P., Lindemaun, R. and Zundel, G. (1976) *FEBS Lett.* **70**, 150-154.
25. Burge, R.E., Adams, R., Balyuzi, H.H. and Reacely, D.A. (1977) *J. Mol. Biol.* **117**, 955-974.
26. Burge, R.E., Fowler, A.G. and Reaveley, D.A. (1977a) *J. Mol. Biol.* **117**, 927-953.
27. Labischinski, H., Barnickel, G., Bradaczeck, H. and Giesbrecht, P. (1979) *Europ. J. Biochem.* **95**, 147-155.
27a Carlstrom, D.J. (1937) *Biophys. Biochem. Cytol.* **3**, 619.
28. Naumann, D., Barnickel, G., Bradaczek, H., Labischinski, H. and Gresbrecht, P. (1982) *Eur. J. Biochem.* **125**, 505-515.
29. Ou, L.T. and Marquis, R.E. (1972) *Can. J. Microbiol.* **18**, 623-629.
30. Black, S.H. and Gerhardt, P. (1962) *J. Bacteriol.* **83**, 960-967.
31. Marquis, R.E. (1968) *J. Bacteriol.* **95**, 775-781.
32. Gerhardt, P. and Judge, J.A. (1964) *J. Bacteriol.* **87**, 945-951.
33. Mendelson, N. (1976) *Proc. Natl. Acad. Sci., USA* **73**, 1740-1748.
34. Mendelson, N. (1978) *Proc. Natl. Acad. Sci., USA* **75**, 2478-2482.
35. Mendelson, N. (1982) *J. Bacteriol.* **151**, 438-447.
36. Fein, J.E. and Rogers, H.J. (1976) *J. Bacteriol.* **127**, 1427-1442.
37. Fein, J.E. (1979).
38. Elliot, T.S.J., Ward, J.B., Wyrick, P.B. and Rogers, H.J. (1975) *J. Bacteriol.* **124**, 905-917.
39. Verwer, R.W.H. and Naninga, N. (1976) *Arch. Mikrobiol.* **109**, 195-197.

40. Verwer, R.W.H., Nanings, N., Keck, W. and Schwarz, U. (1978) *J. Bacteriol.* **136**, 723-729.
41. Beveridge, T.J., Forsberg, C.W. and Doyle, R.J. (1982) *J. Bacteriol.* **150**, 1438-1448.
42. Beveridge, T.J. and Murray, R.G.E. (1976) *J. Bacteriol.* **127**, 1502-1518.
43. Beveridge, T.J. and Murray, R.G.E. (1980) *J. Bacteriol.* **141**, 876-887.
44. Millward, G.R. and Reaveley, D.A. (1974) *J. Ultrastruct. Res.* **46**, 309-326.
45. Forsberg, C.W., Wyrick, P.B., Ward, J.B. and Rogers, H.J. (1973) *J. Bacteriol.* **113**, 969-984.
46. Boylan, J.R., Mendelson, N.H., Brooke, D. and Young, F.E. (1972) *J. Bacteriol.* **110**, 281-290.
47. Rogers, H.J., Thurman, P.F., Taylor, C. and Reeve, J.N. (1974) *J. Gen. Microbiol.* **85**, 335-350.

MICROBIAL BIOENERGETICS AND TRANSPORT

W.A. HAMILTON

Department of Microbiology,
Marischal College,
The University, Aberdeen AB9 1AS, Scotland

SUMMARY

Bioenergetics is described in terms of its pivotal role in directing and integrating the manifold activities of cellular function. The dependence on cell structure and on the physico-chemical state of the intracellular environment is stressed, as is the dynamic nature of the interlinked ion fluxes. The role of endogenous metabolism and the response of the energy-linked reactions to environmental stress are considered. The essential features of spore dormancy, and of sporulation and germination, are considered in the light of this holistic view of bioenergetics.

INTRODUCTION

Specialisation is an inevitable, albeit an unfortunate, consequence of the present extent of our factual knowledge. This is strikingly true of scientific research, and the problem is exacerbated rather than alleviated by the developing facilities for information handling; data and understanding are not necessarily synonymous. Bioenergetics, or rather the rest of biology, has been a particularly notable casualty in this compartmentalisation which is so much a feature of our common experience.

Bioenergetics is a recognised sub-discipline within biology. Although it relates to studies in metabolism, genetics or evolution, it has developed its own techniques and methodology, it employs its own language and dogma, and it has generated its own hierarchy of disciples and high priests. As with any such specialism, it is often extremely difficult for the non-initiate to appreciate the beauty of the individual trees, let alone to comprehend the awesome power of the wood itself.

In this article I hope to distil out the essence of bioenergetics and to set it in the context of cellular function. Having thus established that biological activity and growth are merely expressions at the cellular level of the efficacy and control of the various energy transducing mechanisms, it will be interesting then to consider the biologically inactive spore, and the processes of its formation and germination.

ENERGETICS AND CELLULAR FUNCTION

There are a number of theories and speculations regarding the origins of life and the evolution of the various forms extant at the present time. What these hypotheses have in common is the conviction that of major importance was the formation of the first primitive "cell", with its discrete, and different, intracellular environment. As Haldane (1) has put it, "The critical event that may best be called the origin of life was enclosure of several different self-reproducing polymers within a semi-permeable membrane". The point is that only within the confines of an enclosed cellular environment was it possible for the infrequent and random polymerization reactions associated with the formation of mucleic acids and proteins to become commonplace and co-ordinated. The essential features of this intracellular environment are:- the selection and concentration of nutrients; the retention of essential metabolic intermediates; the removal of unwanted metabolic products; the maintenance of a particular physicochemical milieu in respect of pH, water, activity, ionic strength and composition; and above all, the ability to sense and respond to changes in the external environment such that the intracellular environment is buffered against deleterious fluctuations in these characteristics. The cytoplasmic membrane both physically delimits the intracellular environment, and is responsible for its composition and character. These properties of the membrane stem directly from its unique structural organization; protein associated with and embedded in a bimolecular layer of phosphilipid in its fluid liquid crystalline state (2).

Particularly with regard to the resulting decrease in entropy, it is evident that the synthesis of macromolecules, their organisation into supramolecular structures, and the maintenance of the unique intracellular environment, are all endergonic processes requiring the input of free energy. The cell is nature's way of demonstrating the Second Law of Thermdynamics. Furthermore, often organisms are classified on the basis of their energy sources (phototrophs and chemotrophs), or the particular mechanisms they use to conserve and transduce that energy (oxygenic or non-oxygenic phototrophs; organo- or litho-chemotrophs; aerobes and anaerobes). It was in an effort to promulgate this cellular view of bioenergetics that the Society for General Microbiology held its symposium on "Microbial Energetics" in 1977 (3).

CHEMIOSMOSIS

Although detailed studies of the molecular mechanisms of ATP synthesis in oxidative and photosynthetic phosphorylation are of major importance, clearly to confine bioenergetics to such a narrow specialism is to miss the opportunity of gaining real insight into the totality of cellular function. Interestingly enough it was just such a blinkered view that led to total stalemate prior to Mitchell's proposal of his chemiosmotic hypothesis (4,5). The ideas that lead to chemiosmosis developed from Mitchell's studies of ion and nutrient transport in bacteria, and of the maintenance of the intracellular environment (6). The cornerstone of chemiosmosis is the recognition that all enayme catalysed reactions are vectorial in character; the pathways of diffusion of the reacting species to and from the active site being an integral part of the molecular mechanism. Where an enzyme, or transport carrier, is sited within a membrane and functions anisotropically this vectorial character gives rise directly to the phenomena of group, substrate or ion translocation. In a quite literal sense, these vectorial reactions add a third dimension to Lipmann's powerful concept of the coupling of reactions through group transfer, and provide the necessary link between a casual array of metabolic reactions and the integrated biological functioning of a complete cell. They also provide the mechanism whereby the redox, phosphoryl group and other forms of energy in biological systems can be transformed through the agency of ion translocations (primarily H^+) and the establishment of transmembrane gradients of chemical and electrical potential.

This extremely skeletal description of chemiosmosis can be fleshed out by reference to what is now a vast literature, embracing review articles, specialist monographs and general textbooks. Particularly recommended are the writings of Mitchell (7), Harold (8,9,10), Jones (11) and Nichols (12).

EVOLUTION OF CHEMIOSMOTIC PROCESSES

A fascinating demonstration of the range of energy requiring processes that are directly driven by the ptotonmotive force is provided by considering three proposals that have been made concerning the selective pressure which governed the presumed development of a H^+-translocating ATPase in the earliest anaerobic fermentative bacteria. Mitchell himself (13) suggested that the uptake of nutrients by H^+-symport would confer the necessary advantage on any organism with the ability to transduce part of its energy resources into a transmembrane H^+ gradient. Raven and Smith (14) on the other hand have argued that the need to maintain intracellular pH within certain limits would provide a rather more direct selective pressure for a H^+-translocating ATPase. Interestingly, these authors point out the necessity for an electrophoretic cation (possibly K^+) influx in response to the membrane potential, inside negative, generated by H^+ efflux. Due to the low electric capacitance of biological membranes such a mechanism for dissipating the membrane potential is necessary in order to achieve sufficient H^+ flux to affect the intracellular pH or the transmembrane pH gradient. Wilson and Lin (15) concern themselves rather with the physicochemical consequences of enclosing non-diffusible macromolecules within a semi-permeable membrane and suspending such a cell in dilute aqueous solution in which the principal solute is NaCl. Colloid osmotic pressure would result in entry of salts and water, with, unless controlled, the ultimate swelling and bursting of the cell. Whereas extant microbial, and plant, cells are protected by the constraining effect of their cell walls, the earliest cells would have lacked such structures and would have required some other mechanism of volume control. Wilson and Lin propose that this might have been achieved by a H^+-translocating ATPase in combination with an electroneutral H^+/Na^+ symport allowing H^+ to re-enter the cell and in so doing to drive the efflux of Na^+. In a contemporary study of cell volume regulation in *Mycoplasma gallisepticum*, Wilson and his co-workers (16) have shown that energy-dependent DCCD-inhibited extrusion of Na^+ (followed passively by Cl^-) is indeed the mechanism involved. Although in this case it remains uncertain as to whether a H^+-translocating ATPase plus a H^+/Na^+ symport, or a Na^+-translocating ATPase is the detailed nature of the molecular mechanism, the two components of the former mechanism are in fact found in all present-day bacteria.

CHEMIOSMOSIS AND CELLULAR FUNCTION

Whatever the validity of these hypotheses to the events of some 400 million years ago, they are all firmly based on existing problems in microbial physiology, and the mechanisms for their solution. In an extremely clear and simple experiment Harold and Van Brunt (17) were able to describe the energy-requiring processes within living cells, and to define those directly dependent upon the protonmotive force. Their experimental organism was the non-motile anaerobic fermentative bacterium *Streptococcus faecalis* which obtains the ATP required for anabolism and macromolecular synthesis from substrate phosphorylation. The protonmotive force in this organism is formed by the action of an H^+-translocating ATPase. In the presence of the protonphore gramicidin the protonmotive force was collapsed and yet the organism could grow normally, provided the medium was rich in nutrients, at slightly alkaline pH, and with high K^+ and low Na^+ concentrations. That is to say, the proton circulation maintains the intracellular pH and controls the nutrient and ionic composition of the cytoplasm. It also drives bacterial motility and, of course, in oxidative and photosynthetic organisms, the principal route of ATP formation is by the protonmotive force-driven reversal of the ATPase.

That chemiosmosis is indeed the primary energy transducing mechanism throughout the biological world is evident from the demonstration of a protonmotive force associated with metabolic activity in every eukaryotic and prokaryotic cell type, including mycoplasmas (16). Even the *Archaebacteria* have a proton current, although the mechanism of its generation may be peculiar to the organism in question. For example, the purple membrane of *Halobacterium halobium* is responsible for light-dependent proton pumping by a unique mechanism involving bacteriorhodopsin rather than chlorophyll (18). Although the mechanism remains obscure, Jarrell and Sprott (19) have recently measured the protonmotive force developed in *Methanobacterium thermoautotrophicum*.

It is important to appreciate however that the magnitude and the protonmotive force in any given cell is not constant, and that its two component parts, the membrane potential ($\Psi\Delta$) and the pH gradient (ΔpH), can vary both in absolute terms and relative to each other. Such fluctuations are in response to environmental stimuli and reflect the energetic requirements of the organism in question. Particularly striking, for example, are the demonstrations that ΔpH, and to a lesser extent $\Delta\Psi$, are influenced by the extracellular pH (20-23). Kashket (24,25) has extended this analysis, and has established that during growth of facultative organisms under aerobic conditions the protonmotive force is between 195 mV and 270 mV, depending on organism, stage of growth cycle, and extracellular pH. Anaerobic growth, on the other hand, is associated with a protonmotive force of around 150 mV, or even 75 mV or less. Kashket has concluded that a threshold value of 200 mV (produced by the redox reactions of oxidative metabolism) is required to drive the ATPase reaction in the direction of ATP synthesis; that is, oxidative phosphorylation. Below 200 mV (produced in this case by the action of the H^+-translocating ATPase on ATP arising from substrate level phosphorylation) the protonmotive force is sufficient for the nutrient uptake and pH and ionic control required during anaerobic growth. The lower values of protonmotive force (75 mV or less) found in Gram negative cells may very well relate directly to their lower intracellular concentrations of ions and nutrients, and osmotic pressure (6).

In all such quantitative studies the accuracy of the methodology and the validity of the assumptions on which it is based are critical. These methods include measurement of cytoplasmic volume in terms of sucrose-impermeable space (or other marker); determination of ΔpH by the distribution across the membrane of a weak acid or base; quantitation of $\Delta\Psi$ from the uptake of K^+ or Rb^+ in the presence of the ionophore valinomycin, or of one of a series of synthetic lipid-soluble cations; and modulation of these parameters by changes in, for example, extracellular pH, K^+ concentration in the presence of valinomycin, or oxygen tension. The inherent limitations of such assay procedures are often not fully appreciated and the findings of Ahmed and Booth (26) are salutary in this respect and must be taken into account by anyone seriously pursuing this line of research.

These uses of ionophores and probes either to modulate the protonmotive force or to determine its magnitude, are dependent on the facilitated diffusion of ionic and molecular species down their electrochemical gradient to a position of thermodynamic equilibrium between the intracellular and extracellular aqueous phases. This methodology has consequently been at least partially responsible for fostering the assumption implicit in many studies, of transport for example, that all energy-dependent fluxes are in equilibrium with their driving force. However, the recorded variations in the protonmotive force during growth (see e.g. 24,25) make this inherently unlikely, and in an extended series of experiments Booth and Hamilton and their co-workers (21, 26-30) have clearly demonstrated that the transport of lactose by *Escherichia coli* does not come into thermodynamic equilibrium with the protonmotive force. It is evident therefore that such processes in the cell are in a dynamic steady state, with mechanisms of kinetic rather than simply thermodynamic control.

K^+ AND Na^+ ION CURRENTS

Reference has already been made to K^+ uptake by electrophoretic uniport (14) and to Na^+ efflux by electroneutral H^+/Na^+

antiport (15). Through its response to $\Delta\Psi$, K^+ uptake was proposed as a means of increasing pH by stimulation of H^+ translocation with the development of a more alkaline intracellular pH. Since, on the other hand, the H^+/Na^+ antiport is driven by ΔpH its operation will have the opposite effect of reducing pH and making the intracellular pH more acid. It has been demonstrated that there are in fact K^+ and Na^+ ion fluxes and that these are linked with and secondary to the H^+ circulation. These secondary ion circulations are considered to be crucial to the modulation of $\Delta\Psi$ and ΔpH, and to the control of intracellular pH, but the exact mechanism of these effects is not completely known and is certainly more complex than the illustration just given. For example:- the H^+/Na^+ antiport may be electrogenic with more H^+ entering the cell than there are Na^+ leaving (31,32); the route of Na^+ entry to the cells in the first place is still unknown; it is now known that there is also a H^+/K^+ antiport with putative roles in the control of intracellular pH and in K^+ efflux (33); K^+ uptake in *E. coli* involves at least two systems, Kdp (a high affinity K^+ pump driven directly by ATP) and the TrkA system (lower affinity but higher capacity, and requiring both ATP and the protonmotive force) (34,35), and similar systems have now been identified in *Strep. faecalis* (36,37).

The secondary K^+ and Na^+ circulations also serve other energy-linked functions in cell homeostasis. Bacterial cells maintain an internal osmotic pressure, of the order of 5 atm in Gram positive organisms and 20 atm in Gram positive. In comparison with the generally lower osmotic pressure of the suspending medium, this gives a positive turgor pressure to the membrane-bounded cytoplasm which is balanced and constrained by the structural rigidity of the cell wall. Cells are capable of increasing their intracellular concentrations of so-called compatible solutes to maintain turgor in response to any increase in extracellular osmotic pressure; *E. coli* uses potassium glutamate to maintain turgor, but Gram positives seem to prefer proline or γ-aminobutyric acid for this purpose (38). It seems likely however that the primary response to changes in osmotic pressure and turgor is a stimulation of K^+ uptake (34).

In parallel with mammalian cells, many secondary transport systems in bacteria are driven by the Na^+ rather than the H^+ circulation. This is true not only of marine species (39) and of *H. halobium* (40), but also of alkalophilic organisms (41) and, in the case of such substrates as glutamate and melibose, even *E. coli* (42).

In a thoughtful analysis of the H^+ circulation as the primary convertible energy currency of the cell, Skulachev has suggested that the principal function of the unequal distribution of K^+ and Na^+ between the microbial cell and the medium is to act as a buffering system for the protonmotive force (43,44). That is, the transmembrane electrochemical gradients of K^+ and Na^+ represent an energy store which may either be used by the cell directly, or after conversion back into a H^+ gradient.

The cardinal importance of these H^+, K^+ and Na^+ ionic currents is eloquently demonstrated by the evidence that the gradients are seen to decrease on exhaustion of nutrients or approach to the stationary phase in batch culture, and are rapidly re-established by the addition of fresh nutrient medium (see e.g. 24,25,45-47). In fact it seems reasonable to suggest that a very high proportion of the maintenance energy requirements identified in growing cultures is a direct expression of the need to maintain ionic gradients and to control cytoplasmic pH, ionic strength, volume and turgor pressure. Similarly the major function of endogenous respiration is likely to be the use of reserve polymers laid down in time of plenty (and ultimately on their depletion, the use of the cell's functional and structural polymers) to maintain homeostasis and so prolong viability, although there is as yet insufficient data available to draw too firm conclusions on this point (48). In an interesting paper Jolliffe and colleagues (49) have shown that cell autolysis can arise when the protonmotive force is reduced to values of less than 85 mV by starvation or treatment with ionophores.

CONCLUSION

Thus we have established a picture of the role of bioenergetics in respect of the biological activity and viability of cellular organisms which, while accepting the predominant role of ATP and the reactions of chemical synthesis, has laid particular stress on nutrient transport and the maintenance of the intracellular environment as the necessary foundations on which everything else depends. In pursuance of this view we have noted the absolute requirement for two aqueous environments separated by a semi-permeable phospholipid membrane which allows the specific and controlled dynamic flux of mobile ionic species. All of which neatly describes what a bacterial spore is not.

From the bioenergeticist's point of view therefore the spore is an extreme structural modification which alters fundamentally the physicochemical character of the cell, and in so doing fixes the cytoplasm so that homeostasis and ion flux are neither necessary nor possible. Sporulation can then be seen as the complex series of reactions required to bring about this transition, while in germination we might seek to determine the nature of the initial event that can so rapidly re-establish membrane permeability and dynamic ion flux, with the consequent onset of metabolic activity.

While this conclusion has been reached from a consideration of the life and times of the vegetative cell, its validity is also indicated, perhaps a little more directly, by reference to the spore literature. One might cite, for example, a number of the contributions to the present volume, and such data as the findings from dielectric studies that spore ions are largely immobilized (50); the correlation which can be drawn between heat resistance and spore water content (38, 51); the polycrystalline state of spore membranes, and the reduced macromolecular mobility and increased non-exchangeable water pool (52); the demonstration of ion fluxes as a very early event in germination (53).

It seems certain that bioenergetics is central to the functioning of all living forms, and is equally the key to our understanding of their many faceted character.

REFERENCES

1. Haldane, J.B.S. (1954) *New Biol.* **16**, 12-20.
2. Ellar, D.J. (1978) in *Companion to Microbiology* (Bull, A.T. and Meadows, P.M., eds.), pp.265-295, Longmans, London.
3. Haddock, B.A. and Hamilton, W.A. (1977) *Symp. Soc. Gen. Microbiol.* **25**, 1-442.
4. Mitchell, P. (1961) *Nature* **191**, 144-148.
5. Mitchell, P. (1966) *Biol. Revs.* **41**, 445-502.
6. Mitchell, P. and Moyle, J. (1956) *Symp. Soc. Gen. Microbiol.* **6**, 150-180.
7. Mitchell, P. (1981) in *Of Oxygen, Fuels and Living Matter* Part 1, (Semenza, G., ed.), pp.1-160, John Wiley & Sons Ltd, Chichester.
8. Harold, F.M. (1972) *Bact. Revs.* **36**, 172,230.
9. Harold, F.M. (1977) *Ann. Rev. Microbiol.* **31**, 181-203.
10. Harold, F.M. (1982) *Curr. Top. Memb. Trans.* **16**, 485-516.
11. Jones, C.W. (1982) *Bacterial Respiration and Photosynthesis*, pp.1-106. Thos. Nelson & Sons Ltd, Walton-on-Thames.
12. Nicholls, D. (1982) *Bioenergetics*, pp.1-190. Academic Press, London.
13. Mitchell, P. (1970) *Symp. Soc. Gen. Microbiol.* **20**, 121-166.
14. Raven, J.A. and Smith, F.A. (1976) *J. Theor. Biol.* **57**, 301-312.
15. Wilson, T.H. and Lin, E.C.C. (1980) *J. Supramolec. Struct.* **13**, 421-446.

16. Rottem, S., Linker, C. and Wilson, T.H. (1981) *J. Bact.* **145**, 1299-1304.
17. Harold, F.M. and Van Brunt, J. (1977) *Science* **179**, 372-373.
18. Lanyi, J.K. (1978) *Microbiol. Revs.* **42**, 682-706.
19. Jarrell, K.F. and Sprott, G.D. (1981) *Can. J. Microbiol.* **27**, 720-728.
20. Padan, E., Zilberstein, D. and Rottenberg, H. (1976) *Eur. J. Biochem.* **63**, 533-541.
21. Booth, I.R., Mitchell, W.J. and Hamilton, W.A. (1979) *Biochem.* **182**, 687-696.
22. Mitchell, W.J., Booth, I.R. and Hamilton, W.A. (1979) *Biochem. J.* **184**, 441-449.
23. Zilberstein, D., Schuldiner, S. and Padan, E. (1979) *Biochemistry* **18**, 669-673.
24. Kashket, E.R. (1981) *J. Bact.* **146**, 369-376.
25. Kashket, E.R. (1981) *J. Bact.* **146**, 377-384.
26. Ahmed, S. and Booth, I.R. (1981) *Biochem. J.* **200**, 573-581.
27. Hamilton, W.A. (1980) *Viertel. Nat. Gesell. Zurich* **125**, 33-42.
28. Booth, I.R. (1980) *Biochem. Soc. Trans.* **8**, 276-278.
29. Hamilton, W.A. and Booth, I.R. (1982) in *Membranes and Transport: a Critical Review* (Martonosi, A., ed.), Plenum Publishing Corp., New York, **2**, 41-46.
30. Ahmed, S. and Booth, I.R. (1981) *Biochem. J.* **200**, 581-589.
31. Krulwich, T.A., Guffanti, A.A., Bornstein, R.F. and Hoffstein, J. (1982) *J. Biol. Chem.* **257**, 1885-1889.
32. Padan, E., Zilberstein, D. and Schuldiner, S. (1982) *Biochim. Biophys. Acta* **650**, 151-166.
33. Brey, R.N., Rosen, B.P. and Sorensen, E.N. (1980) *J. Biol. Chem.* **255**, 39-44.
34. Epstein, W. and Laimins, L. (1980) *Trends Biochem. Sci.* **5**, 21-23.
35. Kroll, R.G. and Booth, I.R. (1981) *Biochem. J.* **198**, 691-698.
36. Bakker, E.P. and Harold, F.M. (1980) *J. Biol. Chem.* **255**, 433-440.
37. Kobayaski, H. (1982) *J. Bact.* **150**, 506-511.
38. Gould, G.E. and Measures, J.C. (1977) *Phil. Trans. R. Soc. Lond.* **B278**, 151-166.
39. Niven, D.F. and MacLeod, R.A. (1980) *J. Bact.* **142**, 603-607.
40. Lanyi, J.K. (1979) *Biochim. Biophys. Acta* **559**, 377-397.
41. Guffanti, A.A., Blanco, R., Benenson, R.A. and Krulwich, T.A. (1980) *J. Gen. Microbiol.* **119**, 79-86.
42. Schellenberg, G.D. and Furlong, C.E. (1977) *J. Biol. Chem.* **252**,9055-9064.
43. Skulachev, V.P. (1978) *FEBS Lett.* **87**, 171-179.

44. Skulachev, V.P. (1980) *Can. J. Biochem.* **58**, 161-175.
45. Schultz, S.G., Epstein, W. and Solomon, A.K. (1963) *J. Gen. Physiol.* **47**, 329-346.
46. Harold, F.M., Pablasova, E. and Baarda, J.R. (1970) **196**, 235-244.
47. Harold, F.M. and Papineau, D. (1972) *J. Memb. Biol.* **8**, 27-44.
48. Horan, N.J., Midgley, M. and Dawes, E.A. (1981) *J. Gen. Microbiol.* **127**, 223-230.
49. Jolliffe, L.K., Doyle, R.J. and Streips, U.N. (1981) *Cell*, **25**, 753-763.
50. Ellar, D.J. (1978) *Symp. Soc. Gen. Microbiol.* **28**, 295-325.
51. Beaman, T.C., Greenmyre, J.T., Corner, T.R., Pankratz, H.S. and Gerhardt, P. (1982) *J. Bact.* **150**, 870-877.
52. Stewart, G.S.A.B., Eaton, M.W., Johnstone, K., Barrett, M.D. and Ellar, D.J. (1980) *Biochim. Biophys. Acta* **600**, 270-290.
53. Swedlow, B.M., Setlow, B. and Setlow, P. (1981) *J. Bact.* **148**, 20-29.

TABLE 1 *Biological activities of calcium*[(i-iii)]

1. Imparts structural and functional integrity to the matrix of mineralized tissue and other matrices
2. Extracellular but in solution – activation of hydrolytic enzymes and membrane systems
3. Inside cells but in solution – maintenance of protein structure
4. Flux from outside to inside cells – triggering, the major activity of calcium e.g. differentiation, contraction, hormone release, modulation of various enzymic activities

(i) Biochem. Soc. Symp. (1974), 'Calcium and Cell Regulation', Vol.39.

(ii) 'Calcium Binding proteins: Structure and function' (1980), Developments in Biochemistry, Vol.14, Siegel, F.L. *et al.*, eds., Elsevier, North Holland.

(iii) 'Calcium and Cell Function', Cheung, W.Y., ed., Volumes 1 and 2 (1982), Academic Press

relatively slower timescale for activation of extracellular enzymes). Calcium selectivity of the binding site(s) is also required if the action of calcium is not to be totally antagonized by a variety of cations. Activity is therefore a function of (i) the prevailing calcium concentrations, (ii) the strength and selectivity of binding, and (iii) the structural and kinetic factors of the binding reaction – of importance here are any fluctuations in conformation as well as the conformations themselves since action may demand any one of the possible states.

CALCIUM CHEMISTRY

The diversity of functions promoted by the calcium ion relies on the nature of interaction of calcium with ligands. Central to an analysis of its biological activity is the chemistry of Ca^{2+}. The binding of calcium is essentially electrostatic (1) and its association with a ligand relies on the relative cation to anion size. (This viewpoint leads to the notion of the availability in nature of holes of appropriate size for the cation, cf. macrocyclic ligands, but neglects the structure

TABLE 2 *Examples of calcium dependent proteins*[1]

	ROLE	LIGAND DONORS	log K_b	EXCHANGE RATE AND CONFORMATIONAL EFFECT
EXTRACELLULAR				
Phospholipase A2	Enzyme	non-sequential; 1 CO_2^-; 3 >CO and $2H_2O$	3.5	$\sim 10^3$; small
Dentine phosphoprotein	Structure nucleation	CO_2^- and monoester phosphate	~ 5	Long lived complex, cation diffuses on surface; very small
Prothrombin	Trigger	non-sequential; 3.4 CO_2^-	3.5	$>10^5$; considerable
INTRACELLULAR				
Calmodulin	Trigger	closely sequential		
		2 sites	~ 7	$\sim 10^1$; considerable
		2 sites	~ 5	10^3; considerable
		each of 4 CO_2^- and 2 >CO		
Parvalbumin	Transport/ buffering	closely sequential 4 CO_2^-, 1 CO and 1-OH/H_2O	~ 8	~ 10 ; considerable
Calsequestrin	Vesicular store	?CO_2^-	~ 4	fast

1 "Calcium binding proteins and Calcium Function", Siegel, F.L. *et al.*, eds. Elsevier, North Holland, 1980.

and dynamics (i.e. equilibria) of the water and the ligands (see below). Unlike the strict 6-coordinate stereochemistry displayed by magnesium, the structures of calcium salts reflect no simple apparent radius or fixed coordination geometry (2). The lack of stereochemical demand of calcium upon its ligands together with the easy loss of hydration enables it to form irregular lattices (cross-links). This ability to occupy different types of site in a solid allows calcium to function effectively in the formation of bones and shells. Calcium is used to cross-link structures like polysaccharides and proteins rather like S-S bridges and sugar-peptide bridges form cross-links of proteins and cell wall structures. Calcium cross links can however be readily reversed and are rapidly responsive to change in conditions. This dynamic function of the cation relies on the nature of its ligand interactions.

The main binding groups for calcium are oxyanions (carboxylate, phosphate and sulphonates, Table 2). The binding affinity is related to the number of anionic groups available (cf. acetate to malonate or succinate, Table 3). Discrimination and selectivity of complex formation over magnesium is obtained when a neutral oxygen donor (carbonyl), ether or alcohol group) is involved in binding (e.g. oxydiacetate, Table 3) when at 10^{-3}M ligand and metal, the conditions outside cells, there is relatively no competition from magnesium.

TABLE 3 *Binding affinity (log K_b) of Ca(II) and Mg(II) with carboxylate ligands*[1]

	Ca(II)	Mg(II)
Acetate	0.7	0.8
Malonate	2.5	2.8
Succinate	2.0	1.2
Oxydiacetate	3.4	1.7
Dipicolinate	4.4	2.7
EDTA	10.7	8.9
EGTA	10.7	5.4

1 "Stability constants of metal ion complexes", Sillen, L.G. and Martell, A.E., eds., Chemical Society, London, 1964.

Greater affinity and selectivity is required inside cells where calcium targets (Table 2) possess a group of donor centres similar to EGTA (cf. the selectivity of EDTA, Table 3, lacking the neutral oxygen donor). The smaller size of the

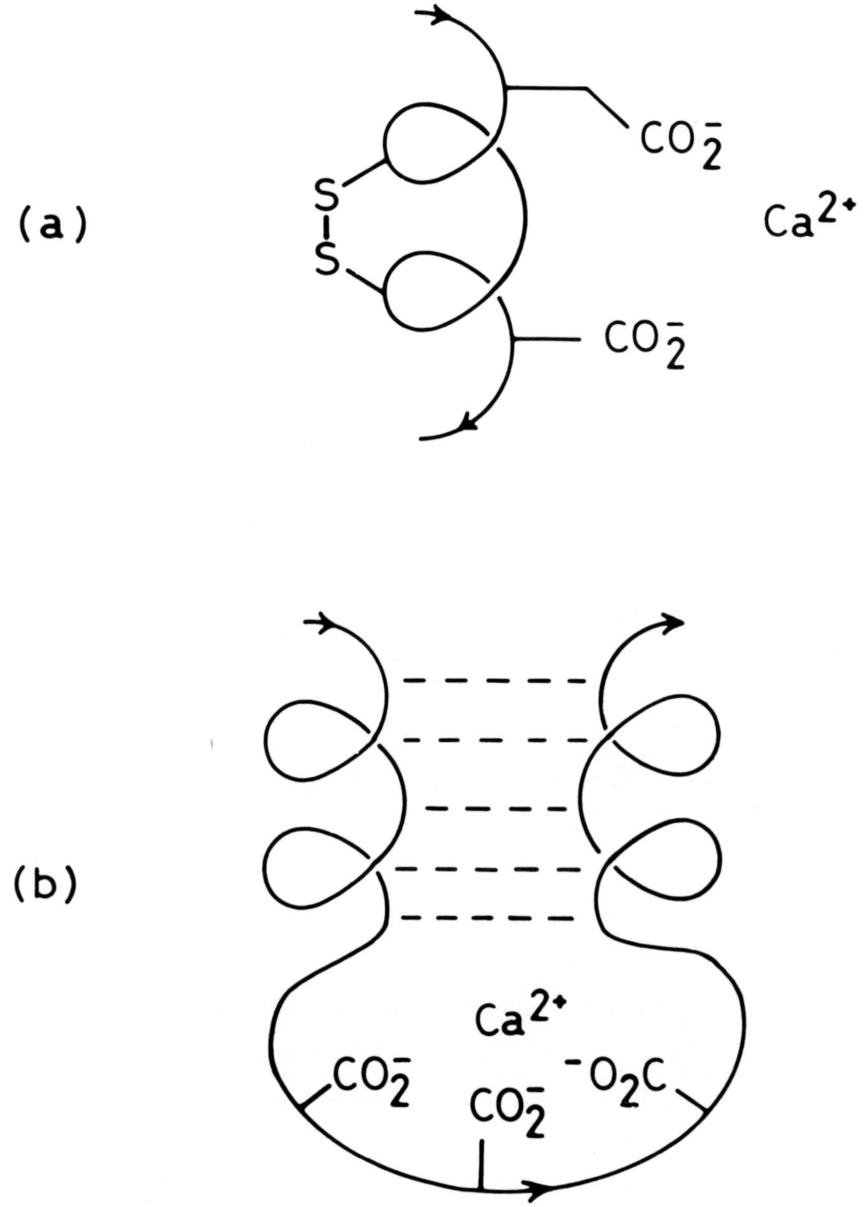

FIG. 1 *Schematic geometry of the calcium ligand framework in a protein. (a) Donor sidechains from different regions of the primary structure constrained by S-S bridges. (b) Closely sequential ligands on a relatively mobile framework.*

magnesium cation is relatively unfavourable for such multidentate binding of the comparatively bulky oxygen donor ligands. The binding strength of a ligand with several anionic donors is however seen to be dependent on the restriction to internal mobility of the ligand i.e. the geometry of the ligand is more confined in the complex (cf. malonate and succinate, Table 3). The configurational entropy of the free ligand opposes complex formation. If a protein is therefore to form a polycarboxylate ligand (Table 2) then the configurational entropy of the ligand framework must be restricted (Fig. 1). (Similarly, constraints on the freedom of motion and thereby organization of binding centres is achieved by arranging them on a surface as in phospholipid headgroups). The protein may well have to adjust its structure on charge neutralization when calcium binds leading to the question of the nature of the energetic and functional consequences of cation binding in biology. Biological processes are further often rate and not equilibrium controlled. It is the kinetic properties of the calcium ion which are outlined next.

A contributing factor in the use of calcium for the regulation of rate controlled events is its inherently flexible coordination requirements which permit complex dissociation with an acceptable rate constant. Eigen and Hammes (3) pointed to the rapid exchange of water molecules around cations (Ca >Mg by a factor of 10^3) and showed that fast on/off reactions of multidentate ligands could be explained on the basis of stepwise association and dissociation — water molecules being bound (or displaced) one at a time instead of in a single, cooperative step. This type of reaction when involving calcium binding to a multidentate ligand in order to achieve considerable affinity (above), requires intrinsic flexibility of the ligands in order to achieve fast on/off reactions, i.e. not only is the calcium ion coordination sphere fluctional- the calcium control protein must possess mutual flexibility thereby matching trigger and its target. Such mortality is indeed observed in the apo-forms of the intracellular control proteins (ref. 4 and below). By way of distinction, the role of magnesium is to maintain structural arrangements on a longer time-scale. On/off rates of ligand-metal reactions also reflect the ability of ion movement in a matrix of polar groups. Calcium, with its faster water dissociation rate (ease of distortion of the hydrate coordination sphere) compared to magnesium, can diffuse rapidly even in a matrix of ligands with high affinity. This point is relevant to the question of how biological systems have evolved calcium channels both ineffective for and not blocked by magnesium thereby enabling calcium currents to become a major control element coupled to a large variety of activities mediated by calcium-specific protein targets (Table 2).

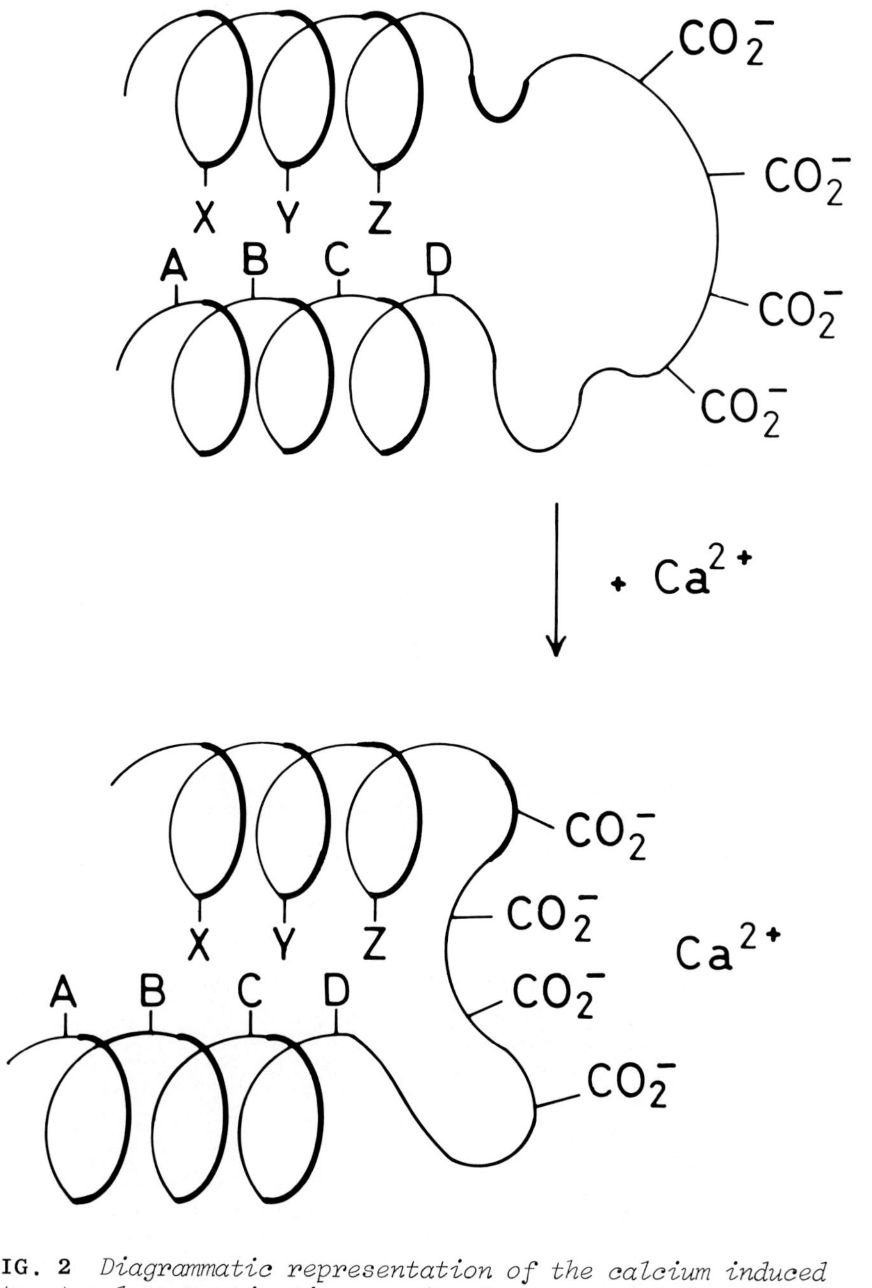

FIG. 2 *Diagrammatic representation of the calcium induced structural reorganization as closely sequential ligands 'wrap around' the cation. Hydration of the cation is under the control of protein sidechains.*

CALCIUM SITES IN PROTEINS

The nature and number of ligand donor atoms which contribute to the coordination site(s) of a protein impart, in the first instance, the ability of a protein to bind calcium selectively and reversibly. The binding affinity of a coordination site is further modulated by the overall protein configuration: the protein can generate an extremely specific interaction with a cation through alteration of structure but, given the consequent change in fold energy, only come to bonding distance with cations of given size. In extracellular enzymes (Table 2) the multidentate calcium binding sites are composed of ligands on a relatively rigid backbone contributed by various groups along the polypeptide chain (e.g. phospholipase A2, or *Stahpylococcus* nuclease). In order to achieve the required disposition of ligands in the coordination sphere, a particular protein fold must be pre-generated (through S-S bridges for example, Fig. 1). Lacking the requirement for rapid triggering, the extracellular enzymes can possess relatively rigid coordination site cavities matching the radius of the calcium ion. Activation is consequently slow since dehydration of the calcium ion is a concerted process and the activation energy is high (e.g. in phospholipase A2).

In the intracellular proteins, the role of the protein fold in generating a coordination site is somewhat reduced. The binding centres are made up from more mobile amino acid sidechains which are closer together in the sequence (Fig. 1; Table 2). The binding of calcium to these anionic sites alters the steric conformation of the protein since interactions within the protein (and between cation/protein and solvent) have to be readjusted upon ligation. Cation selectivity can be generated by the protein fold since different configurations, some energetically unfavourable, are needed to bring the ligand atoms into bonding distance with cations of different size (4). The effect of calcium binding is to organize the ligands about the cation and, by a 'domino effect', the resulting stresses influence regions of the protein structure remote from the binding site(s) (Fig. 2), i.e. the transmission of structural effects essential to the role of the intracellular control proteins (e.g. calmodulin, troponin-C). Unlike the cavity-like site of phospholipase A2, the rate of complex formation of calcium with these more flexible binding site loops is a more rapid process: removal of the first molecules of water from the inner coordination sphere of the hydrated cation destablizies the remaining water and triggers complex formation. Through its fold energy and character of the sidechains in the vicinity of the bound cation the protein can impose hydrophobic or steric interactions which affect the rate of

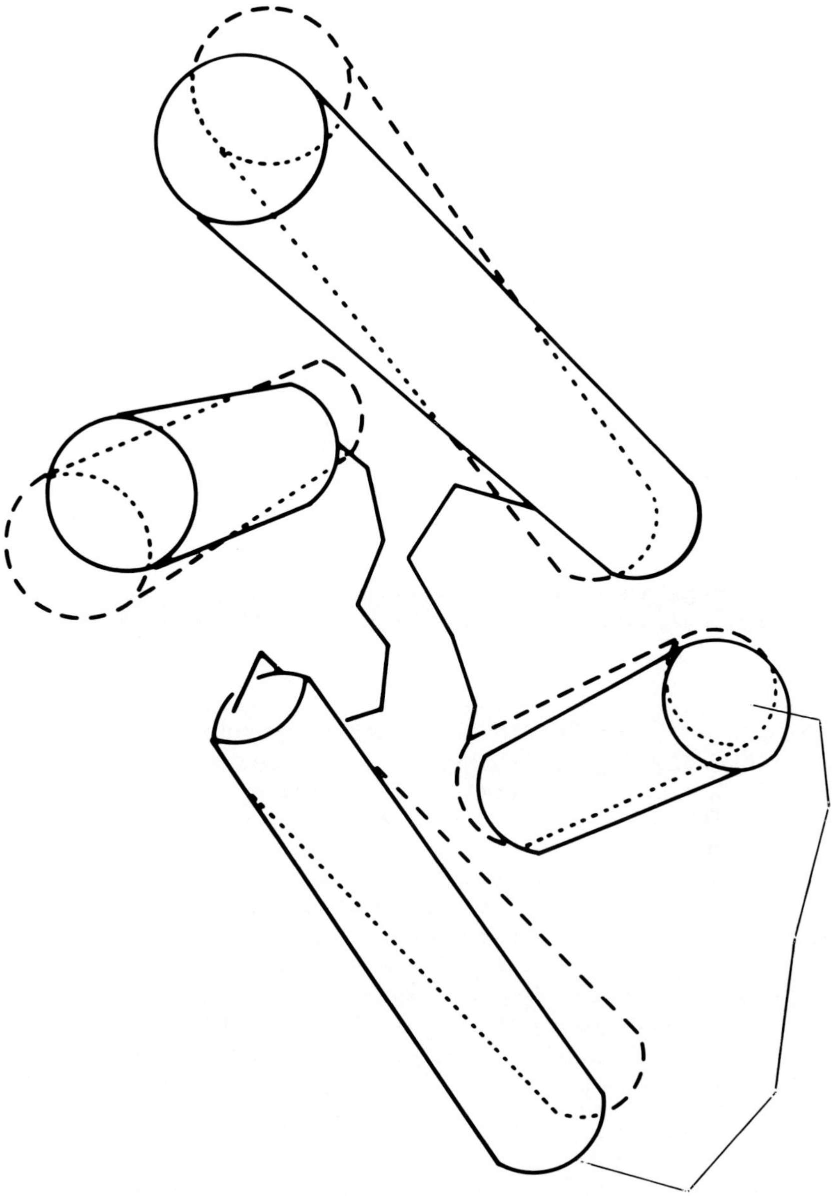

FIG. 3 *Relative motion of the helices in troponin-C and calmodulin which adjusts the linker segments and alters the exposure of helix surfaces.*

reassociation of water with the calcium ion. Binding to a control (trigger) protein therefore involves a succession of steps in complex formation (and dissociation) whose rate constants affect the kinetics of conformational adjustment. Of concern here are fluctuations between conformations of closely similar energy as well as the initial and final states, i.e. a dynamic conformational equilibrium. Concomitant with the functional properties of the protein (e.g. buffering, transport, triggering/modulation) there exists a gradation in the extent and rate of configurational adjustment (Table 2). Directed transmission of information in the calcium trigger proteins, calmodulin and troponin-C, relies on the mode of interaction between the helices which flank the calcium sites — the calcium induced relative motion of the helices (Figs. 2 and 3) is amplified in the relay (reactive) regions of the molecule which connect the protein to its partners in the particular organized system activated (5).

Many extracellular nonemzymatic activities are also triggered by calcium binding to a receptor protein e.g. the role of calcium in the assembly of the macromolecular complexes involved in the activation of the Vit-K-dependent blood coagulation zymogens such as prothrombin (Table 2). This binding site is highly anionic (comprising dicarboxylate sidechains of γ-carboxyglutamate) and relatively disordered before calcium binding. In keeping with the adaptable stereochemistry of the Gla sidechains, binding is fast and results in a considerable conformational change of the Gla-containing portion (6). The selectivity of the action of calcium here lies in its ability (cf. magnesium) to bind further groups after association with prothrombin, cross-linking the protein to phospholipid surfaces. Similar roles for calcium in the assembly of macromolecular complexes are found in complement activation and in the binding of proteins (e.g. concanavalin A) to exposed sugar moieties of cell surfaces. Calcium cross-links within extracellular proteins restrict the protein to a limited set of essential conformations while allowing limited local mobility (i.e. acting as a link about which groups may swivel without dissociation. Such cross links are also found in polysaccharides and lipids as well as forming part of quaternary structures (e.g. in the subunit structure of the tomato bushy stunt virus particle (7)).

The ability of the calcium ion to act as a cross-linking agent results in its ready involvement in the formation of microcrystalline solids. Mineralization of bone is believed to involve a Gla-containing protein which adsorbs on the growing calcium phosphate crystals(8). A second type of solid state protein, involved in the nucleation of dentine, has an unusually high content of physphorylated serine and acidic residues (9) (Table 2). It is a largely extended polymer across whose surface the cation diffuses, the array of ligands acting as a 'runway' for the cation to the site of nucleation. Many small organic molecules (e.g. phosphocitrate, trimetaphosphate) inhibit crystal growth and are similar to the bone and tooth proteins in that they possess tightly packed anions on a molecular framework of the required coordination geometry and affinity. The ability of the nucleating protein to sequester calcium can be modulated by the ionic composition of the surrounding medium as well as its degree of super-saturation — a complicated multiple equilibrium (10,11) involving effects of enzymatic activity, hormonal control, competing cations and energized ion gradients.

CALCIUM BINDING TO MEMBRANE SURFACES

Calcium activation of membrane-linked functions ranges from pinocytosis to exocytosis. Cation interaction with a membrane surface has been treated in terms of a surface potential resulting from the arrangement and density of surface charge on both sides of the membrane interface, as well as the ionic composition of the solution (e.g. refs. 12,13). Biological surfaces are not usually rigid and carry a variety of charged groups from proteins and lipids, as well as sugar hydroxyl groups of glycopeptides. Such intrinsically malleable surfaces can be distorted (shaped) in various ways by calcium binding; calcium's flexible coordination requirements (cf. magnesium) allow many coordination patterns. Fluctuating clusters on the membrane surface can occur with calcium and the state of the membrane is then dependent on the calcium ion and is in communication with the surrounding medium. Calcium pumps or channels can act to alter both the mechanical and electrical stress on the membrane and its constituents while the precise anionic state and calcium binding of a membrane is related to metabolic activity (14) through phosphorylation as a means of control over membrane structures.

The binding of groups which modulate membrane-bound calcium (e.g. inducers or drugs) could allow the membrane to alter its (dynamic) structure, resulting in a possible trigger of further changes such as the selective entry of specific solutes and the transmission of these surface events across the membrane. Transmission and energization of activity is often the result of initial local charge separation. Such events are likely to be involved in the activation/germination phenomenon of bacterial spores.

ACKNOWLEDGEMENTS

Analysis of the relevance of calcium chemistry to biological systems and the ideas outlined above derive from the work of Professor R.J.P. Williams.

REFERENCES

1. Phillips, C.S.G. and Williams, R.J.P. (1966) *Inorganic Chemistry*, Oxford University Press.
2. Williams, R.J.P. (1974) *Biochem. Soc. Symp.* **39**,133-138.
3. Eigen. M. and Hammes, G.G. (1963) *Adv. Enzymol.* **25**, 1-38.
4. Levine, B.A., Thornton, J.M., Fernandes, R., Kelly, C.M. and Mercola, D. (1978) *Biochim. Biophys. Acta* **535**, 11-23.
5. Levine, B.A., Dalgarno, D.C., Klevit, R., Scott, G.M.M. and Williams, R.J.P. (1982) *Ciba Symp.* **93**, 72-90.
6. Pluck, N.D., Esnouf, M.P., Israel, E.A. and Williams, R.J.P. (1980) in *Regulation of Coagulation* (Mann, P. and Taylor, A., eds.), pp.67-74, Elsevier North Holland.
7. Robinson, I.K. and Harrison, S.C. (1982) *Nature* **297**, 563-568.
8. Haushka, P.V. and Carr, S.A. (1982) *Biochemistry* **21**, 2538-2547.
9. Cookson, D.J., Levine, B.A., Williams, R.J.P., de Bernard, B., Jontell, A. and Linde, A. (1980) *Eurp. J. Biochem.* **110**, 273-278.
10. Hastings, A.B. (1980) *Trends Biochem. Sci.* **5**, 84-85.
11. Williams, R.J.P. (1975) *Biochim. Biophys. Acta* **416**, 237-286.
12. Hille, B. (1976) *Ann. Rev. Physiol.* **38**, 139-171.
13. Hauser, H., Levine, B.A. and Williams, R.J.P. (1976) *Trends Biochem. Sci.* **1**, 278-281.
14. Williams, R.J.P. (1979) *Biochem. Soc. Trans.* **7**, 481-509.

Genetics & Gene Manipulation of Spore Formers

MOLECULAR CLONING OF *BACILLUS* GENES AND MOLECULAR CLONING IN *B. SUBTILIS*

D.H. DEAN

Departments of Microbiology, Genetics and Program in Molecular, Cellular and Developmental Biology, The Ohio State University, Columbus, Ohio 43210, USA

SUMMARY

A significant number of genes of *Bacillus subtilis* and related species have been isolated by molecular cloning. The regulatory sequences (promoters, ribosomal binding sites, start codons) have been determined on nine *Bacillus* and *Staphylococcus* genes. These sequences may now be compared to regulatory sequences of *Escherichia coli* in order to address three problems related to cloned *B. subtilis* DNA or using *B. subtilis* for cloning: (1) general failure of foreign genes to be expressed, (2) instability of *B. subtilis* DNA in *E. coli* banks, and (3) deletion of cloned DNA when it is returned to *B. subtilis*.

INTRODUCTION

With the development of recombinant DNA techniques, the isolation of genes from *Bacillus subtilis* has become a reality. It will soon be possible to discuss the organization of these genes, their regulation and expression with a great deal of confidence. Although very few data have been collected, it is currently possible to compare the regulatory sequences of nine *B. subtilis* vegetative genes and sporulation genes and to reflect upon unusual and unexpected results which have already been observed from cloning *B. subtilis* DNA. It is, after all, the unexpected results which are the most interesting.

TABLE 1a *Cloned vegetative genes from* Bacillus subtilis *and related organisms*

GENE	VECTOR	HOST	REFERENCE
amy[a]	plasmid, pUB110	*B. subtilis*	[50]
*amy*E-*amy*R	phage, phi3T	*B. subtilis*	[69]
*aro*D	phage, Charon 4A	*E. coli*	[62]
*aro*I	phage, Charon 4A	*E. coli*	[16]
pen[b]	plasmid	*E. coli* *B. subtilis*	[19]
pen[c]	phage, NM74	*E. coli*	[2]
pen	plasmids	*B. subtilis*	[28]
ctc	plasmid, pMB9	*E. coli*	[48]
dal	phage, Charon, 4A	*E. coli*	[16]
δ-endotoxin[d]	plasmid, pBR322	*E. coli*	[59]
*gln*A	phage, Charon 4A	*E. coli*	[16]
*gly*B	plasmid, pHV33	*E. coli* *B. subtilis*	[53]
*his*A	plasmid, pHV33	*E. coli* *B. subtilis*	[53]
	phage, ρ11	*B. subtilis*	[30]
*leu*A	plasmid, pMB9	*E. coli*	[36]
	plasmid, pHV33	*E. coli*	[53]
	phage, Charon 4A	*E. coli*	[16]
	plasmid, RSF2124	*E. coli*	[48]
	pLS103	*B. subtilis*	[63]
*leu*B	plasmid, RSF2124	*E. coli*	[48]
*leu*C	plasmid, RSF2124	*E. coli*	[48]
lys-1	phage, Charon 4A	*E. coli*	[15]
	phage, ρ11	*B. subtilis*	[30]
*met*B4	phage, Charon 4A	*E. coli*	[16]
*met*C3	phage, Charon 4A	*E. coli*	[16]
*phe*A	phage, Charon 4A	*E. coli*	[16]
*pur*A	phage, Charon 4A	*E. coli*	[16]
*pur*B33	phage, Charon 4A	*E. coli*	[16]
*pyr*D	phage, Charon 4A	*E. coli*	[16]
*pyr*D,B,F,E,	phage λgt	*E. coli*	[4]
sul	plasmid, pUB100	*B. subtilis*	[40]
*trp*C	plasmid, pUB110	*B. subtilis*	[31]
thr-5	plasmid, pHV33	*E. coli*	[53]
tms-26	phage, Charon 4A	*E. coli*	[16]

TABLE 1b *Cloned sporulation genes from* Bacillus subtilis

GENE	VECTOR	HOST	REFERENCE
*spo*0A	phage, Charon 4A	E. coli	[16]
*spo*0B	phage, Charon 4A	E. coli	[16]
*spo*0B-pheA	plasmid, pUB110	B. subtilis	[29]
*spo*0B	phage, rho11 & phi-105 and plasmid pUB110	B. subtilis	[30]
*spo*0H[b]	pBD64	B. subtilis	[9]
*spo*0F	plasmid, pBS161-1 and phage NM607b	B. subtilis E.coli	[55] [54]
*spo*0F	phage, rho-11 & phi-105 and plasmid pUB110	B. subtilis	[30] [32]
*spo*IIA	plasmid pHV33	E. coli B. subtilis	Piggott (pers. comm.)
*spo*IIC	phage, Charon 4A	E. coli	Sonenshein (pers. comm.)
*spo*IIIC	phage, Charon 4A	E. coli	Sonenshein (pers. comm.)
*spo*VC	plasmid, pMB9	E. coli	[47]
*spo*VG(0.4kb gene)	plasmid, pMB9	E. coli	[56]

RESULTS

Cloned *B. subtilis* Genes

A significant number of *B. subtilis* genes have been cloned by one of several molecular cloning systems (Table 1a,b). Subcloning of genes of *B. subtilis* into *E. coli* has led to the following general observations: (1) *B. subtilis* genes are expressed in *E. coli* but the expression is usually weak, (2) *E. coli* genes or foreign DNA in general are not directly expressed in *B. subtilis*, (3) "Banks" of *B. subtilis* DNA maintained in *E. coli* on plasmid vectors are highly unstable, (4) DNA cloned in *E. coli* which is reintroduced into *B. subtilis* is subject to deletions. These observations will be discussed in detail below.

There are several remarkable physiological observations which may be made about the cloned *Bacillus* DNA. Of course, it is important to note that a number of *B. subtilis* genes are expressed in *E. coli* such as: *leu* [4,36], *thy*P (from the phage phi3T) [11,13], *pyr* [4], *pen*P [2,19,28] *pen*I [28], *trp* [31] and delta-endotoxin [59]. Many of these are expressed strongly enough to be selected for by complementation of *E. coli* auxotropic mutations. It was clear from the earliest recombinant DNA experiments [3] that gram positive antibiotic resistance genes could be expressed in *E. coli*.

The cloning of the *B. licheniformis* penicillinase gene (beta-lactamase or *pen*) showed peculiar effects upon the expression in *B. subtilis*. When Imanaka and co-workers [28] cloned the inducable *pen*P-*pen*I operon on both high and low copy plasmids, they found, as expected, that the high copy plasmid allowed the greatest production of beta-lactamase. What was not expected was that when they attempted to clone the constitutive *pen*P gene (mutant Col, *pen*I$^-$) it could only be expressed in the low copy plasmid. It was not clonable on a high copy plasmid. These results indicate that the beta-lactamase, when over expressed, interferes with the normal cellular activities [28]. Their results appear to be in conflict with those of Gray and Chang [19] who cloned the *pen*P constitutive gene from strains 749/C (*pen*I$^-$) on a high copy plasmid (although both groups used complicated hybrid plasmids both used the pUB110 origin) and obtained great overproduction.

Unusual effects have also been observed when attempts to clone sporulation genes were made. Cloning of the *spo*OF gene has been reported by two groups [30,29,54]. Both have noted that when they attempted to clone the gene on a high copy plasmid, sporulation was blocked. The gene could be cloned on the temperate phage, rho11 which integrates as one or a few copies, with no ill effects on sporulation [29]. This gene has been reported to be the *abs* gene which encodes adenosine bis-triphosphate synthetase, the enzyme which makes pppAppp, a highly phosphorylated nucleotide, reportedly involved in the initiation of sporulation [54]. Certain other sporulation genes such as *spo*OA [J. Hoch, pers. comm.] and *lys* [D. Stahly, pers. comm.] have not been able to be cloned on multiple copy plasmids. Perhaps these genes encode repressors or products which are toxic to the recipient cells.

One of the most difficult problems in the genetics of sporulation genes will be the determination of the role or function of spore genes once they have been cloned. Along these lines, Kobayashi and co-workers [32] have shown, through the use of a *B. subtilis* maxi-cell (UV irradiated cell) protein synthesis system that the product of the *spo*OB gene is a 39,000 dalton protein. Dubnau and colleagues [9], have also identified the protein product of the *spo*OH gene as 27,000 daltons, by using minicells.

Sequence Analysis of Cloned *B. subtilis* and Related Genes

Direct expression of foreign DNA in *B. subtilis* has been reported in only two cases. Rubin et al. [58] cloned the *E. coli trp* gene into *B. subtilis* on an integration vector and it is assumed that expression was related to the fact that the gene was integrated into the chromosome. The second example is the expression of the *E. coli* Tet gene (tetracycline resistance) from pBR322 on the hybrid plasmid pJK502. Expression in this case may be read through from a gram positive promoter on the hybrid plasmid [34]. The general lack of expression in all other cases has delayed the use of *B. subtilis* as a host for molecular cloning. The simplest explanations for this phenomenon are: (1) failure to transcribe the foreign gene, and/or (2) failure to translate the foreign gene. The sequence of the regulatory regions of a few *B. subtilis* and related genes are now available and shed some light on these simple views. As will be seen, these simple explanations alone do not account for all of the observations.

Table 2 lists the promoters of some vegetative genes and sporulation genes which are expressed in *B. subtilis*. For comparison, the consensus *E. coli* promoter sequence is also listed. It is apparent from inspection, that the promoters of the vegetative genes are not widely different from the canonical promoter of *E. coli*, yet experiments demonstrate that the *cat* gene of the *E. coli* plasmid paCYC184 is not transcribed in *B. subtilis* [17]. Unfortunately the promoter sequence of this *cat* gene is not available to compare with that of *B. subtilis*. Interestingly the *E. coli* gene for ampicillin resistance on the plasmid pBR322 is transcribed and translated but is not

TABLE 2 *Transcriptional regulatory sequences of cloned* Bacillus subtilis *genes*

GENE	DNA SEQUENCE OF PROMOTOR -35 REGION	-10 REGION	REFERENCE
E. coli consensus sequence	...TTGACA	...TATAAT...	[57]
"Veg" gene (unknown function)	...TTGACA...	...TACAAT	[39,46]
penicillinase (b. licheniformis)	...TTGCAT...	...AATACT...	[35]
tms	...TTGAAA...	...TATATT...	[39,46]
Erm	...TTCATA...	...TATAAT...	[24]
Cam	...TTGATT...	...TCTAAT...	[25]

processed into the active form of beta-lactamase. While it is clear that *B. subtilis* does not express *E. coli* genes it is not obvious from the promoters that certain *E. coli* genes could not be transcribed in *B. subtilis*, yet they fail to be transcribed, *in vitro*, efficiently by *B. subtilis* RNA polymerase in several cases [37,49,61,66].

Expression may also be blocked at the translational step; either at the initiation of translation binding of the *Bacillus* ribosome to the mRNA, or during the elongation process by a substantial difference in the codon preference of the two organisms. It has been observed from *in vitro* studies that *B. subtilis* ribosomes recognize and bind less tightly to *E. coli* mRNAs than *E. coli* ribosomes. Furthermore, *E. coli* ribosomes require elongation factor EF3 which has relatively less effect on gram positive ribosomes [43]. It has been proposed [42] that ribosomes of *B. subtilis* and other gram positive organisms require a more extensive ribosomal binding (Shine-Delgarno) sequence. Inspection of a number of sequenced genes from gram positive organisms (Table 3) shows that the vegetative genes tend to have more 16S RNA complementing sites than usual for *E. coli* (see *trp* sequences Table 3). Four bacteriophage sequences which follow this trend are also cited by McLaughlin *et al.* [42]

E. coli 16S RNA 3'OH AUU*CCUCC*ACUAG....5' [60]

B. subtilis 16S RNA 3'OH UCUUU*CCUCC*ACUAG....5'

TABLE 3 *Ribosomal binding sites on cloned DNA which are expressed in* Bacillus subtilis; *Comparison to* E. coli *sites*

GENE	NUCLEOTIDE SEQUENCE	REFERENCES
Bacillus or Staphylococcus		
alpha-amylase[1]	...*AAA*T*GAGAGGGA*GAGGAAAC[ATG]...	[49]
beta-lactamase[2]	...*AAA*C*GGAGGGA*GACGATTTTG[ATG]...	[19]
beta-lactamase[3]	...TATC*GGAGGG*TTTATT[TTG]...	[42]
Erm[3]	...AT*AAGGAGGAA*AAAAT[ATG]...	[24]
Cam[3]	...TT*AGGAGG*C*A*TATCAA[ATG]...	[25]
spoVG[4]	...*AAAGGTGG*T*GA*ACUACU[GTG]...	[46]
E. coli		
trpL	...A*AAGG*GTATCGACA[ATG]...	[68]
trpE	...ATTA*GAG*AATAACA[ATG]...	[68]
trpD	....C*AGGA*GCTTTCTG[ATG]...	[68]
trpC	...CAC*GAGG*GTAAATG[ATG]...	[68]
trpB	...T*AAGGA*AAGGAACA[ATG]...	[68]
trpA	...C*GAGG*GGAAATCTG[ATG]...	[68]

[1]*B. amyloliquifaciens*, [2]*B. licheniformis*, [3]*S. aureus*, [4]*B. subtilis*
Italicized bases complement *B. subtilis* 16 s rNA. Bracketed bases are the start codons.

TABLE 4 *Codon preference of two* Staphylococcus aureus *genes expressed in* Bacillus subtilis *compared to* E. coli *genes (data from [24,25,51,68])*

	E.c.		S.a.			E.c.		S.a.			E.c.		S.a.			E.c.		S.a.	
	RP	T	Em	Cm		RP	T	Em	Cm		RP	T	Em	Cm		RP	T	Em	Cm
Phe TTT	6	9	12	8	Ser TCT	12	1	2	7	Tyr TAT	1	11	5	8	Cys TGT	0	2	1	1
Phe TTC	10	7	4	1	Ser TCC	9	1	2	0	Tyr TAC	7	5	3	5	Cys TGC	2	5	1	0
Leu TTA	1	10	15	7	Ser TCA	0	4	6	3	Stop TAA	4	1	1	1	Stop TGA	0	0	0	0
Leu TTG	1	8	2	4	Ser TCG	1	8	0	1	Stop TAG	0	0	0	0	Tro TGG	0	3	1	3
Leu CTT	1	4	4	3	Pro CCT	0	1	4	4	His CAT	0	5	6	4	Arg CGT	16	8	0	0
Leu CTC	1	8	0	0	Pro CCC	0	2	0	2	His CAC	3	6	3	0	Arg CGC	12	13	4	0
Leu CTA	0	2	0	0	Pro CCA	2	2	1	2	Gln CAA	5	10	5	4	Arg CGA	0	1	0	1
Leu CTG	43	37	0	1	Pro CCG	21	9	0	1	Gln CAG	15	18	1	2	Arg CGG	0	1	0	0
Ile ATT	9	13	6	6	Thr ACT	18	2	3	4	Asn AAT	2	8	15	13	Ser AGT	1	3	5	2
Ile ATC	22	12	2	2	Thr ACC	15	5	1	1	Asn AAC	18	9	7	5	Ser AGC	5	7	3	1
Ile ATA	0	0	12	0	Thr ACA	1	1	4	7	Lys AAA	55	14	21	7	Arg AGA	0	0	7	2
Met ATG	19	5	4	2	Thr ACG	0	7	1	1	Lys AAG	8	4	5	1	Arg AGG	0	1	2	2
Val GTT	35	9	7	2	Ala GCT	60	10	2	2	Asp GAT	8	19	9	5	Gly GCT	25	10	1	5
Val GTC	3	7	0	0	Ala GCC	7	18	1	0	Asp GAC	24	11	2	3	Gly GGC	24	12	2	0
Val GTA	28	5	2	4	Ala GCA	31	14	2	2	Glu GAA	41	12	10	3	Gly GGA	0	3	1	4
Val GTG	7	15	4	2	Ala GCG	16	16	0	0	Glu GAG	7	11	2	4	Gly GGG	0	6	1	2

RP, totals from nine ribosomal proteins of *E. coli* [51]; T, from *trp*C gene of *E. coli* [68]; EM, from erythromycin resistance gene of pE194 [24]; Cm, from chloramphenicol gene of pC194 [25]

The *B. subtilis* 16 S rRNA sequence is from C. Woese and H. Noller cited in other published works [35,42]. It is also observed (Table 3) that the two sporulation genes which have had their ribosomal binding sites sequenced [44,45,46] follow the McLaughlin *et al.* model [42] for *B. subtilis*.

Finally, there is the possibility that the efficiency of expression is affected by the preference either organism may have for the codons which specify the amino acids. If these two organisms had widely different codon preferences, representing different tRNA pools, each would be equally inefficient on the other organism's genes. If one organism had an even balance of tRNAs it might be more permissive in expressing the genes on the other even if the latter had strong codon preferences. Because very few *Bacillus* or gram-positive genes have been sequenced completely it is not possible to compare, definitely, the codon preferences of *E. coli* and *B. subtilis* but the temptation to analyse the two *S. aureus* genes for erythromycin and kanamycin was too great for the author to resist (Table 4). Again, without placing too great an emphasis on broader ramifications from such preliminary data, it is remarkable that the strong codon preferences for CTG (Leu) and CAG (Gln), noted for almost all *E. coli* genes [51,65,68] are not observed in these two genes.

Stability of Cloned *Bacillus* DNA in *E. coli*

At least three cloned banks of genes have been reported. Rapoport *et al.* [53] have cloned sheared fragments of *B. subtilis* DNA in the Bam HI site of pHV33 via poly AT tails; Hutchison and Halvorson [26] reported a bank of sheared *B. subtilis* (lysogenic for phi-105) DNA into pMB9 via poly AT tails; and, Ferrari *et al.* [16] have cloned partially fragmented *B. subtilis* DNA (previously treated with Eco RI methylase) into Charon 4A [16] and mabda-gt WES.

Hutchison later reported [27] that the bank developed by him had undergone considerable deletions. In a study on a specific clone, he observed patterns of instability which were related to an insertion element identified by heteroduplex analysis to be within the pMB9 cloning vector. Another type of instability was reported by Rapoport *et al.* [53] and by Dedoner *et al.* [7] in their bank. In this case when the cloned DNA was returned to *B. subtilis*, it could integrate into the chromosome. This instability was dependent on the recE4 gene. It was indicated but not discussed that this bank may also undergo structural instability and, indeed, this has been found. Interestingly, certain very important genes remain enriched in this bank (see Piggot, this volume). Other recent reports of genetic instability in cloned DNA indicate that direct repeats are involved. In these cases recA *E. coli* mutants will greatly reduce the structural instability [1]. Even though plasmid borne banks of *B. subtilis* DNA suffer major deletions, such is apparently not the case in banks constructed with bacteriophage vectors.

Instability of foreign DNA in *E. coli* is not confined to *Bacillus* DNA. A. Cohen *et al.* [5] found instability in banks of yeast DNA, and S. Cohen reported deletions of *S. aureus* plasmid DNA in psC101 [6], and suggested that inverted repeats were involved.

Structural Instability of Cloned DNA in *Bacillus subtilis*

The development of "shuttle" or "bridge" plasmids [12,19,21, 34] has allowed the cloning of *B. subtilis* DNA into *E. coli* and returning it to *B. subtilis*. Several groups have reported major deletions during the return of such cloned DNA to *B. subtilis* [14,17,19,20,21,33,34,51,62,64]. It is clear that the deletion (recombination) phenomenon is *rec*E4 independent [15,62,64] and affects both competent cell transformation and protoplast induced transformation [17,64].

In a definitive set of experiments Uhlen *et al.* [64] showed that if plasmids were isolated immediately after the transformation event (i.e. after only a few generations) no deletions could be detected. All of the 100 colonies tested from the subculturing of a primary transformant colony contained the phenotype of the insert DNA. On the contrary, if the primary transformant was allowed to grow for 20-200 generations, all of the subcultured colonies had lost their inserted DNA as well as some of the vector. This indicated that the deletion phenomenon is a rather rare event (less than 1:1000) and that deleted plasmids replicate faster than the non-deleted plasmids. These observations have been independently reported by Kreft and Hughes [34]. Uhlen *et al.* [64] propose that the deletion event is associated with the transformation process, a view shared by Goebel *et al.* [17] and Cryczan and Dubnau [10,22].

The deletion phenomenon is particularly noted in experiments which employ pC194, or constructs developed from it [18,19,64] and Grandi *et al.* [20] have localized a site near a particular Mbo I site where the union of pC194 and pS194 occurred by spontaneous recombination to form pSC194 (also called pCS194 and pSA2100) [38]. This indicates a recombinational hot spot or insertion element in pC194 and it is apparently advisable to avoid this plasmid as a cloning vector. Primrose and Ehrlich [52] take exception to this conclusion, noting that deletions observed in their studies do not have an end point near the designated Mbo I site. The deletion

TABLE 5

Foreign Genes Cloned and Expressed in Bacillus subtilis

GENE	VECTOR	REFERENCE
Hepatitis B Core Antigen	pBD9	[23]
Foot and Mouth Disease Virus Coat Antigen	pBD9pBR322	[23]
E. coli Cam (CAT)	pGFT1	[18]
E. coli trp	pPL608	[67]
Mouse DHFR	pPL608	[67]
E. coli thy	pMB9* or pBR322*	[58]
E. coli Tet (pBR322)	pJK502	[34]
E. coli *lac*Z, Y, A	pCED2	[8]

*integrated into the *B. subtilis* genome

phenomenon is proposed to be "an efficient recombination machinery, active on sequences with a limited amount of homology" [14]. It is true that the deletion phenomenon in *B. subtilis* is observed with other plasmid replicons such as pJK3 (pBC16-1+pBR322) and pJK501 (pBS1+pBR322) [17] and pUB110 [21].

Successful Cloning of Foreign DNA in *B. subtilis*

B. subtilis has been successfully used as a molecular cloning host for a number of foreign genes of economic and academic interest (Table 5). With the expection of the *E. coli thy* and Tet genes (discussed above), the other genes are expressed through the use of expression vectors; plasmids which provide *Bacillus* regulatory sequences upstream to the insertionally inactivated cloning site. *Bacillus* plasmids such as pUB110, pBD9 and pPL606 are useful for such purposes.

CONCLUSIONS

A considerable amount of information has already been gleaned from the few cloned *B. subtilis* genes which have been analysed, but there is so much more to learn. The similarity of *B. subtilis* vegetative promoters and *E. coli* gene promoters suggests a reasonable mechanism for the ease of expression of the former genes in *E. coli*. The lack of expression of *E. coli* genes in *B. subtilis* seems to lie in the failure of the latter to initiate transcription [37,46,61] or translation on mRNA from

E. coli templates. This is reflected in the sequence analysis of the ribosomal binding sites [43]. Successful cloning of foreign DNA is readily achieved through the use of commonly available plasmids.

The lack of stability of *B. subtilis* DNA in *E. coli* banks has been associated with repeated homologous sequences [26] and perhaps may be reduced by use of *rec*A hosts [1]. Preparation of cloned *B. subtilis* DNA may be more stable in bacteriophage banks, but the only reported such bank [16] is only 20-70% complete. It is clear that the continued progress of our knowledge of *B. subtilis* genes is dependent, in part, upon the development of other and more complete gene banks.

ACKNOWLEDGEMENT

The author thanks R. Losick for reading and correcting portions of the manuscript relating to his research.

REFERENCES

1. Albertini, A.M., Hofer, M., Calos, M.P. and Miller, J.H. (1981) *J. Bacteriol.* **147**, 622-632.
2. Brammar, W.J., Muir, S. and Morris, A. (1980) *Mol. Gen. Genet.* **178**, 217-224.
3. Chang, A.C.Y. and Cohen, S.N. (1974) *Proc. Natl. Acad. Sci. USA* **71**, 1030-1034.
4. Chi, N.-Y.W., Ehrlich, S.D. and Lederberg, J. (1978) *J. Bacteriol.* **133**, 820-821.
5. Cohen, A., Ram, D., Halvorson, H.O. and Wensink, P.C. (1976) *Gene* **3**, 135-147.
6. Cohen, S.N., Cabello, F., Chang, A.C.Y. and Timmis, K. (1977) in *Recombinant Molecules: Impact on Science and Society* (Beers, B.F. and Bassett, E.G., eds.), pp.91-105.
7. Dedonder, R., Rapoport, G., Billault, A., Fargette, F. and Klier, A. (1980) in *DNA-Recombination Interactions and Repair* (Zadrazil, S. and Sponar, J., eds.), pp.95-104.
8. Donnelly, C.E. and Sonenshein, A.L. (1982) in *Molecular Cloning and Gene Expression in Bacilli* (Ganesan, A.T., Chang, S. and Hoch, J.A., eds.), pp.63-72, Academic Press.
9. Dubnau, E., Ramakrishna, N., Cabane, K. and Smith, I. (1981) *J. Bacteriol.* **147**, 622-632.
10. Dubnau, D., Cryczan, T., Contende, S. and Shivakumar, A.G. (1981) in *Genetic Engineering* (Setlow, J.K. and Hollaender, A., eds.), pp.115-131.
11. Duncan, C.H., Wilson, G.A. and Young, F.E. (1977) *Gene* **1**, 153-167.
12. Ehrlich, S.D. (1979) *Proc. Natl. Acad. Sci. USA* **75**, 1433-1436.

13. Ehrlich, S.D., Bursztyn-Pettegrew, H., Stroynowski, I. and Lederberg, J. (1977) in *Recombinant Molecules: Impact on Science and Society* (Beers, B.F. and Bassett, E.G., eds.), pp.69-80.
14. Ehrlich, S.D., Niaudet, B. and Michel, B. (1982) *Curr. Top. Micro. Immun.* **96**, 19-29.
15. Fujii, M. and Sakaguchi, K. (1980) *Gene* **12**, 95-102.
16. Ferrari, E., Henner, D.J. and Hoch, J.A. (1981) *J. Bacteriol.* **146**, 430.432.
17. Goebel, W., Kreft, J. and Nurger, K.J. (1979) in *Plasmids of Medical Environmental and Commercial Importance* (Timmis, K. and Puhler, A., eds.), pp.471-480.
18. Goldfarb, D.S., Doi, R.H. and Rodriguez, R.L. (1981) *Nature* **293**, 309-311.
19. Gray, O. and Chang, S. (1981) *J. Bacteriol.* **145**, 422-428.
20. Grandi, G., Mottes, M. and Sgaramella, V. (1981) *Plasmid* **6**, 99-111.
21. Gryczan, T., Shivakumar, A.G. and Dubnau, D. (1980) *J. Bacteriol.* **141**, 246-253.
22. Gryczan, R. and Dubnau, D. (1978) *Proc. Natl. Acad. Sci. USA* **75**, 1422-1428.
23. Hardy, K., Stahl, S. and Kupper, H. (1981) *Nature* **293**, 309-311.
24. Horinouchi, S. and Weisblum, B. (1982a) *J. Bacteriol.* **150**, 804-814.
25. Horinouchi, S. and Weisblum, B. (1982b) *J. Bacteriol.* **150**, 815-824.
26. Hutchison, K.W. and Halvorson, H.O. (1980) *Gene* **8**, 267-278.
27. Hutchison, K.W., Sachter, K. and Halvorson, H.O. (1981) in *Sporulation and Germination* A.L. and Tipper, D.J., eds.), pp.
28. Imanaka, T., Tanaka, T., Tsunekawa, H. and Aiba, S. (1981) *J. Bacteriol.* **147**, 776-786.
29. Jayaraman, K., Keryer, E. and Szulmajster, J. (1981) *FEBS Letts.* **10**, 273-277.
30. Kawamura, F., Saito, H., Hirochika, H.M. and Kobayashi, Y. (1980) *J. Gen. Appl. Microbiol.* **26**, 345-355.
31. Keggins, K.M., Lovett, P.S., Duvall, E. (1978) *Proc. Natl. Acad. Sci. USA* **75**, 1423-1427.
32. Kobayashi, Y., Hirochika, H., Kawamura, R. and Saito, H. (1980) in *Sporulation and Germination* Sonenshein, A.L. and Tipper, D.J., eds.), pp.114-118.
33. Kreft, J., Bernhard, K. and Goebel, W. (1978) *Mol. Gen. Genet.* **162**, 59-67.
34. Kreft, J. and Hughes, C. (1982) *Curr. Topics Microbiol. Immun.* **96**, 1-17.
35. Kroyer, J. and Chang, S. (1981) *Gene* **15**, 343-347.
36. Mahler, I. and Halvorson, H.O. (1977) *J. Bacteriol.* **131**, 374-377.

37. Lee, G., Talkington, C. and Pero, J. (1980) *J. Mol. Biol.* **152**, 407-422.
38. Lofdahl, S., Sjostrom, J.-E. and Philipson, L. (1978) *Gene* **3**, 161-172.
39. Losick, R. and Pero, J. (1981) *Cell* **25**, 582-584.
40. McDonald, K.O. and Burke, W.F. (1982) *J. Bacteriol.* **149**, 391-394.
41. McLaughlin, J.R., Chang, S.-Y. and Chang, S. (1982) *Nuc. Acids Res.* **10**, 3905-3919.
42. McLaughlin, J.R., Murray, C.L. and Rabinowitz, J.C. (1981a) *J. Biol. Chem.* **256**, 11283-11291.
43. McLaughlin, J.R., Murray, C.L. and Rabinowitz, J.C. (1981b) *Proc. Natl. Acad. Sci. USA* **78**, 4912-4916.
44. Moran, C.P., Lang, N. and Losick, R. (1981) *Nucleic Acids Res.* **9**, 5979-5990.
45. Moran, C.P., Lang, N., Banner, C., Haldenwang, W.G. and Losick, R. (1981) *Cell* **25**, 783-791.
46. Moran, C.P. Jr., Lang, N., LeGrice, S.F.J., Lee, G., Stephens, M., Sonenshein, A.L., Pero, J. and Losick, R. (1982) *Mol. Gen. Genet.* **186**, 339-346.
47. Moran, C.P. Jr., Losick, R. and Sonenshein, A.L. (1980) *J. Bacteriol.* **142**, 331-334.
48. Nagahari, K. and Sakaguchi, K. (1978) *Mol. Gen. Genet.* **157**, 263-290.
49. Ollington, J.R., Haldenwang, W.G., Huynh, T.V. and Losick, R. (1982) *J.Bacteriol.* **147**, 432-442.
50. Palva, I., Petterson, R.F., Kalkkinen, N., Lehlovaara, P., Sarvas, M., Soderlund, H., Taddinen, K. and Kaariainen, L. (1981) *Gene* **15**, 43-51.
51. Post, L.E., Strycharz, G.D., Nomura, M., Lewis, H. and Dennis, P.P. (1979) *Proc. Natl. Acad. Sci. USA* **76**, 1697-1701.
52. Primrose, S.B. and Ehrlich, S.D. (1981) *Plasmid* **6**, 193-201.
53. Rapoport, G., Klier, A., Fargette, R. and Dedonder, R. (1979) *Mol. Gen. Genet.* **176**, 239-245.
54. Rhaese, H.J., Groscurth, R., Amann, E., Kuhne, H. and Vetter, H. (1980) in *Sporulation and Germination* (Levinson, H.S., Sonenshein, A.L. and Tipper, D.J., eds.), pp. 134-137.
55. Rhaese, H.J., Groscurth, R., Vetter, H. and Gilbert, H. (1979) in *Regulation of Macromolecular Synthesis by Low Molecular Weight Mediators* (Koch, G. and Richter, D., eds.), pp.145-159.
56. Rosenbluh, A., Banner, C.D.B., Losick, R. and Fitz-James, P.C. (1981) *J. Bacteriol.* **148**, 341-351.
57. Rosenberg, M. and Court, D. (1979) *Ann. Rev. Genet.* **13**, 319-353.
58. Rubin, E.M., Wilson, G.A. and Young, F.E. (1980) *Gene* **10**, 227-235.

59. Schnepf, H. and Whiteley, H.R. (1981) *Proc. Natl. Acad. Sci. USA* **78**, 2893-2897.
60. Shine, J. and Delgarno, L. (1975) *Nature* **254**, 34-38.
61. Shorenshein, R.G. and Losick, R. (1973) *J. Biol. Chem.* **248**, 6170-6173.
62. Tanaka, T. (1979) *J. Bacteriol.* **139**, 775-782.
63. Tanaka, T. and Sakaguchi, K. (1979) *Mol. Gen. Genet.* **165**, 269-276.
64. Uhlen, M., Flock, J.-I. and Philipson, L. (1981) *Plasmid* **5**, 161-169.
65. Wain-Hobson, S., Nussinov, R., Brown, R.J. and Sussman, J.L. (1981) *Gene* **13**, 355-364.
66. Wiggs, J.L., Bush, J.W. and Chamberlain, M.J. (1979) *Cell* **16**, 97-109.
67. Williams, D.M., Schoner, R.G., Duvall, E.J., Preis, L.H. and Lovett, P.S. (1981) *Gene* **16**, 199-206.
68. Yanofsky, D., Platt, T., Crawford, I.P., Nichols, B.P., Cristie, G.E., Horowitz, H., Van Cleemput, M. and Wu, A.M. (1981) *Nuc. Acid Res.* **9**, 6647-6668.
69. Yoneda, Y., Graham, S. and Young, F.E. (1979) *Biochem. Biophys. Res. Commun.* **91**, 1556-1564.

BACILLUS SUBTILIS PROTOPLAST FUSION, NUCLEOIDS, AND CHROMOSOME INACTIVATION

P. SCHAEFFER and L. HIRSCHBEIN

Institut de Microbiologie associé au C.N.R.S., Université de Paris XI, F91405, Orsay, France

SUMMARY

Recombinants, previously considered haploid, are not the only stable products of *B. subtilis* protoplast fusion. Stable diploids, prototrophic by complementation, have been found. Their diploidy was established by analysing the transforming activities of their DNA. In other diploid clones one of the parental chromosomes is inactivated (nor normally transcribed). Consequently these non-complementing clones are parental phenotypically. Even auxotrophic recombinants were found to be diploid. Thus genetic inactivation, extended to a complete chromosome or not, accounts for the various phenotypes of diploid cells presumably identical genetically.

Studying the transforming activities of DNA and nucleoids extracted from the various clones has been essential to unravel this situation. Proteins, which by their association with nucleoids mask their transforming activities for unexpressed markers might also inactivate them *in vitro* as templates for transcription by changing their tertiary structures.

The relevance of bacterial fusion studies in other research fields is discussed.

INTRODUCTION

Polyethyleneglycol (PEG)-induced protoplast fusion is now known as a powerful method to obtain recombinant clones of gram-positive bacteria, particularly *Bacillus* and *Streptomyces* species. A very well documented and provocative review on various aspects of the subject has recently appeared (10). Fusion of *B. subtilis* protoplasts has recently made rapid progress, showing that at least a large fraction of the clones resulting from fusion are diploid and that genetic inactivation, rather than recombination, might account for their various phenotypes. The purpose of this review is to present the main outlines of the information that led to this unexpected new notion.

Chronologically the important steps that have led to this progress have been: the demonstration that among prototrophic clones produced by fusion many are stable diploids; the discovery among the fusion products of an unexpected new type of diploid clones (the biparentals) in which only one of the parental chromosomes is expressed, and the demonstration that, irrespective of their markers, many (at least) of the recombinants also are diploids. These three points are developed below. In addition, a major technical improvement was made when it was realized that, in order to establish the genetic structure of a given fusion product, a careful investigation of the properties of its nucleoids also should be made.

METHODS

As newcomers in fusion research, nucleoids could not be included in Hopwood's review (10), but another review by Hirschbein and Guillen has recently appeared, which covers the isolation, structure and properties of *B. subtilis* nucleoids (9). Among their features, the following are essential to the clarity of this text. From freshly prepared crude lysates of growing cells, bacterial chromosomes can be obtained in highly compact, presumably native forms known as nucleoids. In these complexes (easily purified since in sucrose gradients they sediment as a single peak) folded chromosomes are associated with membrane fragment(s), nascent RNA and proteins. Intact molecules from natural (haploid) strains contain practically all the RNA polymerase, and all the transforming activity of the original cell extract. Their structural integrity, as judged from their sedimentation rate for instance, is destroyed by exposure to RNase, or proteinase, but their transforming activity (sensitive of course to DNase) is unaffected by these treatments.

Concerning the methods previously described to produce fusion, to regenerate bacterial forms, to measure the frequency of recombinants, and of the initial fusion events, the reader is referred to Hopwood's review. Not included in the latter is the recent reference to a minimal regeneration (mR) medium, which unlike the previously used non-selective B medium can, with various supplementations, conveniently be made selective (21).

RESULTS

Stable Prototrophic Clones Produced by Fusion may be Diploid*

In the early work prototrophic clones once reisolated had been found to be stable, since no auxotrophic segregants could be detected in their broth cultures. It was postulated that the primary fusion products were unstable diploids, but that the subclones investigated as a rule were haploid recombinants (24,25). It could also be, however, that stable diploid subclones were present but, being rare, were overlooked. In order to be able to select for them fusions were made of parental strains carrying the same *rec*$^-$ mutation. This approach was a failure since all sorts of recombinants still appeared, including prototrophs, in numbers not significantly different from those obtained in *rec*$^+$ x *rec*$^+$ control crosses (15). When larger numbers of reisolated prototrophic colonies from these crosses were tested, blindly as before, a few (1-2%) diploids were found segregating various kinds of auxotrophic recombinants in their broth cultures. Subclones isolated from these however were often found to have a much greater stability. In order to check the presumed haploidy of these stabilized subclones, the transforming activities of purified DNA prepared from them were tested. By congression it could be shown that the activities of the nutritionally deficient alleles carried by the parents were present (15).

In conclusion heterozygous diploids are produced by fusion, which although unstable are capable of repeatedly going through cell division, and stabilized subclones appear in their progeny. These stabilized prototrophs are still diploid, as shown by the transforming activities of their DNA. The mechanism underlying stabilization is not known, and will be discussed below. The prototrophy of some at least of these clones must be due to functional complementations, which implies that in minimal medium the two apparently complete parental chromosomes carried by the cells are simultaneously expressed. For that reason they are called complementing diploid or Cd clones.

It is clear that, even in bacteria, genetic stability by itself is not sufficient to establish the haploid nature of the strain under study. No prototrophic clone produced by fusion has so far been shown to be truly haploid. Frequent as it is, diploidy may account for the fact that *rec*$^-$ mutations do not significantly affect prototroph formation.

---

* In the present context 'diploid' is used to mean 'containing at least one copy of each parental genome'.

Stable and unstable Cd clones are efficiently selected for when mR medium is used for cell regeneration.

Fusion and Chromosome Inactivation

One apparently minor technical change introduced by Hotchkiss and Gabor in the procedure previously adopted in fusion experiments has had far reaching consequences. The change was simply to introduce, prior to plating on the non-selective regeneration medium, a dilution step high enough (10^5 times or more) to obtain well isolated colonies of regenerated bacteria. The consequence has been the unexpected discovery of a new type of fusion product, initially named biparental (BP) by the authors (11,6). With heterologous fusion occurring nearly at 25% (4, 22), and wall regeneration at 100% (5), the BP clones have been found to reach 10% of the regenerated bacteria (11).

If for the sake of simplicity we call MO the unsupplemented glucose salts medium, A and B the parental types, and MA and MB their respective properly supplemented minimal media, BP cells will grow on MA and MB (and on MAB), but practically never on MO. As a rule they are unstable, however. They breed true to type, but also segregate phenotypically parental cells and various auxotrophic recombinants*. They are interpreted as being heterodiploids, differing from Cd cells by the fact that one (either one) of their parental chromosomes remains unexpressed (it can also be said to be inactive, silent, or turned off). With regard to a particular culture it is convenient to indicate at this point which chromosome is silent. This is shown in parentheses. Although it may not be true of every fusion experiment, in general after fusion as many A(B) colonies appear on MA as there are B(A) colonies appearing on MB. It is not excluded that a selective medium somehow determines which chromosome is inactivated. Whatever the nature of this apparent control the inability of BP cells, whether A(B) or B(A), to grow on MO medium must be ascribed to a lack of functional complementation (11,6). For that reason, and also to oppose BP clones to Cd clones, the designation of the former has been changed to Ncd (for non-complementing diploid).

Comparative studies of the transforming activities of both the purified DNA and the non-purified nucleoids (i.e. the crude lysate) from Ncd clones gave apparently decisive support to these views. Knowing the state, A(B) or B(A) of the Ncd

---

* Of belated appearance, the latter are said to be *secondary* recombinants, as opposed to the early *primary* recombinants, which already appear together with the BP clones on the non-selective regeneration medium.

donor, the experimenter may transform at will an approximate competent culture for markers which in the donor were expressed or unexpressed. When purified DNA is used, transformants are obtained for all markers, whether expressed or unexpressed (11). When the lysate is used however, the transformants for expressed markers outnumber those for unexpressed markers by a factor of nearly 100. Thus, a correlation is clearly revealed between expression *in vivo* on the one hand, and transforming activity *in vitro* on the other hand (7).

Treatments by RNase and protease being the two main steps in DNA purification from a cell lysate, we wondered whether both were required to recover in purified DNA the transforming activities which in lysates were so low. Experimentally, it was found that RNase treatment alone had hardly any effect, but that protease treatment increased the activity for unexpressed markers 20 to 50 times, nearly restoring it to that for expressed markers (1). Comments on this apparently straightforward response are given in the Discussion below.

Diploidy of Recombinants Resulting from Fusion

If recombinations are required after fusion to produce recombinants, they occur normally, as we have seen, in diploid cells that are homozygous for some *rec*$^-$ mutations at least (15). This unexpected observation promoted investigations to see whether other properties of the recombinations occurring in fused bacterial cells were normal or not.

Chromosomal location of the crossovers. Based on transduction data, a fairly detailed chromosomal map of *B. subtilis* is known (8). When the most frequent types of the recombinants after fusion were determined it became clear (6,23), that crossovers were more frequent in two chromosomal regions, one encompassing the origin, and the other the terminus of replication (the fact that chromosome attachment to the wall, and even to the membrane, is impaired in protoplasts has been invoked as possibly accounting for this anomaly).

Reciprocity of recombination. Reciprocal recombinant types have been observed, even in a single clone (6), but one of them is always in 100-fold excess at least (23). Thus, recombination in fused cells seems to be practically non-reciprocal, a finding unexpected in cells carrying at least two complete genomes. Observed with primary and secondary recombinants alike, these multiple oddities led to the suspicion that even stable recombinants might not be haploid.

Ploidy level of recombinants. In this search for a possible diploidy one particular type of secondary recombinants, phenotypically $Spo^{+}Trp^{-}$, has so far been carefully studied. The way such strains were obtained deserves a digression.

In previous work it was claimed that *spoOA* mutations in *spoOA/spoOA*$^{+}$ mero-diploids are dominant over their wild type allele (12), and a subsequent paper (based on arguments which to us seemed irrelevant) challenged this conclusion (27). When fusion was carried out with a pair of triple auxotrophs, one of which carrying a *spoOA* mutation, the majority (70%) of the prototrophic clones subsequently isolated on selective mR medium were *Spo*$^{-}$, and segregated *Spo*$^{+}$ clones (21). This evidence is not sufficient to demonstrate the disputed dominance, however, when the possibility of genetic extinction is taken into account (23). Any *Spo*$^{+}$ segregant appearing in such *Spo*$^{-}$ Cd clone, even very rarely, should be selected from a heated culture in broth. Some were indeed isolated (21), but they no longer were prototrophic: surprisingly, 95% of these secondary recombinants were Trp^{-} monoauxotrophs (23). These were the Spo^{+} Trp^{-} recombinants mentioned above, the presumed haploidy of which was then investigated. DNA extracted from several of them was purified and used in transformation experiments. The conclusion drawn was, in each case and in spite of their phenotype, that the donor cells were heterodiploid for the six genes marked by mutation in the parents. Thus, whether Dc, Ncd or recombinant, all the clones appearing after fusion seem to be diploids. In addition, since the dominant *trp*$^{+}$ allele is present in a Trp^{-} recombinant, it must in these cells be inactivated. Possible genetic structures for the various fusion products, and the currently discussed hypotheses on the molecular mechanism of inactivation are presented below.

DISCUSSION

It has been generally accepted (24,10) that the primary diploid protoplasts contained at least one copy of the parental chromosomes, present as distinct replicons. From these protoplasts to the final bacterial colonies, wall regeneration, chromosome replication and cell division must take place, with or without segregation or recombination of the parental genomes. But the order and precise timing of these events occurring on the plate, as well as their synchrony in the individual cells of clone are not known. Primary colonies are generally heterogeneous but, sooner or later, stable recombinant subclones can be isolated, which were thought to be haploid (24, 10).

Now we see (and this does not say that haploids never form) that the diploid state may be stable at least in Cd and recombinant cells, and surprisingly is compatible with the various phenotypes encountered. This is made possible by genetic inactivation, a process first described in procaryotes by Hotchkiss and Gabor (11). In the frequently occurring Ncd clones discovered by these authors one complete parental chromosome apparently was inactivated, this inactivation preventing any functional complementation between the two genomes and incidentally, the determination of dominance relationships. Simultaneous inactivation of the two chromosomes, if it occurs at all, would obviously be lethal, but both of them may (rarely) escape inactivation, allowing complementation to take place, and the formation of prototrophic Cd clones. Lastly, in diploid recombinants inactivation may be either restricted to a mere chromosomal segment, or else affect a certain type of complete recombinant chromosome (23, and unpublished results of Lévi-Meyrueis and Sanchez-Rivas).

Genetic structures of the various fusion products. The schematic representations of their possible genetic structures, given in Fig. 1, may require some additional comments. The two complete and expressed parental chromosomes in Cd clones (II, centre) are represented as independent replicons. It is possible that this structure is that of unstable prototrophs only, but our ignorance of the nature of the stabilizing event does not permit the drawing of a structure for stabilized clones.

Of the three structures envisaged in III for the $Spo^{+}Trp^{-}$ recombinants, the haploid one (to the left) was not supported by the results of transformation analysis. If they occur at all haploid recombinants, of this type at least, must be rare. Strains are being built which may soon permit the choice to be made between the other two structures (III, centre and right).

Mechanism of chromosomal inactivation in bacteria. Nucleoids from Ncd strains have been studied soon after the strains were described (7). Unlike the nucleoids from the haploid parents or their mixture, which in a sucrose gradient sediment in a single peak, those isolated from Ncd clones give two clearly separated peaks, equivalent in amount of DNA (the different sedimentation rates of these peaks could be due to different amounts of membrane material associated with the chromosomes, or to different levels of compaction of DNA in the latter (7)). Since one chromosome in Ncd cells is unexpressed, its capacity to synthesise mRNA was investigated in the following way. The B parental strain, lysogenized with the phage $\phi105$, was fused with A to produce $A(B_{\phi105})$ and $B_{\phi105}(A)$ cultures. Both before and after prophage induction, mRNA fractions were isolated from both cultures, and hybridized with

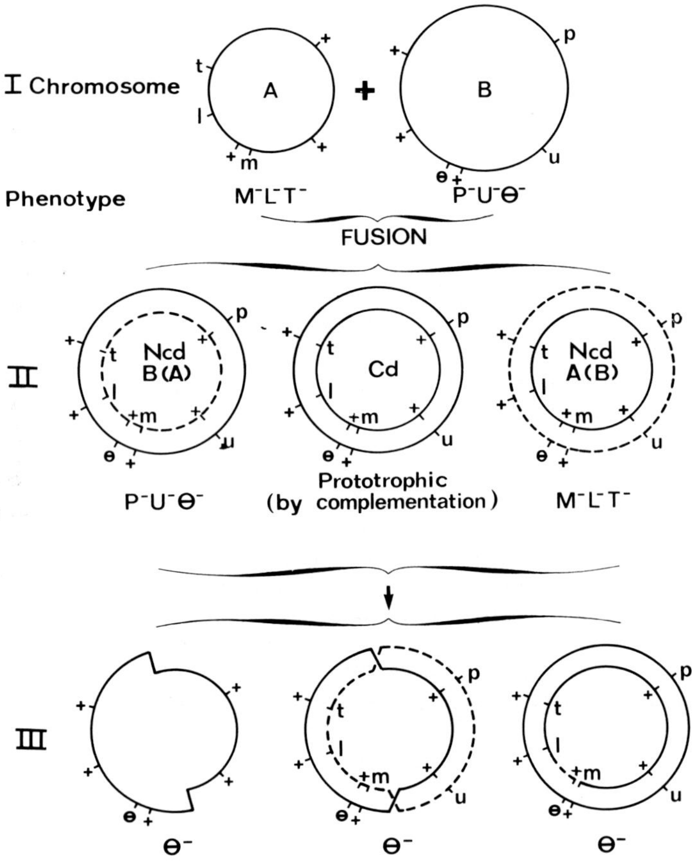

FIG. 1 *Possible genetic structures of the various fusion products. Circles stand for chromosomes (their sizes were made unequal for convenience)*
I. Parental strains. *Parent A is* metB, leu, thr, *parent B* purBm ura, trpF. *The mutant alleles are represented by small letters (m, l, t, p, u and θ), the wild type alleles by +. In capital letters, the corresponding phenotypes.*
II: Diploid fusion products. *Cd, complementing diploid (prototrophic), Ncd, non0complementing diploids (parental phenotypically).*
III: Possible structures for $\bar{\theta}$ recombinants. *Left, the rejected haploid structure. Centre, diploid structure with recombination close to 0 and T (located at 0 and 6 h respectively), and inactivation of a complete recombinant chromosome. Right, diploid structure without recombination, and with inactivation restricted to a segment of chromosome A, encompassing the* $tryF^+$ *allele.*

denatured labelled phage DNA. Significant radioactive hybrid formation was only observed when the prophage was inserted in the active chromosome. Thus in the cell, chromosome inactivation occurs at the level of transcription (Guillen *et al.* 7a). Since the prophage integrated in the silent chromosome does not synthesise repressor, the maintenance of the lysogenic state might be secured by a modified superstructure of the inactivated nucleoid.

Gene expression being correlated with undermethylation of DNA (19), the possibility has been considered, that chromosome inactivation be due to base methylation. The latter being known to lower the buoyant density of DNA (26), it was anticipated that total Ncd DNA might reveal its heterogeneity and band as two peaks in neutral CsCl gradients. This was not the case, however, and inactivation was deemed unlikely to be due to an increased methylation level (7). Since DNA itself does not seem to be modified in inactive chromosomes, what else could be responsible for inactivation? Although the final answer to this question cannot be given yet, an hypothesis can be made based on the fact that the ability of a crude lysate of Ncd cells to transform for unexpressed markers is restored by a protease treatment (1). Clearly, proteins associated with nucleoids are masking this transforming activity. Ascribing the masking effect of this association to a change in the tertiary structure of the nucleoids, we would like to suggest that this changed conformation is also, *in vivo*, inhibiting transcription (1). One difficulty in formulating the hypothesis more precisely at this time is, that two kinds of nucleoids are present in the lysate (7), and the reactivation experiment has so far not been repeated with each kind separately.

Relevance of Bacterial Fusion Studies in Other Research Fields

Applied bacteriology. The industrial applications of bacterial protoplast fusion have recently been reviewed most competently by Hopwood (10). For all we know it does not seem to matter much, as long as they are stable, that performant strains produced in this way result from diploidization and genetic inactivation rather than from genetic recombination.

Spore formation. Fusion has quickly been put to good use as a tool to analyse the sporulation process in Mandelstam's laboratory (2,3). This is no place for us to relate the results, but perhaps it is to say that they could not be obtained without resorting to fusion. Since simultaneous expression of the two parental genomes, and the ensuring complementation seem to be the rule in fused protoplasts that were not allowed to revert to vegetative bacteria (22), the notions on genetic inactivation presented here should be irrelevant in the type of experiments described.

Chromosomal and subchromosomal inactivation in eucaryotes. If studies on bacterial fusion may be helpful in some fields, they may have to borrow from other fields like membrane biology, genetics of somatic cells, tertiary structure of DNA, chromosome inactivation, and biological evolution. Unprepared to comment on the first three we will dwell briefly on the fourth, and make a few remarks on the fifth.

Several cases of inactivation, or better unresponsiveness, of entire chromosomes or their parts are known in various higher organisms such as mammals, insects and plants. The most extensively studied case is that of the X chromosome in somatic cells of female embryos and adults of man and mouse (17). Sometimes called 'lyonisation' in honour of M.F. Lyon who discovered it, this phenomenon affects one (either one) of the two X chromosomes present in those cells. One of its most studied aspects is the molecular mechanism of genetic extinction. In mammals it seems that inactivation of the X chromosome is due to methylation of its DNA (18,24), and that purified DNA of the X chromosome loses, by inactivation, its transforming activity, for some at least of the genes it carries (16,14). Thus on both points the results with mammalian cells seem to be at variance with those obtained with bacteria.

The evidence concerning methylation, however, was with mammalian cells derived exclusively from the use of inhibitors, which may have spurious effects (14), and with bacteria from attempts to detect DNA methylation by its effect on DNA density (7). This method may not discriminate among the various methylation patterns. The best evidence that no chemical change of DNA takes place during chromosome inactivation in bacteria is the restoration of the transforming activities missing in nucleoids by their protease treatment (1). It would thus seem that the inactivation mechanism has changed during the course of cellular evolution, and one would like to know when, perhaps by studying primitive unicellular eucaryotes. It remains that bacteria, the stronghold of the haploid way of life, already have the potential to inactivate a redundant chromosome (just in case?).

As to the differential transforming activity of intact Ncd nucleoids, which we account for by a conformational change in the nucleoid complex due to association with protein(s), it has so far no counterpart with somatic cells. If intact X chromosomes, both in their active and inactive forms, can be purified, they might (before and after proteinase treatment) be compared as donors in the transformation of animal cells for various markers.

Whatever the material, a possible role in genetic inactivation of the transition of DNA from the familiar B form to the left-handed Z form (20) seems to be an interesting speculation. Putting aside the technical difficulties, testing for it seems premature as long as its occurrence in cells can be disputed.

Fusion and cell evolution. For a number of reasons the organelles of eucaryotic cells are suspected to derive from free-living procaryotes which, having entered unknown primitive eucaryotes, lost part of their genetic autonomy and became endosymbionts. If it occurs spontaneously, fusion is *A priori* no more objectionable that phagocytosis as an initial mode of entry. An occasional reader may appreciate the reminder that spontaneous fusion does occur, as it has recently been confirmed (11,13). Artificial fusogenic treatments are presently much in use only because they conveniently increase (some 100 fold) the spontaneous fusion rate. A non-toxic haploidizing agent active on bacteria, when it becomes available, might make it possible to determine the proportion of diploid bacteria in natural populations.

ACKNOWLEDGEMENTS

Our thanks are due to R.D. Hotchkiss and M.H. Gabor for stimulating discussions and exchange of unpublished information, and to the Centre National de la Recherche Scientifique (contrat L.A. no. 136), and the Fondation pour la Recherche Médicale for their financial support.

REFERENCES

1. Bohin, J.P., Ben Khalifa, K., Guillen, N., Schaeffer, P. and Hirschbein, L. (1982) *Mol. Gen. Genet.* **185**, 65-68.
2. Dancer, B.N. (1981) *J. Gen. Microbiol.* **126**, 29-36.
3. Dancer, B.N. and Mandelstam, F. (1981) *J. Gen. Microbiol.* **123**, 17-26.
4. Frehel, C., Lheritier, A.M., Sanchez-Rivas, C. and Schaeffer, P. (1979) *J. Bacteriol.* **137**, 1354-1361.
5. Gabor, M.H. and Hotchkiss, R.D. (1979) *J. Bacteriol.* **137**, 1346-1353.
6. Gabor, M.H. and Hotchkiss, R.D. (1982) in *Genetic Cellular Technology* (Streips, E.N., Goodgal, S. and Guild, W.R., eds.), Vol.1, 283-292. Decker, M., New York.
7. Guillen, N., Gabor, M.H., Hotchkiss, R.D. and Hirschbein, L. (1982) *Mol. Gen. Genet.* **185**, 69-74.

7a Guillen, N., Sanchez-Rivas, C. and Hirschbein, L. (1983) *Mol. Gen. Genet.* **191**, 81-85.

8. Henner, D.J. and Hoch, J.A. (1980) *Microbiol. Rev.* **44**, 57-82.
9. Hirschbein, L. and Guillen, N. (1981) in *Methods of Biol chemical Analysis* (Glick, D., ed.), Vol. 28, pp.297-328. John Wiley and Sons, New York.
10. Hopwood, D.A. (1981) *Annu. Rev. Microbiol.* **35**, 237-272.
11. Hotchkiss, R.D. and Gabor, M.H. (1980) *Proc. Natl. Acad. Sci. USA* **77**, 3553-3557.
12. Karmazyn, C., Anagnostopoulos, C. and Schaeffer, P. (1972) in *Spores V* (Halvorsen, H.O., Hanson, R. and Campbell, L.L., eds.), pp.126-132, Amer. Soc. Microbiol, Washington DC.
13. Landman, O.E. and Pepin, R.A. (1982) in *Molecular Cloning and Gene Regulation in* Bacilli (Ganesan, A.T., Chang, S. and Hoch, J.A., eds.), pp.25-39, Academic Press, New York.
14. Lester, S.C., Korn, N.J. and De Mars, R. (1982) *Somat. Cell Genet.* **8**, 265-288.
15. Levi-Meyrueis, C., Sanchez-Rivas, C. and Schaeffer, P. (1980) *C.R. Acad. Sci. Paris* **291**, (D), 67-70.
16. Liskay, R. and Evans, R. (1980) *Proc. Natl. Acad. Sci. USA* **77**, 4895-4898.
17. Lyon, M.F. (1968) *Annu. Rev. Genet.* **2**, 31-52.
18. Mohandas, T., Sparkes, R.S. and Shapiro, L.J. (1981) *Science* **211**, 393-396.
19. Naveh-Many, T. and Cedar, H. (1981) *Proc. Natl. Acad. Sci. USA* **78**, 4246-4250.
20. Nordheim, A., Pardue, L.M., Laser, E.M., Moller, A., Stollar, B.D. and Rich, A. (1981) *Nature* **294**, 417-422.
21. Sanchez-Rivas, C. (1982) *Mol. Gen. Genet.* **185**, 329-333.
22. Sanchez-Rivas, C. and Garro, A.J. (1979) *J. Bacteriol.* **137**, 1340-1345.
23. Sanchez-Rivas, C., Levi-Meyrueis, C., Lazard-Monier, F. and Schaeffer, P. (1982) *Mol. Gen. Genet.* **188**, 272-278.
24. Schaeffer, P., Cami, B. and Hotchkiss, R.D. (1976) *Proc. Natl. Acad. Sci. USA* **73**, 2151-2155.
25. Schaeffer, P. and Hotchkiss, R.D. (1978) in *Methods in Cell Biology* (Prescott, D., ed.), Vol. 20, pp.149-158. Academic Press, New York.
26. Szybalski, W. and Szybalski, E. (1971) in *Procedures in Nucleic Acid Research* (Cantoni, G.L. and Davies, D.R., eds.), Vol. 2, p.338.
27. Trowsdale, J., Chen, S.M.H. and Hoch, J.A. (1978) *J. Bacteriol.* **135**, 99-113.

THE GENETICS OF SPORE GERMINATION IN *BACILLUS SUBTILIS*

A. MOIR and D.A. SMITH

Microbiology Department, University of Sheffield, U.K.
and Genetics Department, University of Birmingham, U.K.

SUMMARY

Mutants in which spore germination is interrupted have been used to identify loci (*gerA-K*) on the *B. subtilis* chromosome involved in spore germination. In most mutants the recognised defect is only in spore germination but a few possess metabolic defects and other alterations in spore structure and resistance. Cloning *ger* loci could aid identification of the nature, function and regulation of germination gene products. Since the spore germination phenotype is not generally selectable, cloning closely linked selectable markers was attempted. For example, λ vectors which carry the *citG* (fumarase) gene of *B. subtilis*, formed complementing plaques on an *E. coli* fumarase mutant and carried the closely linked *gerA* locus, mutations in which affect specifically the spore germination response to L-alanine and related amino acids. The same procedure is being applied to auxotrophic markers near other *ger* genes.

INTRODUCTION

Much has been learned of the stimuli which will induce spores of *Bacillus spp* to germinate (8,5,35) and many of the changes occurring in the spore during germination have been described at morphological and biochemical levels (7,25,26). The nature of the spore component(s) with which the germinant interacts, and the immediate consequences of this interaction at the molecular level, remain to be identified. Suggested mediators of the germination response have included membrane-associated receptor molecules (30,10,21) or lytic enzymes (27,2,12,14).

but no germination receptor has been unequivocally identified, and the question of how a germinant acts to stimulate germination is still an open one. Other articles in these proceedings examine hypotheses for the mechanism of spore germination in more detail; the aim of this article is to review the continuing development of the genetic analysis of spore germination and to discuss the contribution that such a genetic study might make to our understanding of the biological processes which occur during the breakage of dormancy in bacterial spores.

The term germination is used here to include events occurring from the addition of germinant up to, but not including, the resumption of major biosynthetic activity. Germination is part of an integrated cycle of differentiation, and cannot be considered entirely in isolation.

The Role of Genetics in Studying Germination

Mutants altered in spore germination properties have been obtained in a variety of species of *Bacilli* and *Clostridia* (29,34,28,17,36), and they have extended the description of germination identifying, for example, possible related groups of germinants (24) and stages in germination (15,33). With a few notable exceptions, discussed later, no gross changes in spore structure or composition have been detected in these mutants. Comparisons between mutant and wild type spores need to be extended to the identification of gene products altered or absent in the mutant. It may not always be easy to establish whether any altered product observed is the primary protein product of the mutated gene and whether it has a direct or indirect role in germination.

The choice of *B. subtilis* 168 as an experimental organism permits use of its well-characterised systems of genetic exchange to classify mutations according to their position on the genetic map as well as by phenotype. Such mapping reveals the relationship, if any, between the mutation responsible for a particular defect and those affecting the same or different parts of the differentiation cycle, locating genes or groups of genes whose products are necessary to progress through particular stages of the cycle. Using the techniques of bacterial genetics and molecular cloning, the genes concerned can be isolated perhaps permitting *in vitro* and *in vivo* identification of their protein products.

Types of Mutants Available - Their Interpretation and Potential Exploitation

Enrichment procedures for spores of *B. subtilis* which fail to lose resistance properties in response to concentrations of germinant sufficient to germinate those of wild type, have led to the isolation of a variety of types of germination mutant. To those described by Piggot *et al.* (19), several new groups can be added, and their phenotypes and map locations compared with the earlier mutants.

Mutants in some *ger* loci affect germination in some germinants without altering the response in others. These include *gerA* and *gerC* loci, important for germination in L-alanine and analogous amino acids, and *gerB* and *fruB*, for response to a combination of amino acid and sugars. A recent addition to the latter category is *gerK* (9); mutants in this locus, which maps between *aroI* and *dal*, behave similarly to *gerB* mutants.

Amongst the types of proteins whose function would be likely to be specific to the germinant employed would be those which act as receptors for the primary and critical interaction with germinants which triggers the subsequent sequence of germination events. While *gerB* and/or *K* products might interact with components of the amino acid + sugars mixture, *gerA* might encode, say, an alanine-sensitive receptor in the spore. Detailed studies of *gerA* mutants are consistent with this possible interpretation (24).

The *gerD, E, F, H* and *J* and *spoVIA* and *spoVIB* mutants comprise a heterogeneous group whose response both to amino acids and to amino acid + sugar combinations is altered. Genes in which these mutations occur have a role important in the successful establishment or functioning of the germination apparatus. Their products could act directly in germination, either as common elements of the different triggering systems, or as common mediators, transducing the initial specific response into the subsequent sequence of germination changes or be themselves components of this sequence. Alternatively, their products might act indirectly, regulating either the expression or the assembly of germination-associated proteins in the sporulating cell.

Structural defects have been identified in spores of *gerE, spoVIA* and *spoVIB* and *gerJ* mutants. *SpoVIA* and *B* mutants (H.F. Jenkinson and J. Mandelstam, this volume) are germination-defective and lysozyme-sensitive and their spore coats are altered. At least for the *spoVIA* mutant, an intermediate level of loss of absorbance reflects germination of only a proportion of the spores in L-alanine; there is no evidence of any loss of resistance properties in the proportion of spores which does not germinate (11). This behaviour contrasts with that of the *gerE* mutant, grossly defective in coat structure

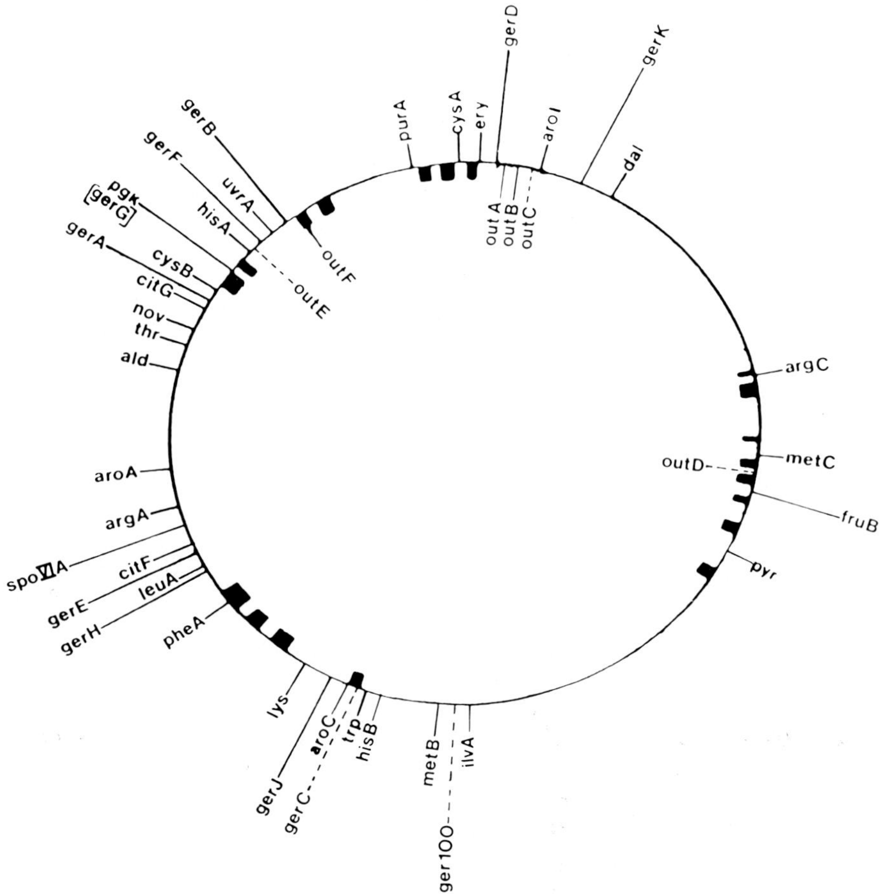

FIG. 1 *Abbreviated genetic map of* B. subtilis *168 to indicate the location of* ger *mutations. The blocks on the inside of the circle indicate the main location of sporulation (*spo*) mutations; out = outgrowth mutations; --- = approximate positions.*

(15), spores of which proceed through loss of heat resistance but are blocked at the phase-grey stage.

The *gerJ* mutations, like *gerE*, block germination at an intermediate phase-grey stage (33). No alteration in spore coat structure has been detected, but the spores are abnormally heat-sensitive, and the progress of development of resistances during sporulation is delayed (32).

It may be very difficult to identify the particular characteristic of these abnormal spores whose absence or malfunction leads to a germination defect, as the mutations may well have pleiotropic effects. They are equally validly described as late sporulation mutations, and their study is relevant to both sporulation and germination stages of the differentiation cycle.

Finally, a new class of germination mutant, in which germination of L-alanine is less sensitive to barbital has recently been described (R. Morse and D.A. Smith, these proceedings). This mutant, *ger-100*, establishes another category of altered germination response, and might provide a way to examine membrane-associated events. Whether the increased resistance is specific to germination remains to be determined.

The *ger* loci do not appear to be clustered near recognised sporulation (*spo*) or outgrowth (*out*) loci (Fig. 1). Until the results of complementation tests are available, it is not possible to determine whether a locus comprises one or more genes. The current list of loci is by no means exhaustive — different mutant isolation strategies will no doubt extend it further.

The dormant spore already possesses the mechanism which awaits stimulation to terminate dormancy. Developing spores during the late stages of sporulation have already acquired particular germination characteristics (4), while inhibitors of macromolecular synthesis do not interfere with the initiation or progress of germination (13). Thus proteins required for germination must be present in the mature spore and are almost certainly elaborated during sporulation. The *ger* genes are therefore probably under sporulation control; study of their regulation may contribute to the understanding of the orchestration of gene expression during spore development, just as identification of their gene products could direct the focus of biochemical studies on triggering and germination.

Cloning of the Fumarase Gene and *gerA* Locus

The isolation of DNA including the recognised *ger* loci is currently being attempted. The fact that the germination phenotype cannot be selected has hindered the direct screening of DNA in libraries for the presence of *ger* genes. The approach adopted instead has been to isolate genes represented by auxotrophic markers that map close to *ger* genes; the nearby *ger* gene may be obtained simultaneously, or may be sought in gene libraries using the adjacent cloned DNA as a hybridisation probe. Genes which could be used include *citG*, 60-80% cotransformed with *gerA* mutations (24), *citF*, approximately 90% cotransduced with *gerE* by SPP1 (15), those of the *ilv-leu* cluster, cotransduced at approximately 20-30% with flanking *gerE* and *gerH* loci (A. Moir and I. Roberts, unpublished) and *hisA* approximately 55% cotransduced with *gerF* by SPP1 (31).

We have sought these genes in a library of *B. subtilis* DNA, partially digested with EcoRI and cloned in the λ phage vector λ*gtWES* (the library supplied by Drs. Ferrari, Henner and Hoch is described in more detail in reference 16). λ*gtWES* is a disabled derivative of λ, which can neither form plaques on a suppressor-free strain nor integrate into the host chromosome to form a lysogen. A helper phage, $\lambda imm^{434}R^-$, can complement the defects in essential genes and, itself integration-

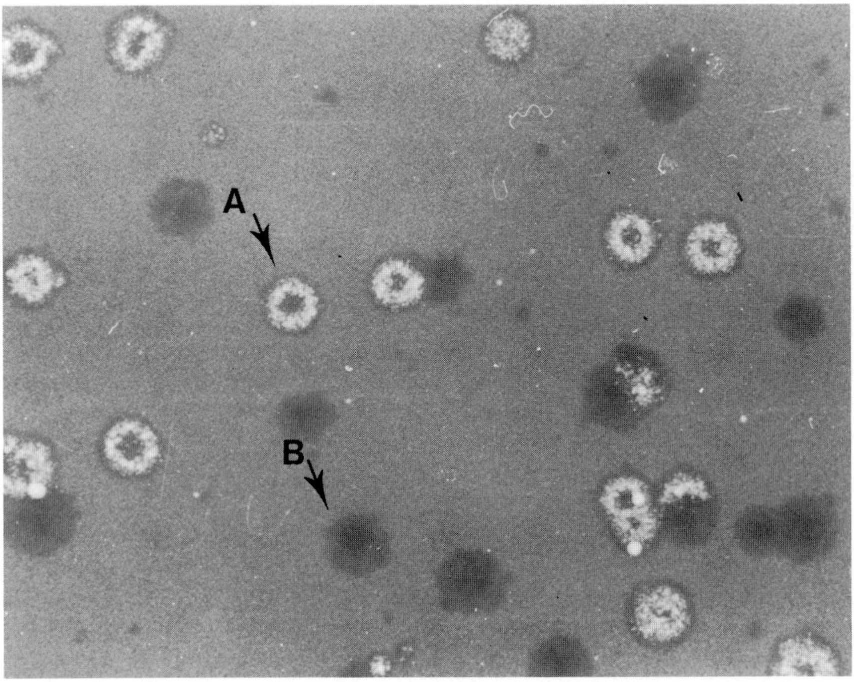

FIG. 2 *Complementation of an* E. coli *fumarase defect by phages carrying* B. subtilis citG. *The photograph shows the surface of an agar plate which represents a stage in the purification of* λBScitG *phages. The plaques were generated in a lawn of a fumarase*$^-$ *strain by cells coinfected with* $\lambda imm^{434}R^-$ *and a partially purified suspension of the* λcit *transducing phage. Examples of complementing (1) and non-complementing (B) plaques are arrowed.*

proficient, provide homology for *rec*-promoted integration of λ*gtWES* to form a double lysogen.

Complementation in λ plaques was described by Franklin (6) and has been used extensively to detect cloned *E. coli* genes (1). Attempts to select clones carrying the *B. subtilis* genes were made on the basis of their putative ability to complement equivalent defects in *E. coli* (as already established for *leuC*, 18,3).

Using this approach, phages carrying *citG*, the *B. subtilis* structural gene encoding fumarase, have been isolated (16). Complementing plaques containing double lysogens of helper and isolate λ*BScitGI* are seen in Fig. 2. λ*BScitGI* contained EcoRI fragments of approximately 1.5 and 5 kilobase pairs, whereas λ*BScitGII* carried these and two additional small EcoRI fragments of 0.9 and 0.8 kbp (16).

A very large number of *gerA* mutants have been isolated. SPPI cotransduction frequencies with *citG4* (70-90% (24)) suggest that the *gerA* locus could comprise one or two genes separated by one or two genes at most from the *citG* locus. The *gerA1* mutation was chosen as a test marker because it was amongst those which gave the lowest of the range of cotransduction frequencies, and therefore, might be amongst the furthest from *citG*. Transformation of a *citG4 gerA1* double mutant showed that *gerA1*$^+$ is cotransduced with *citG4*$^+$ using either of the two phages as donor, although the linkage of *gerA* and *citG4* using these phages as donor is lower than that observed using chromosomal DNA (Table 1). A third genetic

TABLE 1 *Cotransformation of* gerA *with* citG *using cloned DNA as donor.*

	TRANSFORMANTS			
DONOR	Nos./ml	Nos. Cit$^+$Ger$^+$	Total tested	%Cit$^+$Ger$^+$
Wild-type chromosomal DNA (0.2μg)	1.2×10^3	32	46	70
λ*BScitGI* (2×10^9pfu≡0.1μg)	4×10^3	22	53	41
λ*BScitGII* (2×10^9pfu≡0.1μg)	6×10^3	19	41	46

Chromosomal DNA or λ*cit* phage preparations were used as donors to transform a *citG4 gerA1 trpC2* strain to Cit$^+$ on minimal lactate medium (22). Germination phenotype was scored using the tetrazolium overlay method (17).

marker on the cloned DNA is a site of chromosomal rearrangement in *trpE26* mutants shown to reside in the 1.5 kbp fragment (23).

The 5 and 1.5 kbp fragments have been recloned (A. Moir, unpublished) from *λBScitGI* into plasmid pHV33 which replicates in *E. coli* or *B. subtilis* (20). Only plasmids carrying both the 5 and 1.5 kpb fragments can complement a fumarase defect in either *B. subtilis* or *E. coli* although a plasmid carrying the 5 kbp fragment could repair *citG4* (and *gerA1*) in transformation of a *rec*$^+$ recipient. Thus the sequences required for expression of a functional fumarase protein span the junction of the two EcoRI fragments. Cleavage of the plasmid carrying the 5 kbp fragment with restriction endonuclease *Hind III* destroys the linkage of *gerA1* to *citG4*, this suggests that *gerA1* lies near the second *Hind III* target or on the opposite side of this target from *citG4* (Fig. 3). Preliminary experiments suggest that *gerA1* cannot be complemented in *trans* by the plasmid carrying the 5 kbp fragment. The exact location of *gerA* alleles, the number of genes comprising the *gerA* locus and whether the gene or gene cluster is completely contained in the cloned DNA can now be examined.

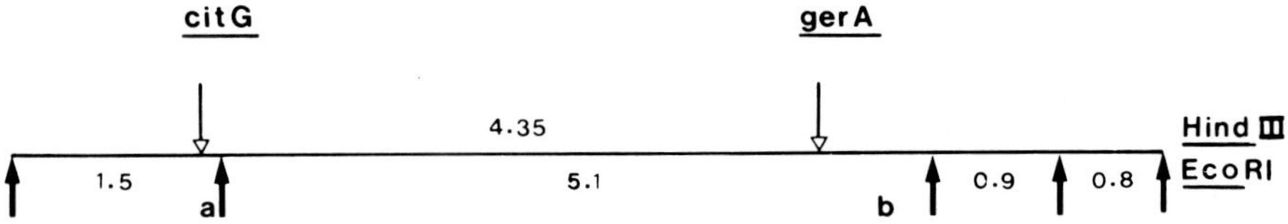

FIG. 3 *Correlation of genetic and physical map of the* citG – gerA *region. Arrows indicate restriction sites. Numbers refer to fragment sizes in kilobase pairs (kbp). Distances between* Hind III *and nearest* EcoRI *sites a = 0.15 and b = 0.9 kbp.*

Cloning of Genes near other *ger* Loci

Phages complementing an *E. coli hisB* mutation were isolated and shown to carry a single EcoRI fragment of approximately 8.2 kbp (D. Walton and R. Morse poster communication at this meeting). These complimented *E. coli hisA, B* and *D* defects therefore carrying genes encoding at least these functions, and repaired all six *B. subtilis hisA* alleles tested (*1,3, 10,11*). *GerF21*$^+$ could not, however, be detected in experiments seeking its cotransfer with *hisA1* from the cloned DNA and is probably not present in the fragment (D. Walton unpublished). Similarly, λ phages with inserts which together encompass the *ilv-leu* gene cluster have been isolated by

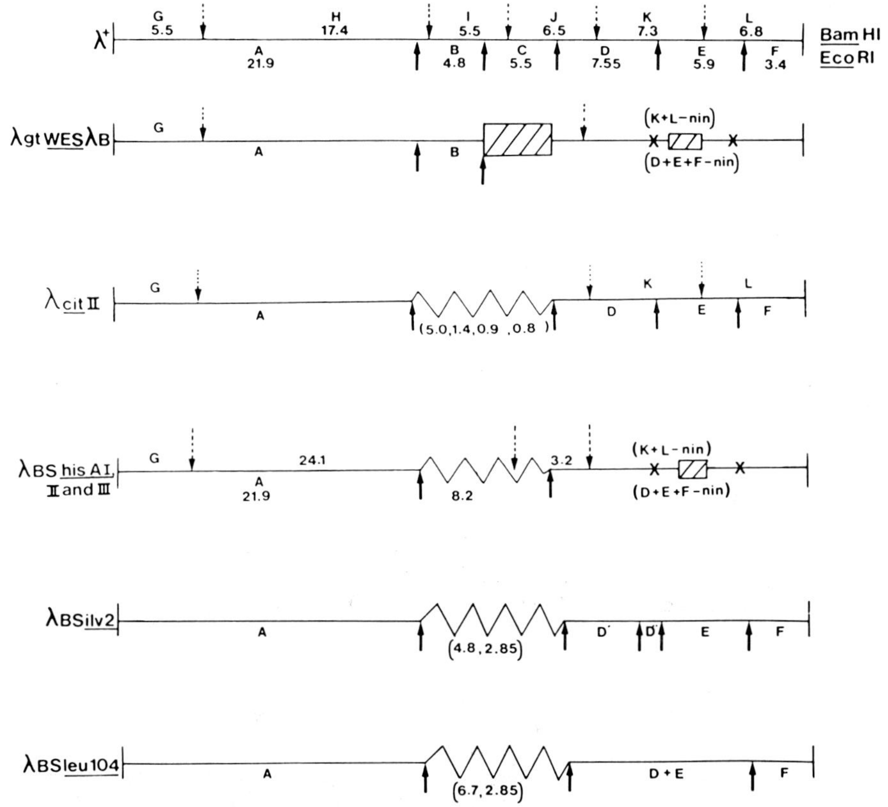

FIG. 4 *Approximate restriction maps of representative phages carrying* B. subtilis *DNA isolated by complementation of* E. coli *defects.*

▧ = *deleted region.* ⩘ = *inserted* B. subtilis *DNA*

complementation of *E. coli ilvC* and/or *leuB* functions. None of the phages so far tested appear to carry *gerE* or *gerH*, at least when examined for cotransformation of *ger* with *ilv* or *leu* genetic markers (A. Moir and I. Roberts, poster communication at this meeting and unpublished). Figure 4 summarises the restriction maps of a selection of the complementing phages isolated. As well as the study of *ger* genes, they provide material for the study of enzymes of metabolic pathways in *B. subtilis* and their regulation. *His* and *leu* phage DNAs are being used as hybridisation probes to locate *gerF*-, *E*- and *H*-carrying clones (R. Sammons, personal communication).

The complementation observed for cloned *B. subtilis* genes

in λ vectors varies from extremely to fairly weak. This may depend upon the promoters that can be used either from the phage or internally in the cloned DNA, and upon the potential level of function of the enzyme in *E. coli*, which would be expected to be higher for a free cytoplasmic enzyme than for say an enzyme component, which had to form part of a multi-enzyme complex, (*citF*, for example, encoding components of membrane-bound succinate dehydrogenase, has not been isolated by the complementation procedure). However, isolation of interesting regions of *Bacillus* DNA is now clearly possible. The next stage, identifying and locating *ger* genes, has begun. The third stage, examination of their products and regulation may, we hope, provide some insight into the mechanisms operating in establishment and operation of the germination apparatus.

ACKNOWLEDGEMENTS

The work described has been supported by the Science and Engineering Research Council in the form of grants (DAS) and a Research Fellowship (AM) and research studentships from both the SERC and the Medical Research Council.

REFERENCES

1. Borck, K., Beggs, J.D., Brammar, W.J., Hopkins, A.S. and Murray, N.E. (1976) *Mol. Gen. Genet.* **146**, 199-207.
2. Brown, W.C., Vellom, D., Ho, I., Mitchell, N. and McVay, P. (1982) *J. Bacteriol.* **149**, 969-976.
3. Chi, N-Y.W., Ehrlich, S.D. and Lederberg, J. (1978) *J. Bacteriol.* **133**, 816-821.
4. Dion, P. and Mandelstam, J. (1980) *J. Bacteriol.* **141**, 786-792.
5. Foerster, H.F. and Foster, J.W. (1966) *J. Bacteriol.* **91**, 1168-1177.
6. Franklin, N.C. (1971) in *The Bacteriophage Lambda* (Hershey, A.D., ed.), pp.621-638, Cold Spring Harbor Laboratory, New York.
7. Gould, G.W. (1969) in *The Bacterial Spore* (Gould, G.W. and Hurst, A., eds.), pp.397-444, Academic Press, London.
8. Hills, G.M. (1950) *J. Gen. Microbiol.* **4**, 38-47.
9. Irie, R., Okamoto, T. and Fujita, Y. (1982) *J. Gen. Appl. Microbiol.* **28**, 345-354.
10. Janoff, A.S., Coughlin, R.T., Racine, F.M., McGroarty, E.J. and Vary, J.C. (1979) *Biochem. Biophys. Res. Comm.* **89**, 565-570.
11. Jenkinson, H.F. (1981) *J. Gen. Microbiol.* **127**, 81-91.
12. Johnstone, K. and Ellar, D.J. (1982) *Biochim. Biophys.*

Acta **714**, 185-191.
13. Keynan, A. and Halvorson, H. (1965) in *Spores III* (Campbell, L.L. and Halvorson, H.O., eds.), pp.174-179. American Society for Microbiology, Washington D.C.
14. Labbe, R.G., Tang, S.S. and Francheschini, T.J. (1981) *Biochim. Biophys. Acta* **678**, 329.333.
15. Moir, A. (1981) *J. Bacteriol.* **146**, 1106-1116.
16. Moir, A. (1983) *J. Gen. Microbiol.* **129**,303-310.
17. Moir, A., Lafferty, E. and Smith, D.A. (1979) *J. Gen. Microbiol.* **111**, 165-180.
18. Nagahari, K. and Sakaguchi, K. (1978) *Mol. Gen. Genet.* **158**, 263-270.
19. Piggot, P.J., Moir, A. and Smith, D.A. (1981) in *Sporulation and Germination* (Levinson, H.S., Sonenshein, A.L. and Tipper, D.J., eds.), pp.29-39, American Society for Microbiology, Washington D.C.
20. Primrose, S.B. and Ehrlich, S.D. (1981) *Plasmid* **6**, 193-201.
21. Racine, F.M., Skomurski, J.F. and Vary, J.C. (1981) in *Sporulation and Germination* (Levinson, H.S., Sonenshein, A.L. and Tipper, D.J., eds.), pp.224-227, American Society for Microbiology, Washington D.C.
22. Rutberg, B. and Hoch, J.A. (1970) *J. Bacteriol.* **104**, 826-833.
23. Sammons, R.L. and Anagnostopoulos, C. (1982) *FEMS Lett.* **15**, 265-268.
24. Sammons, R.L., Moir, A. and Smith, D.A. (1981) *J. Gen. Microbiol.* **124**, 229-241.
25. Santo, L.Y. and Doi, R.H. (1974) *J. Bacteriol.* **120**, 475-481.
26. Setlow, P. (1981) in *Sporulation and Germination* (Levinson, H.S., Sonenshein, A.L. and Tipper, D.J., eds.), pp.13-28, American Society for Microbiology, Washington D.C.
27. Strange, R.E. and Dark, F.A. (1956) *Biochem. J.* **62**, 459-465.
28. Trowsdale, J. and Smith, D.A. (1975) *J. Bacteriol.* **123**, 83-95.
29. Vary, J.C. (1975) *J. Bacteriol.* **121**, 197-203.
30. Vary, J.C. (1978) in *Spores VII* (Chambliss, G. and Vary, J.C., eds.), pp.104-108, American Society for Microbiology, Washington D.C.
31. Warburg, R.J. (1981) Ph.D. Thesis, University of Birmingham.
32. Warburg, R.J. (1981) in *Sporulation and Germination* (Levinson, H.S., Sonenshein, A.L. and Tipper, D.J., eds.), pp. 98-100, American Society for Microbiology, Washington D.C.
33. Warburg, R.J. and Moir, A. (1981) *J. Gen. Microbiol.* **124**, 243-253.
34. Warren, S.C. (1969) *J. Gen. Microbiol.* **55**, xviii.

35. Woese, C.R., Morowitz, H.J. and Hutchinson, C.A. (1958) *J. Bacteriol.* **76**, 578-588.
36. Wyatt, L.R. and Waites, W.M. (1971) in *Spore Research* (Barker, A.N., Gould, G.W. and Wolf, J., eds.), Academic Press, London.

MUTATIONS IN *BACILLUS SUBTILIS* 168 AFFECTING THE INHIBITION OF SPORE GERMINATION BY A BARBITURATE

R. MORSE and D.A. SMITH

Genetics Department, University of Birmingham, P.O. Box 363, Birmingham B15 2TT, UK

SUMMARY

The barbiturate 5,5'-diethyl barbituric acid (barbital) at concentrations of 1-10 mM is a specific inhibitor of the germination of wild type *Bacillus subtilis* spores. Altering the temperature or decreasing the pH of the germination mixture increases the degree of inhibition as does raising the inhibitor concentration. However, raising the temperature of sporulation or including barbital in the sporulation medium both decrease the susceptibility of the resulting spores to inhibition. These results suggest that inhibition results from disruption of the spore membranes. Two mutants whose spore germination in both L-alanine and nutrient broth as well as on nutrient agar is less sensitive to barbital have been isolated. The site of mutation of one has been mapped close to *ilvA* and *metB* by phage SPP1 transduction and transformation.

INTRODUCTION

The transition from a dormant bacterial spore to an actively metabolizing vegetative cell encompasses both germination (12) and outgrowth (29). The former is a degradative process mainly characterized by the breakdown of peripheral spore layers, being induced by various compounds (14) including amino acids, sugars and nucleosides (or a combination of these). Outgrowth involves the synthesis of new macromolecules such as nucleic acids, protein and carbohydrate to produce a vegetative cell. Each of these processes is inhibited by a structurally diverse group of compounds (20,30). Those that inhibit outgrowth such as 2,4-dinitrophenol, iodoacetate and azide also usually affect vegetative growth (probably inhibiting both in similar ways). Germination inhibitors include 8-hydroxy-quinoline (20), 1,3-dimethylxanthine (8) methyl anthranilate (21) and barbital (26,27). Their mode of action is unknown although analysis of partition coefficients by Yasuda-Yasaki *et al.* (33) using regression analysis has implicated the hydrophobic regions of some of these compounds in the inhibition. To probe the underlying mechanism of spore germination we have used barbital and have isolated mutants whose spores are altered in a function(s) affected by it. Their phenotype contrasts with that of previously isolated spore germination (*ger*) mutants whose spores respond abnormally to specific germinants (17).

METHODS

Media

Spores were normally prepared on potato glucose yeast extract (10) solidified with 1.5% (w/v) Difco agar (PGYE agar). For the barbital resistance and tetrazolium test they were prepared on Difco nutrient agar (NA). Bacterial cultures were grown in Difco Penassay broth (PAB). Transductants and transformants were selected on minimal agar (MA) (1) containing 5 mg ml^{-1} glucose and supplemented with amino acids to 50 μg ml^{-1} as required. For the isolation of barbital resistant mutants and mapping experiments, barbital was added to NA before autoclaving to give a final concentration of 50 or 70 mM (at higher concentrations the barbital crystallized out).

Bacterial Strains

These are listed in Table 1. Strain 1604 was used as the wild-type.

Spore Preparation

Unless otherwise states spores were prepared at 37°C as previously described (16). For determining the effect on spores of temperature variation or presence of barbital during sporulation the resuspension method of Sterlini and Mandelstam (28) was used.

Germination Studies

Difco-nutrient broth (NB), or 1 mM alanine (ALA) in 10 mM-Tris/HCl (pH 7.0-9.0) were used as germinants, spores being added

TABLE 1 *Strains of* Bacillus subtilis *168*

Designation	Genotype	Origin
1506	*lys-1*	Spontaneous *trp*$^{+}$ revertant of GSY254 (C. Anagnostopoulos)
1604	*trpC2*	Moir *et al.* (17)
BD11	*thr-5 trpC2 cysB3*	D. Dubnau
BD34	*leuA8 thr-5 metB5*	P. Piggot
BD92	*hisA1 cysB3 trpC2*	C. Anagnostopoulos
BD112	*cysA14*	D. Dubnau
CU371	*trpC2 ilvBΔ1*	S. Zahler
GSY111	*ilvA1 trpC2*	C. Anagnostopoulos
MB21	*metC3 leuA8 tal-1*	P. Piggot
RW1	*aroB2 hisB2*	*trp*$^{+}$ transductant from SPP1 cross 1604 x SB137
SB120	*trpC2 aroD120*	J. Hoch

to give a 3 ml suspension containing 10^8 spores ml^{-1}. Germination was monitored by the fall in absorbance at 580 nm (A_{580}) in a UPM Vitatron spectrophotometer (MSE Fisons). Readings were taken at 15 min intervals for the first 60 min and then at 30 or 60 min intervals. The spore suspension was vortexed for 2 sec before each reading. Both the temperature and the pH of the suspension were varied. Spores were routinely activated by heating at 70°C for 30 min before exposure to germinants and observed by phase contrast microscopy before and after each experiment. All experiments were carried out three times and representative data from one experiment appear in figures.

Barbital was dissolved in the germinant solution either by raising the pH to approximately 11 by the addition of NaOH pellets and then obtaining the required pH by adding 1M HCl, or by heating the mixture at 70°C for 30 min.

For measuring loss of heat resistance during germination, 20 μl samples of spore suspension were taken at 5 min intervals, added to 2 ml of distilled water, heated to 70°C for 30 min and the number of surviving spores determined by plating appropriate dilutions onto NA. A_{580} measurements were made whenever samples were taken.

The germination phenotype of spores on plates was determined by the tetrazolium overlay method (17).

Isolation of Barbital Resistant Mutants

For this 10 ml of an exponential PAB culture of strain 1604 was centrifuged, resuspended in the same volume of 100 mM $MgSO_4$, and ultra-violet irradiated for 60 sec using a Hanovia low-pressure lamp at a distance of 20 cm (0.1% survival). Spores were then prepared on PGYE agar and harvested to give a final suspension of 10^8 spores ml^{-1}. 0.4 ml samples were heat activated at 70°C for 30 min, the spores spread onto NA containing either 50 or 70 mM barbital and incubated for 48 h at 37°C. Colonies were subcultured on NA and spore preparations made from single colonies. Germinability was then tested in 10 mM-Tris/HCl at pH 8.4 + 1 mM alanine + 50 mM barbital over a period of 3 h 37°C.

Mapping

Lysate preparations and transduction crosses using phages PBS-1 and SPP1 were as described by Jamet and Anagnostopoulos (15) and Ferrari *et al.* (11) respectively. The methods of Anagnostopoulos and Spizizen (1) were used to prepare DNA from vegetative cells and carry out transformations except that recipients were initially grown overnight on MA + glucose (0.5%) + casein hydrolysate (0.02%) + $MgSO_4$ (5 mM) + amino acids (50 µg/ml) as required. Also incubation time in the medium before stepdown was increased from 4 to 6 h.

To determine the barbital resistance or sensitivity of recombinant colonies they were subcultured on the appropriate selective medium, single colonies stabbed out into NA in glass Petri dishes and incubated at 37° for 3 days to form spores. These inocula were then exposed to chloroform vapour (3 ml chloroform in the lid of inverted plates which were wrapped in aluminium foil and kept at 18-20°C for 16 h). Spore inocula were then transferred onto NA + 50 mM barbital and incubated for 48-72 h at 37°C. Colonies appearing as quickly as those from mutant control inocula were scored as barbital resistant; those failing to grow and behaving as wild-type as barbital sensitive.

RESULTS

Effect of Barbital on Spore Germination

The pH and concentration dependence of the inhibition of spore germination by barbital was investigated to determine the

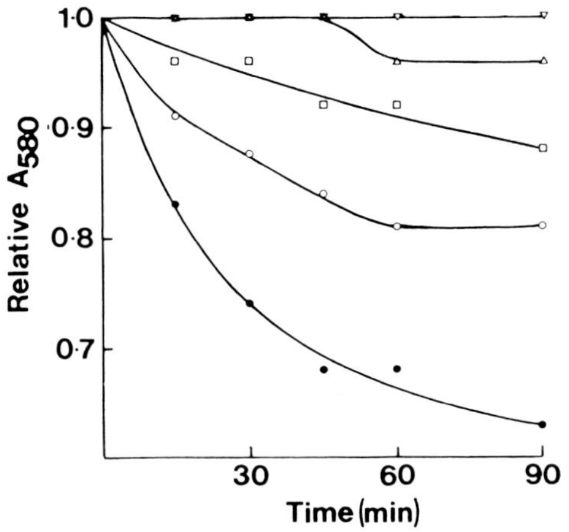

FIG. 1 *Germination of wild-type spores in NB at pH 9.0 containing different concentrations of barbital at 37°C.*
●, no barbital; ○, 10 mM; □, 25 mM; Δ, 50 mM; ∇, 75 mM.

concentration of barbital at which inhibition would be just maximal but at the same time employing a pH not far from the optimum for germination. This should have facilitated the isolation of mutant spores whose sensitivity to barbital inhibition might only differ slightly from that of wild-type spores. In NB at pH 9.0 inhibition of germination of spores of strain 1604 was 50% at 10 mM barbital, 70% at 25 mM and total at 75 mM (Fig. 1) whereas at pH 8.4 it was 70% at 10 mM and total at 25 mM (Fig. 3b). With the pH reduced further to 7.0 inhibition was total at the lower concentration of 10 mM (data not shown). Spore suspensions showing incomplete inhibition by barbital comprised mixtures of phase dark and bright spores.

These results are in line with those of Sierra and Bowman (27) who obtained a hundredfold change in percentage inhibition over the range pH 7-9 with spores of another strain of *B. subtilis*. In addition at a given pH, the percentage inhibition of germination increased with barbital concentration (Figs. 1 and 3) as also reported by the same authors. However, it has been shown (23) that the spore germination kinetics of strain 1604 do not accord with Lineweaver-Burke kinetics and thus we have not been able to demonstrate non-competitive inhibition (27).

As well as the inhibitory effects of barbital on change of A_{580} of spore suspensions, it was also shown that in NB at pH 7.0 10 mM barbital completely suppressed loss of heat

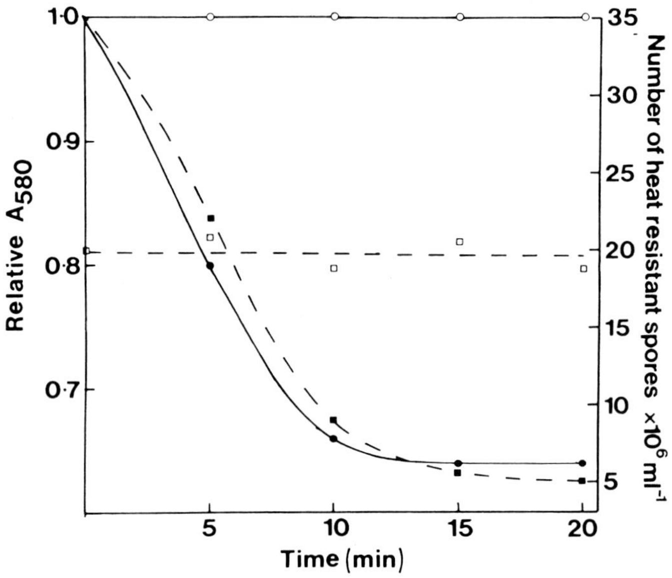

FIG. 2 *Germination and loss of heat resistance of wild-type spores in NB at pH 8.4 containing 10 mM barbital.* A_{580} *(solid line):* ●, *no barbital;* ○, *barbital. Heat resistance (broken line):* ■, *no barbital;* □, *barbital.*

resistance by spores (Fig. 2), a concentration shown to have no effect on spore outgrowth or vegetative growth (30 and D.A. Walton unpublished).

Effect of Varying Sporulation Conditions on Barbital Inhibition

Altering the membrane environment of the receptor might produce a change in the inhibition pattern especially if, as is thought to happen, barbital exerts its effect via the membrane (see Discussion). Increasing the growth temperature of *Bacillus spp* has been shown to change the pattern of branched chain fatty acids in membrane lipid (9,25), altering the structure of the membrane and affecting membrane-linked functions (18,32). Such alterations are likely to remain fixed in the spore (a metabolically inactive body) if induced during sporulation, irrespective of the germination temperature (7). The effect of temperature variation and the presence of barbital during sporulation on the barbital sensitivity of the

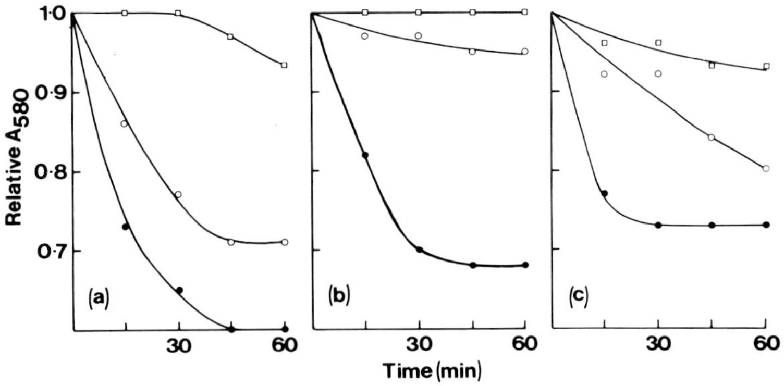

FIG. 3 *Germination of wild-type spores prepared at (a) 26°, (b) 37° and (c) 42°C in NB at pH 8.4 containing different concentrations of barbital at 37°C. μ, no barbital; ○, 10 mM; □, 25 mM.*

resulting spores was therefore tested.

Spores were prepared by the resuspension method at 26,37 and 42°C. Their germination at 37°C in NB at pH 8.4 containing barbital at 7.5, 10 and 25 mM was then measured (Fig. 3). At each concentration of barbital, the germination of spores produced at 37° was clearly more sensitive to inhibition by barbital than the germination of those produced at 26 or 42°C.

The presence of barbital during sporulation also had an effect on the subsequent germination of the spores in barbital. Three different wild-type spore preparations were made by the resuspension method at 26°C with both the growth and resuspension media containing 80 mM barbital. Their germination at 37° in NB and in NB + 10 mM barbital at pH 8.4 was compared with the germination under similar conditions of spores prepared from a sub-culture of the same colony in the absence of barbital (Table 2). Inclusion of barbital in the sporulation medium consistently improved the resistance of the resulting spores to barbital inhibition, although results were variable. Using inocula from the same colony, spore preparation made in the presence or absence of barbital differed in their germinability even in the absence of barbital although no overall pattern was discernable.

Isolation and Germination Characteristics of Barbital Resistant Mutants

After mutagenesis, sporulation and selection on NA + barbital,

TABLE 2 *Effect of Sporulation in Barbital on Subsequent Germination in Barbital*

	Experiments					
	1		2		3	
Absence (−) or presence (+) of barbital during sporulation	−	+	−	+	−	+
Relative $\underline{A}_{580}$* after 60 min. germination in absence of barbital	0.60	0.43	0.54	0.62	0.52	0.73
Relative $\underline{A}_{580}$ after 60 min. germination in presence of barbital	0.975	0.72	1.0	0.89	0.77	0.845
% Inhibition of germination†	93.8	50.8	100	71.0	52	42.6

$$^{*}\text{Relative } \underline{A}_{580} = \frac{(\text{Optical Density at } t=0)-(\text{Optical Density at } t)}{\text{Optical Density at } t=0}$$

$$^{\dagger}\%\text{Inhibition} = 100 \times \frac{(\text{Rel.}A_{580} \text{ in pres. of barbital})-(\text{Rel.}A_{580} \text{ in abs. of barbital})}{1-\text{Rel.}\underline{A}_{580} \text{ in absence of barbital}}$$

each sample of the suspension yielded on a plate approximately 10 colonies of varying size and morphology. Twenty-nine of these colonies were subcultured and the germination of the spores tested under optimal conditions for germination i.e. in ALA at pH 8.4 (23) but in the presence or absence of 50 mM barbital. Two, which appeared as large, translucent colonies on the selection plate, were putative barbital resistant mutants since they yielded spores which germinated better over a period of 60 min (25% fall in $\underline{A}_{580}$) than wild-type spores (14% fall in $\underline{A}_{580}$). The effects of varying pH and barbital concentration on the germination of spores of one of these mutants (*ger-100*) was tested.

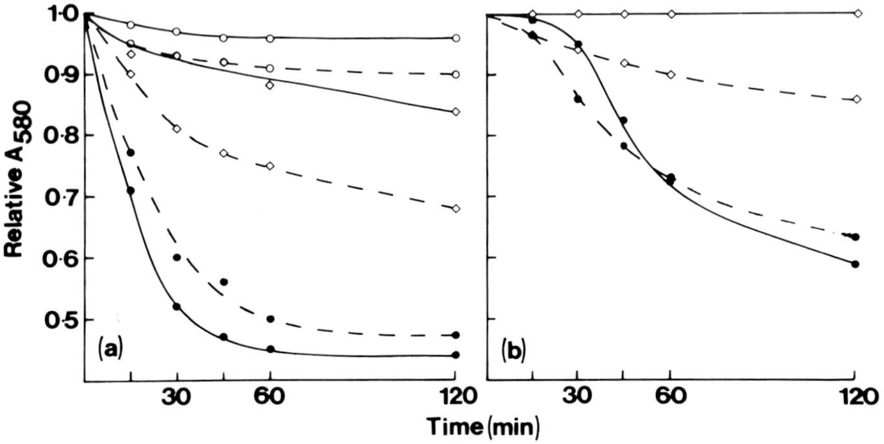

FIG. 4 *Germination of wild-type and* ger-100 *spores in Tris HCl + ALA containing barbital at 37°C.* A_{580} *of wild-type (solid line) and* ger-100 *(broken line) at (1) pH 7.6 and (b) pH 7.0.* ●, *no barbital;* ◊, *5 mM;* μ, *10 mM.*

In 1 mM ALA + 5 mM barbital at pH 7.0 over an extended period of 2 h at 37 C a wild-type spore suspension showed no fall in A_{580} whereas that of a *ger-100* suspension fell by 15% (Fig. 4b) which confirmed the barbital resistance of the mutant spores. In the absence of barbital, the A_{580} of the wild-type and mutant spore suspensions each fell by approximately 40% so that *ger-100* spores are not just better germinators *per se*. The effects of 5 and 10 mM barbital in the same germinant at pH 7.6 were then tested. Both wild-type and mutant spores were less susceptible to inhibition by 5 mM barbital than at pH 7.0 the A_{580} falling by 17 and 32% respectively (Fig. 4a). At this higher pH the A_{580} of both spore preparations in the absence of barbital fell by approximately 50% of their original value. At double the barbital concentration (10 mM) germination inhibition was similar

TABLE 3 *Location of* ger-100 *on the* B. subtilis *Chromosome by Cotransfer*

Recipient strain	Selected marker	Number of recombinants screened	Cotransfer of *ger-100* with selected marker (%)	Type of cross
1506	lys^+	208	14*	PBS-1 transduction
RW1	$hisB^+$	80	35	PBS-1 transduction
GSY111	$ilvA^+$	240	92**	PBS-1 transduction
GSY111	$ilvA^+$	80	58	SPP1 transduction
BD34	$metB^+$	160	52	Transformation

In all crosses *ger-100* was the donor
Data pooled from 2* and 3** experiments.

to that obtained with 5 mM barbital at pH 7.0 the A_{580} falling by 4 and 16% for wild-type and mutant respectively. Both wild type and mutant spores showed similar patterns of inhibition in NB + either 10 or 50 mM barbital at pH 7.0 (data not shown). Finally, after 3 h germination in ALA + 5 mM barbital at pH 7.0 (see Fig. 4) spores of *ger-100* comprised a mixture of phase dark (60%) and phase bright (40%) spores.

The Map Location of a Mutation Conferring Barbital Resistance

The *ger-100* mutation could well be located within a gene coding for a 'germination protein' at such a position as to confer barbital resistance but not a germination deficient phenotype, or in some other component that indirectly negates the inhibitory action of barbital. The possibility that it mapped within one of the previously defined germination loci *gerA,B,C,D, F* or *J* was tested by using *ger-100* as a donor seeking cotransduction with auxotrophic markers close to the loci. Phage PBS-1 lysates of *ger-100* were used to transduce 1506 (*lys-1*), BD11 (*thr-5*), BD92 (*hisA1*), CU371 (*ilvBΔ41*), GSY111 (*ilvA1*), MB21 (*metC3*), SB120 (*aroD120*) and RW1 (*hisB2*) to prototrophy and transductants screened for their ability to germinate on NA + barbital (Table 3). Colonies from mutant control inocula usually appeared after 48 h incubation as did those of barbital resistant transductants: wild-type colonies appeared after a further 24-48 h. Occasionally, barbital sensitive transductants yielded colonies after 48 h incubation but these were much smaller and easily distinguishable from resistant colonies. Control inocula of wild-type and *ger-100* vegetative cells both resulted in colonies after 48 h incubation confirming that the test specifically measured a difference in sensitivity of germination of inhibition by barbital. Cotransduction of the *ger-100* mutation with *hisB2* (35%), *lys1* (14%) and *ilvA1* (92%) was obtained (Table 3). These frequencies were consistent since variation between experiments was <5%.

Finer mapping using phage SPP1 transduction and transformation in similar crosses with strains GSY111 and BD34 respectively as recipients showed *ger-100* to be 58% cotransduced with *ilvA1* and 52% cotransformed with *metB3* (Table 3). Since *metB* and *ilvA* are 39-61% cotransformed dependent upon the mutations used (2) *ger-100* may be located between *ilvA* and *metB* (Fig. 5).

DISCUSSION

The extent of inhibition of wild-type spore germination by different concentrations of barbital (Figs. 1 and 3) corresponded well with that found by Sierra and Bowman (27). These

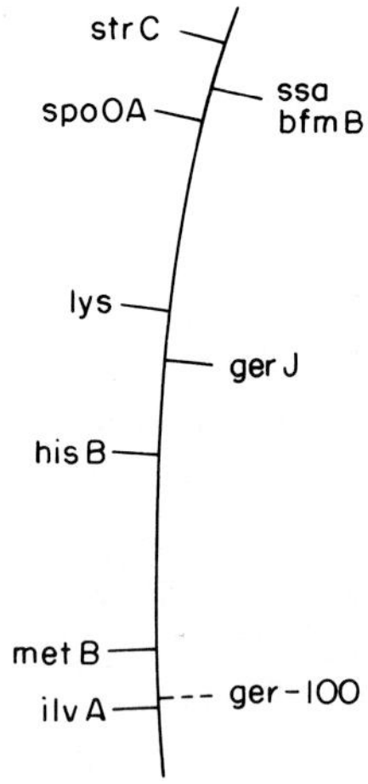

FIG. 5 *Part of the genetic map of* B. subtilis *168 indicating the probable location of* ger-100.

results, together with those of the experiments which indicated highest sensitivity to barbital was achieved at power pH, support their hypothesis that barbital has to be in the non-dissociated form to exert its inhibitory effect. One possibility is that barbital must either pass across a membrane to its site of action or that this site is located in one of the spore membranes. That this inhibition occurs very early in the germination process (probably affecting the trigger reaction and/or any immediate changes that this brings about) is confirmed by inhibition of loss of heat resistance as well as A_{580} of spore suspensions (Fig. 2), since the former is the earliest known event in germination (13).

Barbital is thought to decrease the Na^+/K^+ conductance across nerve cell membranes (4) the non-polar part of the molecule penetrating the lipid bilayer causing an expansion and subsequent increase in the electrical resistance of the membrane. Similar changes in the structure of spore membranes could certainly affect the germination process. However, whether such disruption alters the binding of germinants to the postulated germinant receptor (22) or influences germination events thereafter must remain uncertain.

The greater sensitivity to inhibition by barbital of spore germination compared to vegetative growth could result from the higher branched-chain fatty acid content of spore membranes (7), making them more susceptible to disruption. This is consistent with the effects of varying temperature (Fig. 3) or exposure to barbital (Table 2) during sporulation, both of which would be expected to alter the structure of the resulting spore membranes and thus lead to increased resistance of spore germination to barbital inhibition. Analogous results were found for chlorocresol inhibition of *B. subtilis* spore outgrowth (3). The germination of spores of *ger-100* was approximately half as sensitive as those of wild-type to inhibition by barbital (Fig. 4) but their "pattern" of inhibition with respect to varying inhibitor concentration and pH was unaltered. The *ger-100* mutation could either affect the germinant receptor molecule itself, making it less sensitive to changes in the membrane surrounding it, or be in a gene affecting the composition and hence susceptibility of the membrane to disruption by barbital. The latter is favoured by the mapping data (Table 3) which places *ger-100* close to *ilvA* a position different from any previously mapped *ger* mutations (19), particularly *gerA* mutations which may well result in a defect in a very early event in germination (24).

Partial loss of A_{580} or spore suspensions during germination is a characteristic of *gerJ* mutants (31) initiation taking place normally but germination being blocked at a late stage giving phase grey spores. The partial germination of spores of wild-type (Figs. 1 and 3) and *ger-100* at limiting concentrations of barbital (Fig. 4) suggests similarity to *gerJ* mutants but the absence of phase grey spores from wild type and *ger-100* suspensions does not support this. Nevertheless *ger-100* and *gerJ* sites of mutation map in the same region (Fig. 5). However *gerJ* mutations were 79% cotransduced with *lys1* using phage PBS1 (31) in contrast to *ger-100* (14% Table 3) so that they are quite separate (Fig. 5). Two mutations affecting the membranes of *Bacillus spp* are also linked to *lys-1*, although on the opposite side from *ger-100*. The *bfmB* mutant has a requirement for branched chain fatty acids and the *ssa* mutant has altered phospholipid composition and decreased sensitivity of sporulation to alcohol inhibition whereas *ger-100* could be a third pleiotropic mutation in this region affecting membrane composition.

ACKNOWLEDGEMENTS

We are grateful to Mr G.J. Dring (Unilever Research Laboratories) for suggesting the use of barbital, to Dr D.A. Walton for access to unpublished data and to Dr A. Moir (Department of Microbiology, Sheffield University) for comments on the manuscript. Excellent technical assistance was given by Mrs C. Davies, Mrs S. Haley, Dr H. Howell and Mrs J. Yeomans. The work was supported by the Science and Engineering Research Council in the form of a postgraduate studentship (R.M.) and research grants (D.A.S.).

REFERENCES

1. Anagnostopoulos, C. and Spizizen, J. (1961) *J. Bacteriol.* **81**, 741-746.
2. Barat, M., Anagnostopoulos, C. and Schneider, A.M. (1965) *J. Bacteriol.* **90**, 357-369.
3. Bell, N.D.S. and Parker, M.S. (1975) *J. Appl. Bacteriol.* **38**, 295-299.
4. Blaustein, M.P. (1968) *J. Gen. Physiol.* **51**, 293-307.
5. Bohin, J.P. and Lubochinsky, B. (1982) *J. Bacteriol.* **130**, 944-955.
6. Boudreaux, D.P. and Freese, E. (1981) *J. Bacteriol.* **148**, 480-486.
7. Bulla, L.A., Nickerson, K.W., Mounts, T.L. and Iandolo, J.J. (1975) in *Spores VI* (Gerhardt, P., Costilow, R.N. and Sadoff, H.L., eds.), pp.520-525, American Society for Microbiology, Washington D.C.
8. Cassone, A. and Simonetti, N. (1971) *Experienta* **27**, 981-983.
9. Daron, H.H. (1970) *J. Bacteriol.* **101**, 145-151.
10. Dring, G.J. and Gould, G.W. (1971) in *Spore Research 1971* (Barker, A.N., Gould, G.W. and Wolf, J., eds.), pp.131-141, Academic Press, London.
11. Ferrari, E., Canosi, U., Galizzi, A. and Mazza, G. (1978) *J. Gen. Virol.* **41**, 563-572.
12. Gould, G.W. (1969) in *The Bacterial Spore* (Gould, G.W. and Hurst, A., eds.), pp.397-444, Academic Press, London.
13. Gould, G.W. and Dring, G.J. (1972) in *Spores V* (Halvorson, H.O., Hansen, R. and Campbell, L.L., eds.), pp.401-408. American Society for Microbiology, Washington D.C.
14. Halvorson, H.O. (1959) *Bacteriol. Rev.* **23**,267-272.
15. Jamet, C. and Anagnostopoulos, C. (1969) *Mol. Gen. Genet.* **105**, 225-242.
16. Lafferty, E. and Moir, A. (1977) in *Spore Research 1976* (Barker, A.N., Wolf, J., Ellar, D.J., Dring, G.J. and Gould, G.W., eds.), pp.87-105, Academic Press, London.
17. Moir, A., Lafferty, E. and Smith, D.A. (1979) *J. Gen. Microbiol.* **111**, 165-180.
18. Overath, P., Schairer, H.U. and Stoffel, W. (1970) *Proc. Natl. Acad. Sci., U.S.A.* **67**, 606-612.
19. Piggot, P.J., Moir, A. and Smith, D.A. (1981) in *Sporulation and Germination* (Levinson, H.S., Sonenshein, A.L. and Tipper, D.J., eds.), American Society for Microbiology, Washington D.C.
20. Powell, J.F. (1950) *J. Gen. Microbiol.* **4**, 330-338.
21. Prasad, C. (1974) *J. Bacteriol.* **119**, 805-810.
22. Rossignol, D.P. and Vary, J.C. (1979) *Biochem. Biophys. Res. Comm.* **89**, 547-551.

23. Sammons, R.L. (1980) Ph.D. Thesis, University of Birmingham.
24. Sammons, R.L., Moir, A. and Smith, D.A. (1981) *J. Gen. Microbiol.* **124**, 229-241.
25. Shen, P.Y., Coles, E., Foote, J.L. and Stenesh, J. (1970) *J. Bacteriol.* **103**, 479-481.
26. Sierra, G. (1968) *Appl. Microbiol.* **16**, 801-802.
27. Sierra, G. and Bowman, C. (1969) *Appl. Microbiol.* **17**, 372-378.
28. Sterlini, J.M. and Mandelstam, J. (1969) *Biochem. J.* **113**, 29-37.
29. Strange, R.E. and Hunter, J.R. (1969) in *The Bacterial Spore* (Gould, G.W. and Hurst, A., eds.), pp.45-83, Academic Press, London and New York.
30. Vinter, V. (1970) *Appl. Microbiol.* **33**, 50-59.
31. Warburg, R.J. and Moir, A. (1981) *J. Gen. Microbiol.* **124**, 243-253.
32. Wilson, G. and Fox, C.F. (1971) *J. Mol. Biol.* **55**, 49-60.
33. Yasuda-Yasaki, Y., Namiki-Kanie, S. and Hachisuka, Y. (1978) in *Spores VII* (Chambliss, G. and Vary, J.C., eds.), pp.113-116, American Society for Microbiology, Washington D.C.

ISOLATION OF GENES IN *BACILLUS SUBTILIS* INVOLVED IN SPORE OUTGROWTH

E. FERRARI*, F. SCOFFONE**
M. GIANNI* and A. GALIZZI*

**Istituto di Genetica, Universita di Pavia, Italy*
***Istituto di Genetica Biochimica ed Evoluzionistica del C.N.R., Pavia, Italy*

SUMMARY

We report the isolation of four genes involved in *Bacillus subtilis* spore outgrowth. For this purpose we used a *B. subtilis* DNA library in λ charon 4A (1). A λ clone carrying the gene *out B* was isolated using a mutant temperature sensitive in the outgrowth phase as recipient in transformation crosses with DNA obtained from the library. The cloned DNA fragment is about 14 Kb long and, in addition to *out B*, carries the *aro I* and *amy E* genes (or part of them). The same library was also screened by hybridization to ^{32}P-labelled RNA prepared from outgrowing spores in the presence of a 500-fold excess of competitor vegetative RNA. By this method three clones were identified that encode genes specifically expressed during spore outgrowth.

INTRODUCTION

The process of outgrowth of *B. subtilis* spores has been studied at the biochemical and genetic level. Both approaches have failed thus far to give an answer to the main problem, i.e. at what level and through which mechanisms, is gene expression regulated during this stage of the cell cycle. The genetic system has suffered from a lack of functional analysis and the molecular basis of the elements involved in the process still eludes us.

With the aim of overcoming at least some of the obstacles

and of determining the role of different genes in the outgrowth process, we have cloned four of these genes in bacteriophage λ. To single out the desired clones from a library of *B. subtilis* DNA in phage λ we followed two lines of research. First we analysed the library for the ability to transform outgrowth temperature sensitive (ts) mutants to ts^+. With this procedure we isolated a clone with the *out B* marker. The other procedure consisted of screening the library by hybridization to ^{32}P-labelled RNA prepared from outgrowing spores in the presence of a large excess of competitor vegetative RNA. Three additional clones were isolated by the latter method.

RESULTS

Isolation of the Out B Gene

The *out B* gene (previously named *gsp 81*) is involved in *B. subtilis* spore outgrowth. A mutant, *OUTB81*, is temperature sensitive during outgrowth and the earliest defect observed ia a block of RNA synthesis (2). The only phenotype associated with the mutation is the temperature sensitivity of spore outgrowth. Since we lack any information about the altered function and do not have any molecular probe to screen for the desired gene, we had to resort to DNA mediated transformation. A library of *B. subtilis* DNA in λcharon 4A (1) was screened for the ability to transform the temperature sensitive *out B* marker of strain PB 2427 to ts^+. The library consists of 19 pools of about 100 individual plaques. Each pool was plated to confluent lysis on agarose plates, the phage collected and DNA extracted. The DNA was used to transform competent cells to the strain PB 2427; selection was for ability of the spores to give rise to colonies when plated at 47°. The pool with ts^+ *out B* transforming activity was replated to single plaques; from 200 single plaques, plate lysates were prepared, DNA was extracted and used again to transform strain PB 2427. One clone (λC4 BsG40) gave a high frequency of transformation to ts^+ *out B*. Since the size of the average insertion in λch4 is 15 kb, we looked for the presence of additional markers linked to *out B* in transformation. The DNA extracted from the clone was capable of transforming *aro I* to prototrophy, *tsc B*, *tsc G* and *tsc H* to ts^+ (3) and *amy E* bearing strains to ability to hydrolyse starch. The relationship between DNA concentration and frequency of aro^+ and ts^+ *out B* transformants, from DNA extracted from λC4 BsG40 and *B. subtilis* strain PB 19, is reported in Fig. 1. Following treatment with a chelating agent a deletion mutant of λC4BsG40 was isolated (4). The mutant phage DNA was still effective in transformation experiments involving the *amy E*, *out B*, *tsc B*, *tsc G*,

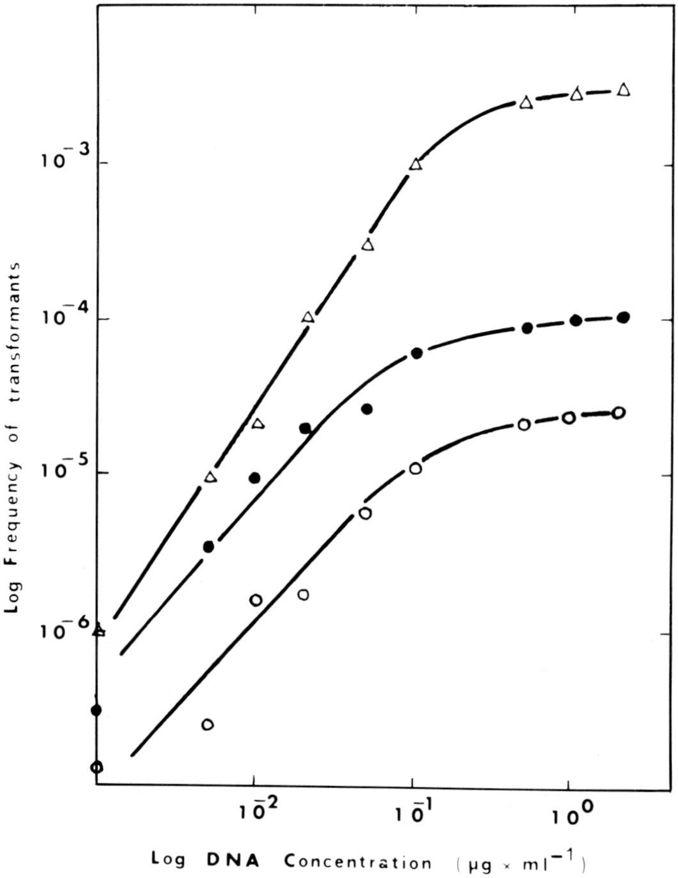

FIG. 1 *Relationship between DNA concentration and frequency of transformants. Frequency of Aro$^+$ transformants at different DNA concentrations using PB 19 (○) and λC4BsG40 (●) as a donor and PB 1715 (*met*B5* dal aro*I906* sac*A321) as a recipient. Frequency of Out$^+$ transformants at different DNA concentrations using λC4BsG40 (Δ) as a donor and PB 2427 (*his*H2* trp*C2* out *B81) as a recipient.*

and *tsc H* markers, but was completely devoid of transforming activity when used to select for aro$^+$.

The DNA of the λC4BsG40 phage and its deletion mutant derivative (λC4BsG40 Δ 20), was subjected to physical analysis using a variety of restriction endonucleases. The entire phage DNA was digested, in parallel with the DNA from the λ*ch4A* vector, and subjected to agarose or acrylamide gel electrophoresis.

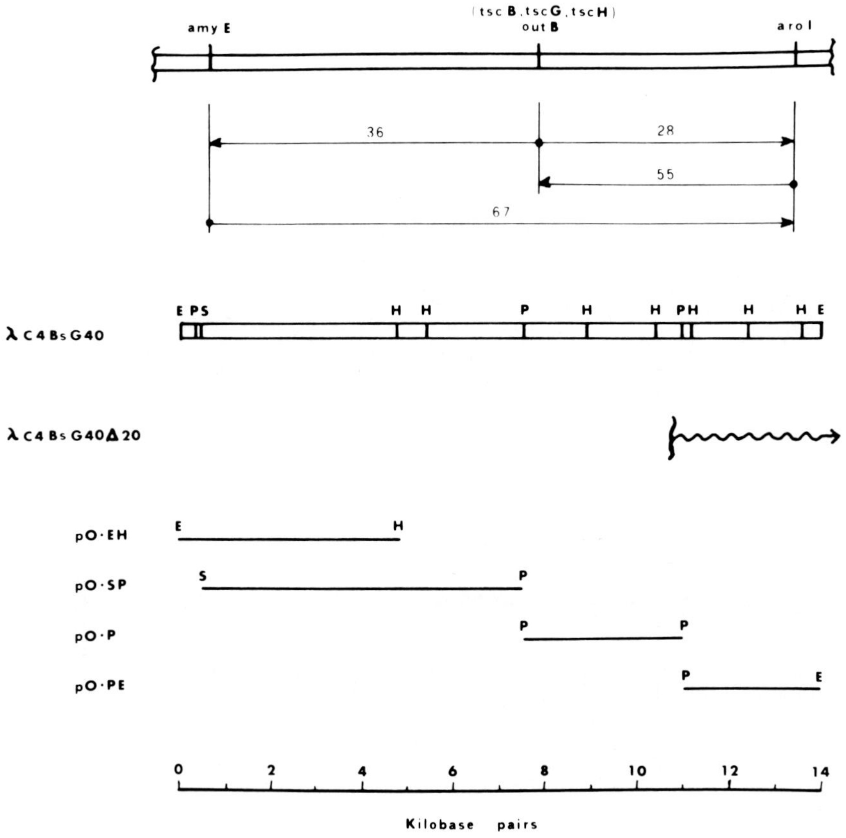

FIG. 2 *Correlation of the physical map of the* outB *region of the chromosome to the genetic map of the region.*
(a) Genetic map of the region. The order of markers has been obtained by reciprocal three-factor transformation crosses (Ferrari, E., Scoffone, F. and Galizzi, A., in preparation). The map distances are expressed as % recombination (1-cotransfer of donor marker) x *100. The head of the arrow points to the donor selected marker.*
(b) Physical map of the restriction endonuclease sites in the insert of chromosomal DNA cloned in bacteriophage lambda λ(C4BsG40). Symbols for the restriction endonuclease used are: E, Eco *RI; H,* Hind *III; P,* Pst *I; S,* Sal *I. The outermost* Eco *RI sites define the ends of the 14 kb* B. subtilis *DNA insert within λCh4. The wavy line represents the extent of the deletion of the mutant phage λC4BsG40 Δ 20. The positions on the physical map of the* B. subtilis *DNA fragments subcloned in pBR 322 and pBR 325 are shown at the bottom of the figure.*

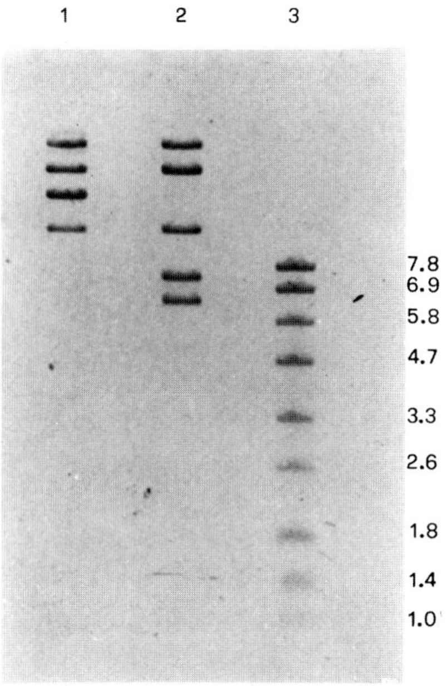

FIG. 3 *Agarose gel electrophoresis of* Eco *RI restricted DNA. Lane 1; DNA of λC4BsG40: lane 2; DNA of λCh4A: lane 3; DNA of SPP1 phage. A single insert approximately 4 kb long is present in the DNA of λC4BsG40. The numbers refer to the length of phage SPP1 DNA fragments, expressed in kb (5).*

The DNA from phage SPP1, digested with *Eco*R1, was run as a molecular weight standard. The results of the analysis are reported in Fig. 2. Digestion of the DNA extracted from clone λC4BsG40, with restriction endonuclease *Eco* RI yielded a single fragment of inserted chromosomal DNA, approximately 14 Kb long (Fig. 3). The *B. subtilis* insert does not have any restriction site for *Bam* HI, *Bgl* II and *Sst* I. The deletion mutant λC4BsG40 Δ 20, unable to transform the *aroI* allele to prototrophy, lacked about 4,000 base pairs. The right end of the insert from the region of 10.7 Kb to the right arm of λ DNA was missing from λC4BsG40 Δ 20. In particular three *Hind* III fragments (1,480, 1,200 and 680 base pairs) and the small 240 bp *Hind* III - *Eco*R1 fragment were missing. A new *Hind* III fragment of 4.4 kb was present at the junction of the insert to the λ right arm. The deletion abolished a *Pst* I site as well. From this we conclude that the deletion encompass about 2.6 kb of the insert and about 1.4 kb of the right arm of the λch4 DNA.

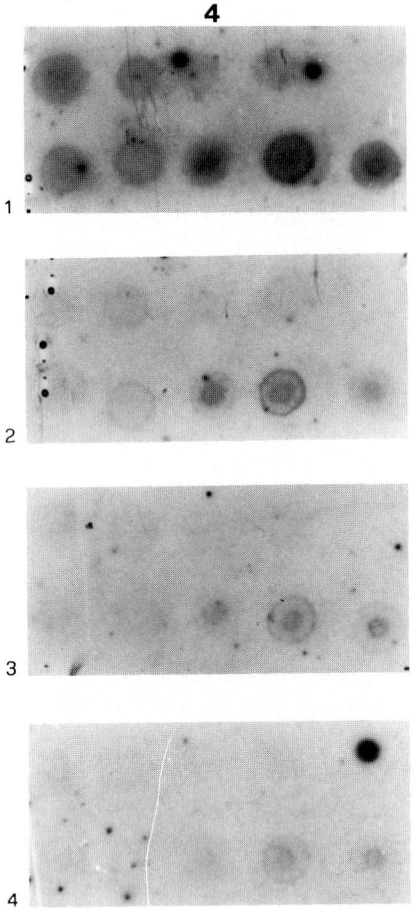

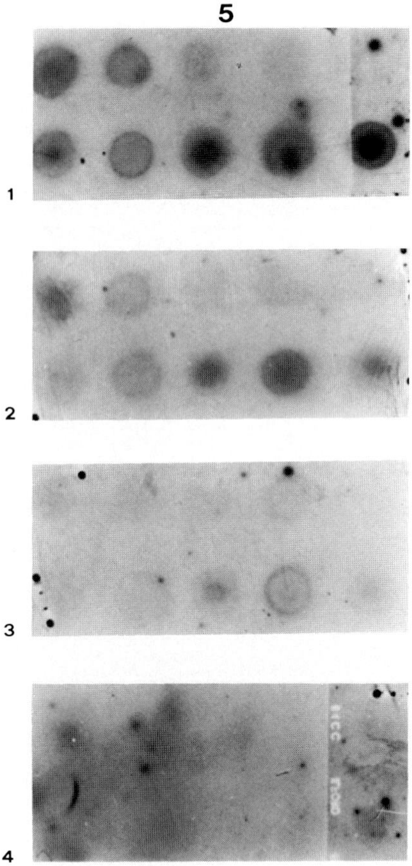

FIG. 5 *See caption opposite.*

FIG. 4 *Analysis of cloned DNA fragments by hybridization-competition. DNA was extracted from ten clones. Five μl aliquots of a solution (about 1.7 μg of DNA) were spotted on nitrocellulose sheets. After denaturation and fixation, the sheets were hybridized to 1 μg of RNA from 12 min outgrowing spores, labeled with γ-^{32}P ATP and polynucleotide kinase (specific activity 1 x 10^7 cpm μg^{-1}). To sheet 1, no competitor RNA was added; to sheet 2, 50 μg of total RNA extracted from vegetative cells; to sheet 3, and 4, 500 μg and 1,500 μg of the same unlabeled RNA were added respectively. The order of the spotted clones is, upper row: UG2, GE6, UC10, IA8, IC12; lower row: IA7, 10E12, IG9, UF7, BA7. On the upper right corner λCh4A DNA was spotted as a control.*

Comparison Between Genetic and Physical Maps

The deletion of about 2.5 kb at the right end of the insert, abolished completely the ability to transform for the *aro I* marker. This fact allows the orientation of the genetic map with the restriction map. More data have been obtained by subcloning several fragments of the insert into suitable plasmids (pBR 322 and pBR 325). The only plasmid that gave transformation to aro^+ was pO.PE (Fig. 2) carrying the *Pst*I - *Eco* RI fragment of 2.8 kb to the right of the insert. In the same way the *out B* gene could be localized in the 3.6 kb *Pst*I fragment present on plasmid pO.P. The same plasmid was effective in transformation experiments using recipient strains with the *tsc B*, *tsc C* or *tsc H* marker. The four genes (*out B*, *tsc B*, *tsc G* and *tsc H*) or part of them, must be present on the same 3.6 kb fragment.

Screening the Library by Hybridization-competition

Our strategy for selection of clones which encode outgrowth specific mRNAs was based on the method described by Mangiarotti *et al.* (6). It consists of the hybridization of a set of phage plaques to a filter replica of labelled RNA isolated from outgrowing spores in the presence of a large excess of cold RNA prepared from vegetative cells. For this purpose, from the original λ ch4 *B. subtilis* DNA library, an ordered collection was prepared in microtiter trays. Each set of plaques was replicated into square Petri dishes using a sterile plastic replicator (transfer plates, microtiter). The replica plates were used to blot DNA onto nitrocellulose filters. Hybridization was with ^{32}P-RNA extracted from outgrowing spores (12 min) and labelled *in vitro* with polynucleotide kinase in the presence of a 500-fold excess of RNA extracted from vegetative cells.

A number of clones consistently gave a positive result. The phage of each clone was purified through three single plaque isolates, grown in liquid culture, run in CsCl density gradients and the DNA extracted. The DNA was spotted onto nitrocellulose filters and hybridized to ^{32}P-RNA from outgrowing spores in the presence of various amounts of competitor vegetative RNA. Of the ten clones examined, seven were unable to

FIG. 5 *Analysis of cloned DNA fragments by hybridization and self competition. To sheet 1, no competitor RNA was added; to sheet 2, 5 μg of cold RNA extracted from outgrowing spores (12 min); to sheet 3 and 4, 50 μg and 250 μg of the same unlabeled RNA were added respectively. Other details are described in the legend to Fig. 4.*

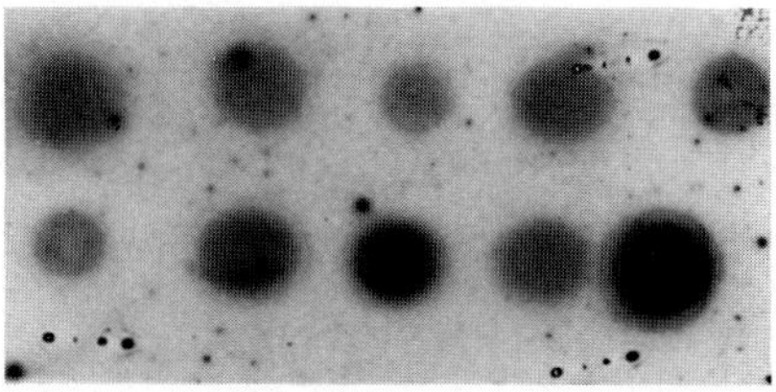

1

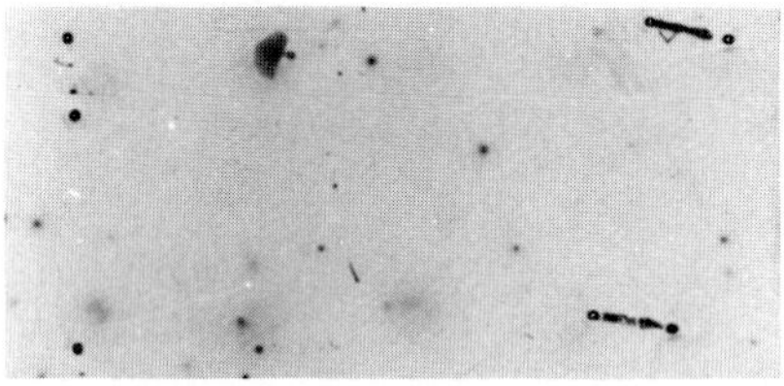

2

FIG. 6 *Mixed competitors experiment. The spotted DNA was hybridized to ^{32}P-RNA from outgrowing spores (12 min). To sheet 1, no competitor RNA was added; to sheet 2, 5 μg of unlabeled RNA from outgrowing spores (12 min) and 500 μg of total RNA from vegetative cells. All other details are as described in the legend to Figure 4.*

give hybrids in the presence of cold vegetative RNA, while three (BA7, IG9 and UF7) gave a positive signal even in the presence of a 1500-fold excess of vegetative RNA (Fig. 4). The hybrids were sensitive to self competition, when the competing RNA (i.e. RNA extracted from outgrowing spores after 12 min of incubation) was present in a 250-fold excess (Fig. 5). As a further control the hybridization was performed in the presence of cold vegetative RNA (500 fold) and cold RNA from outgrowing spores (five fold): with these mixed competitors no hybrid to ^{32}P-RNA from outgrowing spores was detected (Fig. 6). We conclude that the three clones NA7, IG9 and UF7, contain genes expressed during vegetative growth and genes specifically expressed during the outgrowth phase of the cell cycle.

The *B. subtilis* DNA insert of the three clones was about 15 kb (BA7), 11 kb (IG9) and 8 kb (UF7) long.

DISCUSSION

Our purpose was to clone some of the genes involved in *B. subtilis* spore outgrowth. To this goal we used a mutant, temperature sensitive in the outgrowth phase, as recipient in transformation crosses with DNA obtained from a *B. subtilis* library prepared in λcharon 4A. The screening, with DNA extracted from pools and single plaque lysates, has allowed us to identify a clone (λC4BsG40) with a 14 kb *B. subtilis* DNA insert, carrying the *out B* locus. The same phage isolate also carries the *Aro I* and *amy E* genes, or part of them. Isolation of a deletion mutant of the λC4BsG40 phage and subcloning of the insert into pBR322, allowed the localization of the *out B* gene on a 3.6 kb *Pst* I fragment. Obviously the ability to transform to a wild-type phenotype does not guarantee that the entire gene is present in the DNA fragment under scrutiny. A portion of the gene, comprising the site of mutation, will give the same result. In order to define the *out B* gene in a functional way, we are trying to perform a complementation test using plasmid pFH7 integrated into the *B. subtilis* chromosome at the SPβ site, thus providing a region of homology for plasmids derived from pBR 322 and carrying relevant sections of the cloned fragment (7). The analysis should also give an indication of the local ization of the regulatory region.

The same *B. subtilis* DNA library was also screened by hybridization to ^{32}P-labelled RNA prepared from outgrowing spores in the presence of a large excess of competitor vegetative RNA. Three clones have been identified containing genes specifically expressed during outgrowth in addition to genes expressed during vegetative growth. At present we do not know the chromosomal region from which the cloned DNA was derived.

In the four clones isolated, the genes developmentally regulated are interspersed with "vegetative" genes, in agreement with the finding of Losick and co-workers for genes involved in *B. subtilis* spore formation (8).

ACKNOWLEDGEMENTS

This research was supported, in part, by C.N.R. grants CP 81.00213.04 and CT 80.00650.04.

REFERENCES

1. Ferrari, E., Henner, D.J. and Hoch, J.A. (1981) *J. Bacteriol.* **146**, 430-432.
2. Albertini, A.M. and Galizzi, A. (1975) *J. Bacteriol.* **124**, 14-25.
3. Galizzi, A., Siccardi, A.G., Mazza, G., Canosi, U. and Polsinelli, M. (1976) *Genet. Res.* **27**, 47-58.
4. Parkinson, J.S. and Huskey, R.J. (1971) *J. Mol. Biol.* **56**, 369-384.
5. Ratcliff, S.W., Luh, J., Ganesan, A.T., Behrens, B., Thompson, R., Montenegro, M.A., Morelli, G. and Trautner,

T.A. (1979)
6. Mangiarotti, G., Chung, S., Zuker, C. and Lodish, H.F. (1981) *Nucl. Acids Res.* **9**, 947-963.
7. Ferrari, E. and Hoch, J.A. (1982) in *Molecular Cloning and Gene Regulation in Bacilli* (Ganesan, A.T., Chang, S. and Hoch, J.A., eds.), pp.53-61, Academic Press, New York.
8. Ollington, J.F., Haldenwang, W.G., Huynh, T.V. and Losick, R. (1981) *J. Bacteriol.* **147**, 432-442.

Biochemistry & Control of Sporulation

TWO NEW SPORULATION LOCI AFFECTING COAT ASSEMBLY AND LATE PROPERTIES OF *B. SUBTILIS* SPORES

H.F. JENKINSON and J. MANDELSTAM

Microbiology Unit, University of Oxford, Oxford, England

SUMMARY

Two sporulation mutants, selected as forming spores which are slow to germinate, are also abnormal in their development of resistance properties and are altered in coat protein assembly. The first of these mutants forms spores that are lysozyme-sensitive and lack a single coat polypeptide (M.Wt. 36 Kd) normally found in the outer layers of wild-type spores. The mutation defines the locus *spoVIA*. The second mutant forms spores that are slow to develop their resistance properties in sporulation and in which a coat polypeptide (12 Kd) is mis-assembled. This mutation defines a new locus *spoVIB*. Both loci seem to be involved in coat protein assembly.

INTRODUCTION

During stage IV-V of sporulation the germination properties (1) and resistance properties (2) of the spores develop concomitantly with the laying-down of the spore coat proteins (3). The development of these properties and the formation of the coat occur at least in part through the self-assembly of proteins, the genes for which have been expressed much earlier. The evidence for this is that (a) the resistance and germination properties, and the assembly of the coat can occur in the presence of chloramphenicol (1,2,3) and (b) at least 9 out of the 14 proteins identified as being components of the spore coat layers begin to be synthesized during stages II and III (3). If the characteristic properties of the spore are in fact determined by the assembly of the coat, then it should be possible to isolate mutants altered in coat assembly which also display abnormal resistance or germination properties.

By selecting for mutants which form spores that are slow to germinate, we have isolated two strains that are altered in coat protein assembly. Each strain carries a single mutation which defines a new sporulation locus in each case.

METHODS

Bacteria. Bacillus subtilis 168 *trpC2* which sporulates normally was used and is referred to as the wild-type strain. Strains used for genetic mapping included CU267 *leuB16 ilvB2 trpC2* (from S. Zahler), JH326 *leu-2 citF78* (from J. Hoch) and 4673 *gerE36* (from A. Moir).

Induction of sporulation. Cells were induced to sporulate at 37°C by the resuspension method of Sterlini and Mandelstam (4). The time of resuspension is denoted t_0 and subsequent times (in hours) t_1, t_2, etc.

Measurement of spore resistance properties. Resistances to toluene and to heat were determined as previously described (2); the numbers of spores surviving incubation with lysozyme (ED 3.2.1.17, 250 μg/ml, 37°C) were determined by removing samples at intervals and plating suitable dilutions on nutrient agar (Oxoid).

Preparation of spores. For germination experiments, spores were harvested from cultures at t_{20} by centrifugation, washed and purified by centrifugation through 40% (w/v) Urografin (5). Spores from cultures at t_{10} were separated from sporangia by passage through a French pressure cell, and then washed.

Spore coat proteins. The coat proteins were extracted with SDS-dithioerythritol pH 9.8 (3) from washed spore integuments generated by breakage of t_{10} spores in a Braun cell homogenizer (3).

Germination properties. Cleaned spores from cultures at t_{20} were suspended to a density of about 5×10^7 spores/ml in germination medium. The course of germination was followed by measuring the loss of absorbance of the resuspension at 550 nm or by measuring the loss in toluene-resistance.

SDS-polyacrylamide gel electrophoresis (SDS-PAGE). Proteins were fractionated on SDS-acrylamide slab gels (6) and stained and scanned as previously described (3).

Surface protein iodination. Cleaned spores (5 mg dry wt) were surface-labelled with ^{125}I (0.15 mCi) using lactoperoxidase (ED 1.11.1.7, Sigma) as described by Jenkins *et al.* (3). The coat proteins were then extracted as described above, separated by SDS-PAGE and the radioactivity in each band was determined by counting 1 mm slices of the gel lane in a gamma counter.

Genetic mapping. Transduction was carried out using phage PBS1 as described by Karamata and Gross (7). Transformation of competent cells with DNA prepared by the method of Marmur (8) was done in the medium of Young and Spizizen (9). Transforming DNA was used at a concentration of 0.05 μg/ml for mapping. Recombinants were selected on lactate/glutamate minimal agar (1) with appropriate supplements, colonies were picked onto

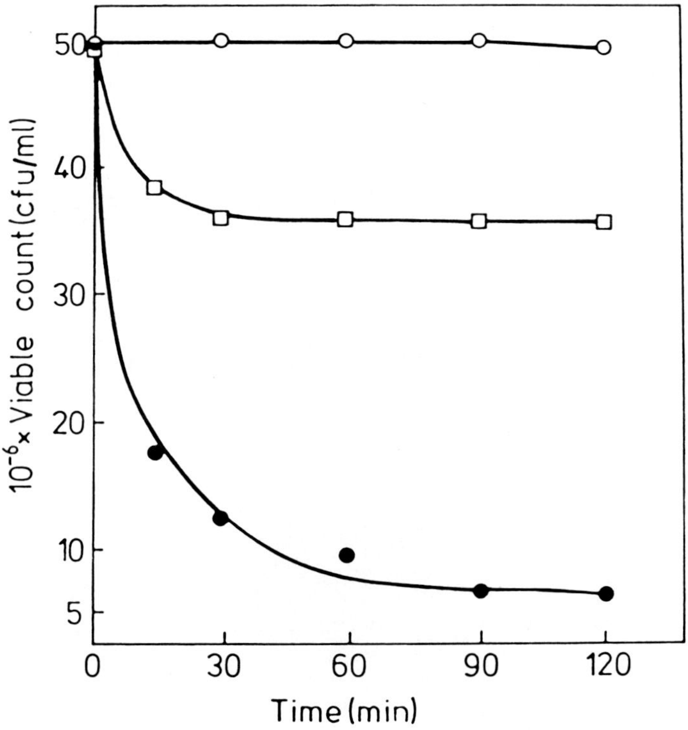

FIG. 1 *Effect of lysozyme on viability of spores of strain 168 (○), 513 (●) and 520 (□). Cleaned spores at t_{10} were suspended at a density of 5 x 10 c.f.u./ml in Na_2HPO_4/NaOH buffer (100 mM, pH 7) containing lysozyme (250 μg/ml) and incubated at 37°C. Samples were removed at intervals for determination of viable count.*

selective medium, and then scored for the Spo⁻ (Ger⁻) phenotype using the method of Lafferty and Moir (11).

RESULTS AND DISCUSSION

Phenotypic Properties of Strains 513 and 520

Both strains were isolated after NTG-mutagenesis of strain 168 and selection for spores that were slow to germinate (5). The strains sporulated normally until stage IV when the spores began to acquire their germination and resistance properties (1,2). In strain 513, resistance to toluene and heat developed normally but >90% of the spores formed by t_{10} were lysed by lysozyme (Fig. 1). In strain 520, the development of all the spore resistance properties was delayed by 40-60 min, and about 30% of the spores formed at t_{10} were sensitive to lysozyme (Fig. 1).

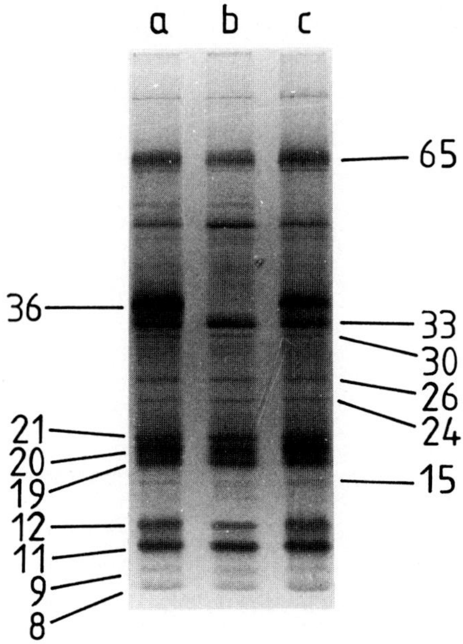

FIG. 2 *SDS-PAGE patterns (stained with Coomassie blue) of proteins extracted from the spore coat fraction of t_{10} spores of strain 168 (lane a), strain 513 (lane b) and strain 520 (lane c). The coat protein bands are numbered by their appropriate M.Wt (Kd).*

To see if these effects were due to an aberration in coat formation, the spore coat proteins from t_{10} spores of each strain were extracted as described in Methods and separated by SDS-PAGE. The spore coat of strain 513 lacked a major component of the normal spore coat, the 36K polypeptide (Fig. 2b). This protein begins to be synthesized late (after t_6) in normal sporulation (3) and its synthesis is prevented by the addition of chloramphenicol at t_6. Spores formed under these conditions are sensitive to lysozyme (2) so the sensitivity of the spores of strain 513 to lysozyme is apparently due to the lack of this protein. By contrast, analysis of the coat extracts from spores of strain 520 showed that all the normal proteins were present (Fig. 2c) except for a small difference in the amount of 36K protein. Since about a third of the spores of this strain at t_{10} are sensitive to lysozyme (see (Fig. 1) and the 36K protein is required for lysozyme resistance, it is reasonable to assume that there is a mixed population of spores at t_{10} which contains some that have failed to incorporate the 36K protein.

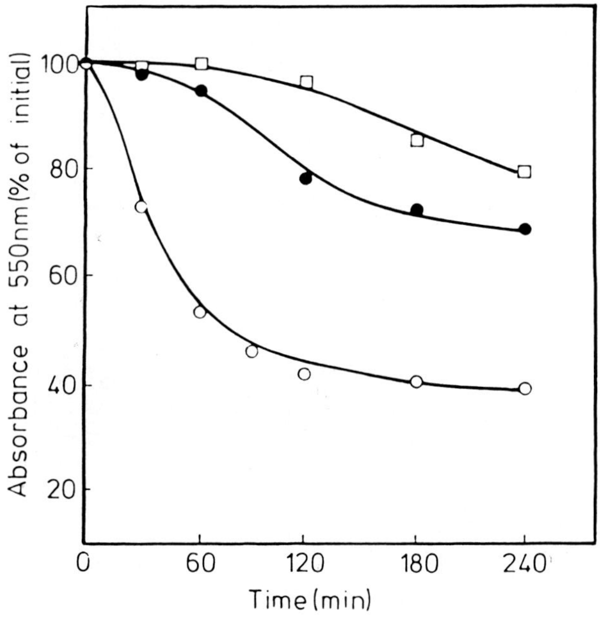

FIG. 3 *Germination of t_{20} spores of strain 168 (○), 513 (●) and 520 (□) in Tris/HCl buffer (10 mM, pH 8.2) containing L-alanine (10 mM) at 30°C.*

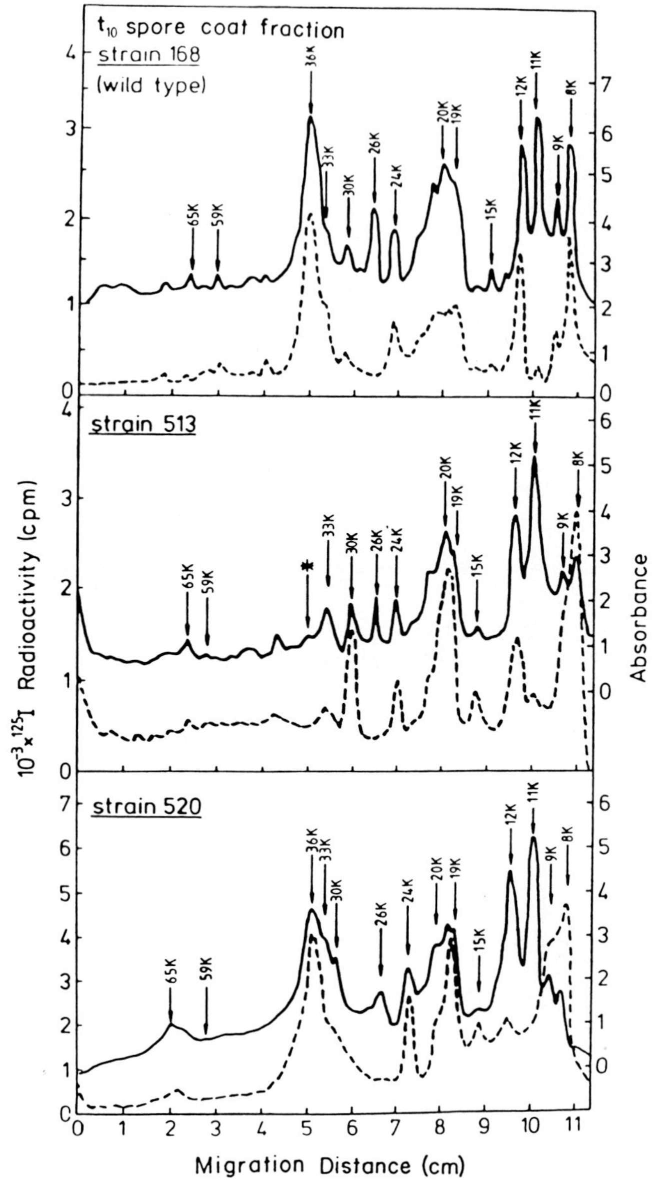

FIG. 4 *Proteins exposed on the surface of t_{10} spores of wild-type strain (a), strain 513 (b), and strain 520 (c). Spores were separated from sporangia, surface-labelled with ^{125}I using lactoperoxidase, and the spore coat proteins were*

Spores of both strains germinated more slowly in L-alanine than those of the wild type (Fig. 3). The spores of strain 520 were slower to lose absorbance (and toluene resistance) than those of strain 513, and in both strains the spores that germinated became phase-dark.

In vitro Labelling of Surface Proteins with ^{125}I

We have shown previously by surface labelling mature spores of the wild-type strain with ^{125}I that four proteins (M.Wt 36K, 12K, 9K and 8K) are the major surface components of the coat (3). Several other proteins can be partly iodinated (e.g. 11K, 24K, 20K and 19K proteins), and the fact that these, too, acquire label in the mature spore indicates that these are coat components and not membrane proteins as has been recently suggested (12).

Since the 36K protein, normally the most abundant component of the outermost layer, is absent from spores of strain 513, treatment of these with ^{125}I should identify those proteins that are normally overlaid by it. In spores of strain 513, the 12K, 9K and 8K polypeptides are highly-labelled as they are in those of the wild-type (compare Fig. 4a and 4b). However, in the mutant there is much higher labelling of the 30K, 24K, 20K, 19K and 15K polypeptides (Fig. 4b). This suggests that these proteins, not significantly exposed in wild-type spores, are normally covered by the 36K polypeptide.

The coat protein extract of strain 520, unlike that of strain 513, showed no qualitative difference from that of the wild-type when examined by SDS-PAGE. However, the surface-labelling technique when applied to the mature spores on this mutant showed an alteration in arrangement of the proteins in the coat. Thus the 12K polypeptide in the coat of the mutant was virtually non-iodinated, i.e. it was not surface exposed (Fig. 4c). In addition the 24K, 19K and 15K proteins were more highly-labelled than they were in wild-type preparations. Since these proteins are among those that are normally overlaid by the 36K protein (see Fig. 4b) the most likely explanation for their being labelled in strain 520 is that they are iodinated on that fraction of the spores that are lysozyme-sensitive (Fig. 1), i.e. which presumably lack the 36K protein.

FIG. 4 (continuation)
separated by SDS-PAGE. The stained gels were scanned with a microdensitometer and then the gel lanes were excised and sliced into 1 mm sections which were counted for radioactivity. (———), Densitometer scan; (-----), ^{125}I radioactivity.

We considered the possibility that the failure to obtain surface-labelling of the 12K polypeptide in strain 520 may have been due to mistiming in either its synthesis or its deposition, or both. However, further experiments (results not shown) showed that it began to be synthesized in strain 520 at the same time as it does in the wild-type strain (at stage II, (3)), and to be deposited in the outer layers of the spore at approximately the same time as in the wild-type (after t_6).

Genetic Mapping

The mutation in strain 513 is 45% cotransducible with *argA3* and 68% cotransducible with *leuA8* (5); it defines a new sporulation locus *spoVIA*. The mutation in strain 520 is 47% cotransformable with *leuB16*, and the order established from three-factor crosses was *ilvB2-leuB16-spo520*. It is separate from the mutation *ger-36* described by Moir (13) which also affects coat formation and which is closely linked to *citF*. The mutation *spo520* therefore also defines a new sporulation locus *spoVIB*. The map in Fig. 5 shows the position of the three loci which are known to be involved in coat assembly. Each locus is defined by a single mutation which in each case causes germination, resistance and coat defects that are always cotransformable.

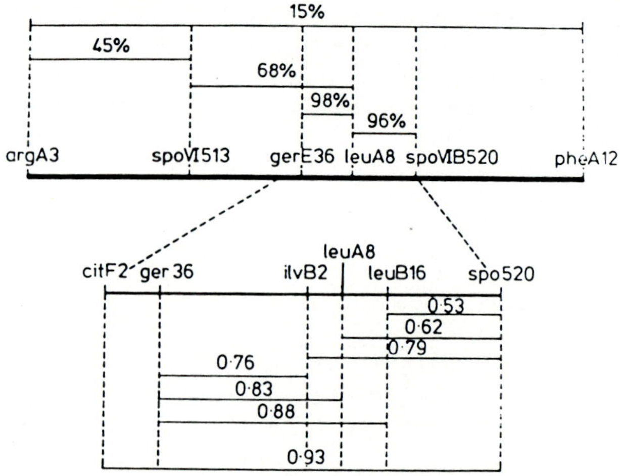

FIG. 5 *Map of* spo *and* ger *loci in the* leu *region on the* B. subtilis *chromosome. The figures above the chromosome are percentage cotransduction frequencies using phage PBS1. The expanded region below the chromosome shows the approximate genetic distances in transformation units (100-% cotransformation) ÷ 100.*

In summary, strain 513 and strain 520, which were both selected as germination mutants, also show defects in their resistance properties and their coat assembly. One (strain 513) showed how the absence of the 36K protein affects the properties of the spore (essentially confirming earlier experiments using chloramphenicol, see Introduction). The other (strain 520) shows that, even when all the component polypeptides of the coat are present, a single mutation can cause a substantial rearrangement in their assembly. These results support the contention that resistance, germination and coat protein assembly are integrally associated, and that a mutation affecting any of these is likely to affect one or both of the others.

ACKNOWLEDGEMENTS

We are grateful to Angela Maunder and Mary Blower for their technical assistance. This work was supported by the Science and Engineering Research Council.

REFERENCES

1. Dion, P. and Mandelstam, J. (1980) *J. Bacteriol.* **141**, 786-792.
2. Jenkinson, H.F., Kay, D. and Mandelstam, J. (1980) *J. Bacteriol.* **141**, 793-805.
3. Jenkinson, H.F., Sawyer, W.D. and Mandelstam, J. (1981) *J. Gen. Microbiol.* **123**, 1-16.
4. Sterlini, J.M. and Mandelstam, J. (1969) *Biochem. J.* **113**, 29-37.
5. Jenkinson, H.F. (1981) *J. Gen. Microbiol.* **127**, 81-91.
6. Laemmli, U.K. and Favre, M. (1973) *J. Mol. Biol.* **80**, 575-599.
7. Karamata, D. and Gross, J.D. (1970) *Molec. Gen. Genet.* **108**, 277-287.
8. Marmur, J. (1961) *J. Mol. Biol.* **3**,
9. Young, F.E. and Spizizen, J. (1961) *J. Bacteriol.* **81**, 823-829.
10. Piggot, P.J. (1973) *J. Bacteriol.* **114**, 1241-1253.
11. Lafferty, E. and Moir, A. (1977) in *Spore Research 1976* (Barker, A.N., Wolf, J., Ellar, D.J., Dring, G.J. and Gould, G.W., eds.), pp.87-105, Academic Press, London.
12. Stewart, G.S.A.B. and Ellar, D.J. (1982) *Biochem. J.* **202**, 231-241.
13. Moir, A. (1981) *J. Bacteriol.* **146**, 1106-1116.

AN EXAMINATION OF THE DEPENDENT-SEQUENCE HYPOTHESIS BY TWO-DIMENSIONAL GEL ELECTROPHORESIS OF PULSE-LABELLED PROTEINS FROM *BACILLUS SUBTILIS spo* MUTANTS

M.D. YUDKIN*, H. BOSCHWITZ and A. KEYNAN

Department of Biological Chemistry, Hebrew University of Jerusalem, Givat-Ram, Jerusalem, Israel
**Permanent address: Microbiology Unit, Department of Biochemistry, South Parks Road, Oxford OX1 3QU, U.K.*

SUMMARY

By labelling wild-type *B. subtilis* for 5 min with [^{35}S]-methionine either on resuspension in starvation medium or 1, 2 or 3 h later and subjecting cell-extracts to two-dimensional get electrophoresis, Yudkin *et al.* (4) detected 74 proteins whose synthesis started or stopped within the first 3 hours of sporulation. Similar experiments have been done with isogenic strains carrying a *spoOA* or a *spoIIA* mutation. The results permit 72 of the changes in protein synthesis to be placed in 4 classes: I - those found in the wild-type and in both mutants; II - those found in the wild-type and the *spoIIA*; III - those found in the wild-type only; and IV - those found in the wild type and the *spoOA* mutant but not in the *spoIIA* mutant. Class I and Class II changes predominated in the first hour but Class III changes predominated in the third hour; only two changes fell into Class IV. These results are in good agreement with the 'dependent-sequence' hypothesis which states that each event that occurs in sporulation depends on the successful completion of all preceding events.

INTRODUCTION

During sporulation in *Bacillus subtilis* a sequence of changes occurs by which the vegetative cell is converted to a spore.

Studies both with inhibitors and with mutations that block sporulation at various points suggested to Mandelstam (1) that the sequence of events is dependent: that is to say, a given event will occur only if all previous events have been successfully completed. More recent findings have cast doubt on the view that the later events of sporulation belong to a simple linear sequence (2, see also 3). Nonetheless, it is still important to know whether the changes that occur within the first few hours of sporulation form a dependent sequence, because the answer to this question will limit the models that can be constructed to explain the control of sporulation. Although the dependent-sequence hypothesis accounts well for the events that have so far been observed in stages 0 - II of sporulation, the number of these events is quite small, and it is possible that if a larger number were studied they would no longer fall into a dependent sequence.

By using two-dimensional gel electrophoresis of pulse-labelled cells, we have detected more than 70 changes in the pattern of protein synthesis in the first 3 hours after *spo*$^+$ cells are transferred from growth medium to resuspention medium (4). We now describe the effect of mutations in *spoOA* or *spoIIA* on these changes, and discuss the results in terms of the dependent-sequence hypothesis.

THEORY AND PREDICTIONS

In Fig. 1, the events of the dependent sequence are symbolized by Roman letters linked by arrows; a given event can occur only if all the previous events in the sequence have occurred.

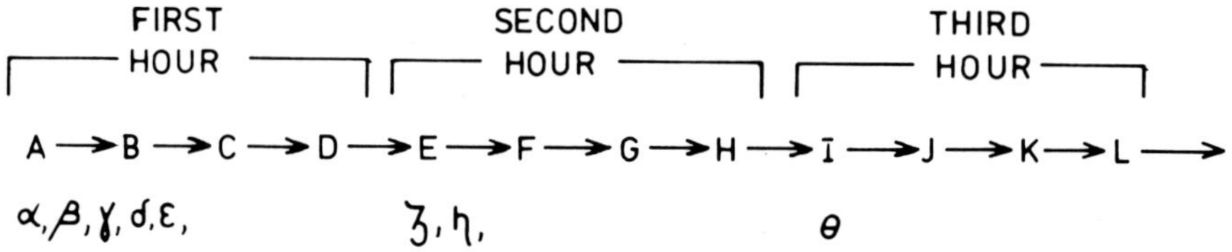

FIG. 1 *The dependence sequence of sporulation and other, non-specific, events. For explanation see text.*

For the sake of discussion we have made the arbitrary and oversimplified assumption that four events occur in each hour after the bacteria have been resuspended in sporulation medium. In addition to these, other changes will occur in the cells after resuspension. These latter changes will be due to the step-down, will not be sporulation-specific, and are symbolized in Fig. 1 by Greek letters separated by commas: we think it reasonable to suppose that many of these changes will occur soon after resuspension but that their number will decline as time goes on.

We now discuss the effect that *spo* mutations will be expected to have. First, consider a very early (*spo0*) mutation, which blocks the dependent sequence immediately after event A. In a strain carrying this mutation event A will occur and so will all the Greek-letter events α to θ; events from B onwards will not occur. Next consider a mutant (let us say a *spoII* mutant) in which the dependent sequence is blocked immediately after E. In this strain events A, B, C, D and E will occur (but not F to L), and so, once again, will all the Greek-letter events.

By contrast with these mutants, a wild-type strain when transferred to resuspension medium will carry out all the events shown in Fig. 1, that is A to L and α to θ. Let us suppose that we can detect and describe these events. Let us further suppose that we resuspend, under identical conditions, the *spo0* mutant, detect the changes that occur and compare them with those in the wild-type; and that we do the same with the *spoII* mutant. We can then classify all the changes that occur in the wild-type according to whether they do or do not also occur in each of the mutants, thus:

	Wild-type	*spo0* mutant	*spoII* mutant
Class I	+	+	+
Class II	+	−	+
Class III	+	−	−
Class IV	+	+	−

What predictions can we make about the distribution of the events that fall into the different classes? Class I (events that occur in all strains) will comprise A, α, β, γ, δ, ε, ζ, η and θ. By referring to Fig. 1, we can see that most of these will occur early; at later times the number of events in this class will decrease. Class II (events that occur in the wild-type and in the *spoII* strain but not in the *spo0* strain) will comprise B, C, D and E; thus most of the changes in this class too will occur early. Class III (events that occur in the wild-type only) will comprise F, G, H, I, J, K and L; unlike the first two classes, the number of changes in this class will *increase* with time. Class IV (events that occur in the wild-type and the *spo0* mutant but not in the *spoII* mutant) will be empty; if the dependent-sequence hypothesis is correct, there can be no change that is blocked by the *spoII* mutation but which occurs in the *spo0* mutant.

We have summarized these predictions in Table 1. They follow from the assumptions: (a) that some events that occur

TABLE 1 *Predicted distribution of changes expected in the different classes defined in the text*

	FIRST HOUR	SECOND HOUR	THIRD HOUR
Class I	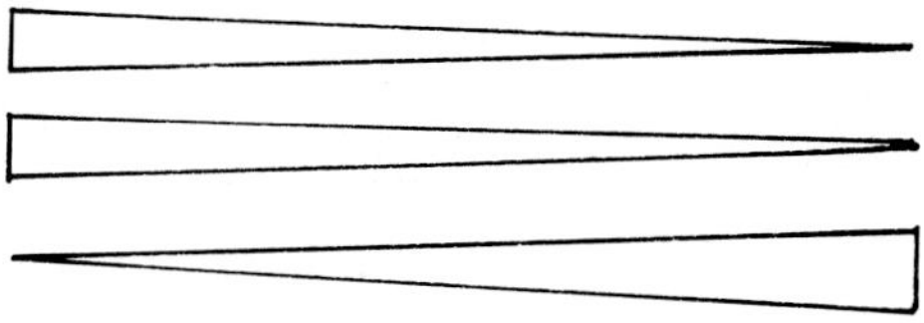		
Class II			
Class III			
Class IV	0	0	0

after *B. subtilis* is resuspended in sporulation medium fall into a linear, dependent, temporal sequence; and (b) that the changes prevented by the *spo0* and by the *spoII* mutation block this sequence at different points, with the *spo0* block preceding the *spoII* block.

EXPERIMENTAL AND RESULTS

By labelling spo^+ cells with [^{35}S]-methionine immediately after resuspension in sporulation medium or one, two or three hours later, displaying the cells' proteins by two-dimensional gel electrophoresis (5) and detecting radioactive proteins by fluorography, we have previously found 74 proteins whose synthesis either starts or stops within the first three hours of sporulation (4). We have now done exactly similar experiments

TABLE 2 *Actual distribution of changes found in the different classes defined in the text*

	FIRST HOUR	SECOND HOUR	THIRD HOUR
Class I	28	4	2
Class II	12	3	1
Class III	3	8	9
Class IV	1	1	0

with isogenic strains carrying either a *spo0A* (*spo-43*) or a *spoIIA* (*spo-69*) mutation. These experiments have allowed us to assign 72 of the changes to the four classes mentioned above, according to whether the change also occurs, or whether it is prevented, in the mutant strains. These assigments are set out in Table 2.

The actual distribution of changes given in Table 2 is in close agreement with the distribution predicted in Table 1. As expected, Class I and Class II changes occur predominantly at early times, with Class III appearing largely in the second and third hour. Class IV, which was expected to be empty, contains only two out of the 72 changes; interestingly enough, one of these is the appearance of a protein whose synthesis is variable even in wild-type strains (unpublished data of M.D. Yudkin).

CONCLUSION

Of the large number of events that occur during sporulation, only a few (those for which biochemical assays are available) have previously been studied in detail. The dependent-sequence hypothesis was designed to account for the relatively few events that had been discovered at the time it was proposed. Our results show that, even when we take a much larger number of events into account, the two mutations *spo0A-43* and *spoIIA-69* affect steps that are joined by a linear dependent sequence. The method (two-dimensional gel electrophoresis of proteins) by which we have detected these events makes no assumptions about the function or activity of the proteins whose synthesis we are studying; we believe, therefore, that the changes we have found represent a large fraction of all the changes in the pattern of protein synthesis that actually occur during the period we are investigating (4). We find it striking that the dependent-sequence hypothesis describes these many changes so well.

ACKNOWLEDGEMENTS

This work was supported by a grant from the Thyssen Foundation to A. Kenyan. Most of it was done during the tenure by M.D. Yudkin of a Fellowship under the Royal Society-Israel Academy Programme.

REFERENCES

1. Mandelstam, J. (1969) *Symp. Soc. Gen. Microbiol.* **19**, 377-403.
2. Jenkinson, H.F., Kay, D. and Mandelstam, J. (1980) *J.*

Bacteriol. **141**, 793-805.
3. Piggot, P.J. and Coote, J.G. (1976) *Bacteriol. Rev.* **40**, 908-962.
4. Yudkin, M.D., Boschwitz, H., Lorch, Y. and Keynan, A. (1982) *J. Gen. Microbiol.* **128**, 2165-2177.
5. O'Farrell, P.H. (1975) *J. Biol. Chem.* **250**, 4007-4021.

POSSIBLE ROLE OF POLYADENYLATED RNA IN THE REGULATION OF SPORE DEVELOPMENT IN *BACILLUS SUBTILIS* AND *BACILLUS POLYMYXA*

J. SZULMAJSTER, P. KERJAN, K. JAYARAMAN* and S. MURTHY*

Laboratoire d'Enzymologie du C.N.R.S., 91190 Gif-sur-Yvette, France
**Department of Molecular Biology, School of Biological Sciences, Madurai Kamaraj University, Madurai 625021, India*

SUMMARY

During sporulation of *B. subtilis* and *B. polymyxa* there is formation of a polyadenylated RNA population. Very little poly(A)RNA was found in vegetative cells or in stationary-phase cells of mutants blocked at the zero stage of sporulation. The antibiotic netropsin, known to inhibit sporulation at a very early stage, inhibits poly(A)RNA synthesis. Based on the property of netropsin to bind to AT sequences in DNA it is suggested that early spore genes might be enriched in d(AT) sequences. Hybridization studies with cDNA probes synthesized on poly(A)RNA and RNA isolated from different stages of growth and sporulation have further confirmed the association of poly(A)RNA and the sporulation process. Double stranded cDNA has been cloned into the pHV33 plasmid vector and the analysis of the inserts in the hybrid plasmids isolated from the selected clones are in progress. The possible role of poly(A)RNA in sporulation is discussed.

INTRODUCTION

Studies on the control mechanisms operating during sporulation at the level of transcription are mainly pursued by two approaches: one is concerned with the structure and mode of action of DNA dependent RNA polymerase, the other deals with the cloning and expression of spore genes.

An appealing feature in regulation of gene expression during sporulation emerged recently from studies on a possible role of polyadenylated RNA in the sporulation processes of *B. polymyxa* and *B. subtilis*, thus adding a new dimension to the control mechanisms involved in bacterial spore development.

METHODS

Bacterial strains, plasmid and growth media: *B. subtilis* 168M (trpC2), its asporogenic derivative 12A and *E. coli* C600 strain carrying the plasmid pHV33 were grown as described previously (1).

Isolation of poly(A)RNA and its labelling by $\gamma(^{32}P)ATP$ were described previously (1,3).

Single and double-stranded cDNA were essentially synthesized according to the procedure of Savage *et al.* (4).

Poly(A)RNA-cDNA hybridization was described elsewhere (1).

Detection of complementary sequences for poly(A)RNA in the *B. subtilis* chromosome was carried out by Southern blots (5).

Isolation of plasmid DNA, endonuclease digestion and ligation were described previously (1).

Agarose gel electrophoresis: DNA was analysed routinely on horizontal 0.8% agarose slab gels in TEB buffer (Tris 89 mM, pH 6; EDTA 2.5 mM; boric acid (89 mM). The samples were mixed with one-tenth volume of glycerol containing 0.8% bromophenol blue and applied to the gel (40 V, 15 h). Gels were stained for 30 min in electrophoretic buffer with ethidium bromide (0.5 μg/ml) and photographed under short-wave UV light.

RESULTS AND DISCUSSION

We have previously shown that during sporulation of *B. subtilis* there is an increase of a poly(A)RNA population that is very low in vegetative cells or in stationary-phase cells of a zero-stage asporogenic mutant (2). Similar results were obtained with *B. polymyxa* (6,7). In the latter case it was also shown that a zero-stage mutant did not show any increase in poly(A)RNA synthesis while in two other mutants blocked at later stages of sporulation (stage II and stage IV) a net increase was observed, although their maximal levels of poly(A)RNA were lower than those observed in the wild-type (7). These results indicated that in *B. polymyxa*, the ability to form poly(A)RNA during sporulation is manifested between stages 0 and II, while in *B. subtilis* the maximum of poly(A)RNA synthesis was reached between t_3 and t_6 (2).

Location and Length of Poly(A) Sequences

Analysis by sucrose gradient centrifugation followed by hybridization with (^{3}H)poly(U) have shown that the poly(A)RNAs of *B. subtilis* sediment with a peak at 6s and a broad shoulder between 5 and 15s reflecting the heterogeneous poly(A)RNA populations, comparable to that found for eukaryotic polyadenylated mRNAs.

The location of the poly(A) sequences in the RNA chains of *B. subtilis* was determined by post-labelling of the 3'-OH terminal with (^{3}H)-Na-borohydride (8), followed by adsorption on oligo(dT)-cellulose. The radioactivity bound thus measures the amount of poly(A) tracts located at the 3' end. By combining this procedure with the treatment by ribonucleases A and T_1 followed by alkaline phosphatase action we have been able to determine the relative amounts of the poly(A) stretches located at the 3' terminus and those located inside the RNA chains (2). The lengths of the poly(A) sequences was estimated by polyacrylamide gel electrophoresis after digestion of the poly(A)RNA by RNAses A and T_1 followed by post-labelling of 3'OH terminal with (^{3}H)Na-borohydride. Synthetic poly(A) of different lengths were used as markers. From the electrophoretic mobility of the radioactively labelled poly(A) samples we have shown that in *B. subtilis* sporulating cells about 65% of the total poly(A) sequences located at the 3'-termini of the RNA chains are about 120-160 nucleotides long, whereas about 35% of poly(A) stretches are located in the internal regions of the RNA chains and contain sequences of about 10-15 nucleotides not retained on oligo(dT)-cellulose.

In *B. polymyxa* the length of the poly(A) tracts at the 3' end was found not to exceed 30 nucleotides and RNA species with even shorter stretches appear during early hours of sporulation (7).

Poly(A)RNA-cDNA Hybridization

For more information on the nature and role of the poly(A)RNA in sporulation of *B. subtilis*, we synthesised complementary DNA (cDNA) probes for poly(A)RNA and determined their hybridization specificity with RNAs isolated from various sporulation stages or from stationary phase cells of a zero stage sporulation mutant or from vegetative cells. The average size of the cDNA was determined by alkaline sucrose gradient centrifugattion and shown to be of about 6 - 9.5s, that is about 400-1300 nucleotides. This cDNA was used to hybridize with RNA species at different stages of growth and sporulation. It can be seen in Fig. 1 that no hybridization was observed when total RNA isolated from either vegetative or from t_4 stationary-phase

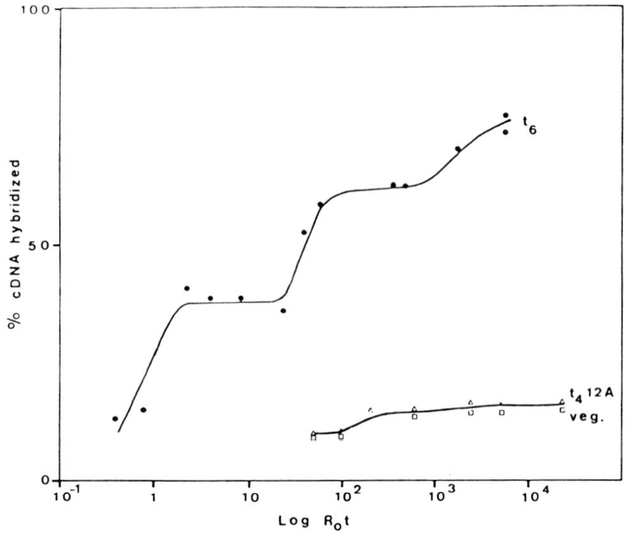

FIG. 1 *Hybridization of single stranded cDNA probes with different poly(A)RNA preparations:* ●——● *t_6 purified poly(A)RNA,* △——△ *t_4 total RNA from a zero stage sporulation mutant,* □——□ *total RNA from vegetative cells.*

cells of a zero stage asporogenic mutant of *B. subtilis* (12A) was used. However when purified poly(A)RNA from t_6-sporulating cells was hybridized to its homologous cDNA probe in the presence of an excess of vegetative RNA, three distinct kinetic classes of abundancies were observed and their R_ot values are shown in Table 1.

The hybridization efficiency of t_4 cDNA with homologous and heterologous poly(A)RNA in the presence of an excess of vegetative RNA is shown in Table 2. It can be seen that, if

TABLE 1 *Analysis of hybridization of t_6 cDNA to t_6 poly(A)RNA shown in Fig. 1.*

	% hybridizable cDNA	R_ot 1/2 (moles x s/l)
Component I	40.7	0.54
Component II	26	30
Component III	14	973

TABLE 2 *Hybridization of t_4-^3H-cDNA with poly(A)$^+$RNA isolated from different stages of sporulation*

cDNA from stage:	poly(A)RNA from stage:	cpm hybridized	%
t_4	t_4	675	28
t_4	t_4	2336	100
t_4	t_6	1859	79

the hybridization efficiency of t_4 cDNA with t_4 poly(A)RNA is taken as 100%, the per cent hybridization of the same cDNA with the poly(A)RNA isolated from t_2 cells or t_6 cells was 28 and 79% respectively. These figures most likely reflect the differences in the population of poly(A)RNA present at the two specific sporulation stages. This is in agreement with the generally accepted idea that at each stage of sporulation different classes of mRNA are expressed in a sequential manner (9,10,11).

Detection of Sequences Complementary to Poly(A)RNA in the *B. subtilis* Chromosome

The next step in these studies was the detection, on the *B. subtilis* chromosome, of sequences complementary to poly(A)RNA. This was achieved by submitting *B. subtilis* DNA to BamHI endonuclease restriction, followed by separation of the resulting fragments by agarose gel electrophoresis. These fragments were then denatured and transferred by blotting on strips of nitrocellulose paper (5) and used for hybridization with radioactive probes prepared by labelling (with ^{32}P-γ-ATP in presence of polynucleotide kinase) poly(A)RNA isolated from t_4 cells. Figure 2 shows the microdensitometer tracing of the radioautograph of the hybrids detected by the ^{32}P-labelled poly(A)RNA. It can be seen that there are five major sets of fragments of 2.3 - 6.1 Kb which retained the highest radioactivity.

Molecular Cloning of *B. subtilis* DNA fragments Hybridizable to poly(A)RNA

The five BamHI endonuclease restriction fragments which were found to hybridize efficiently to poly(A)RNA were ligated to

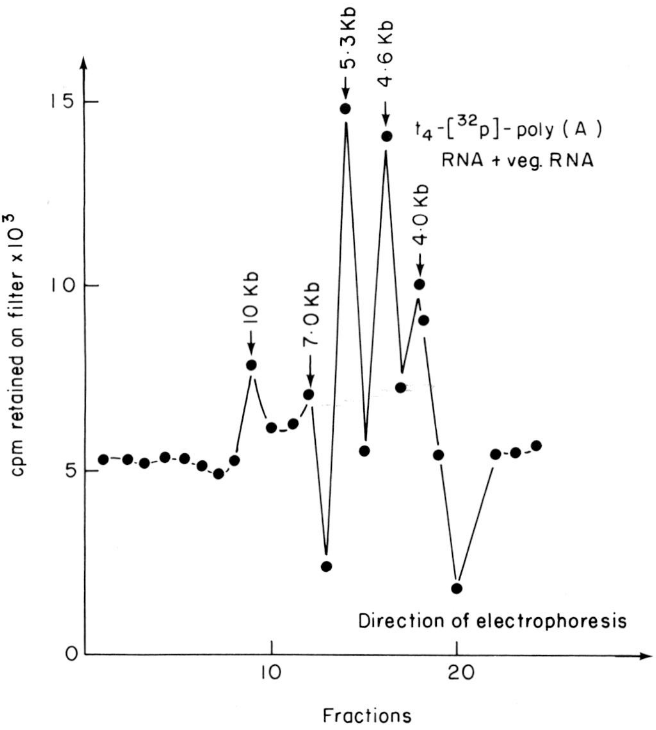

FIG. 2 *Microdensitometer tracing of a radioautograph obtained from Southern blot of BamHI endonuclease restricted fragments of* B. subtilis *DNA detected by* ^{32}*P-labelled poly(A)RNA.*

the dephosphorylated plasmid vector pHV33 (12). This plasmid (MW 4.7 Md) is a bifunctional vector which can replicate and express its resistance markers in *E. coli* (Ap^r, Tc^r, Cm^r) and *B. subtilis* (Cm^r). Among the Amp^r Tc^s clones (insertional inactivation of the Tc^r gene) about 20 were found to hybridize *in situ* (colony hybridization) with labelled poly(A)RNA. Analysis of clear lysates (13) of these clones by agarose gel electrophoresis revealed that some of them were carrying plasmids with DNA inserts of 1.5 - 2.0 Kb, some others contained inserts of smaller size, most likely generated by spontaneous deletions (14).

Cloning of Double Stranded cDNA Transcribed from Poly(A)RNA

Using the same plasmid vector we have also cloned double stranded cDNA transcribed from poly(a)RNA. For this purpose single-stranded (SS) cDNA was synthesized from t_5 poly$(A)^+$RNA using reverse transcriptase and $(dT)_{12-18}$ according to the method described by Savage *et al.* (4). The ss cDNA was extended at its 3' end by a poly(dA) homopolymeric tail (20-30 nucleotides) using (^{3}H)dATP and terminal transferase (15). The tailed cDNA was made double stranded with reverse transcriptase and $(dT)_{12-18}$ followed by treatment with DNA polymerase I (see legend to Fig. 2). The blunt-ended ds cDNA was then extended at its 3' end with dATP and terminal transferase.

Similarly, poly(dT) tails of about 20 nucleotides were added to the PstI cleaved pHV33 plasmid DNA that had been purified through phenol extraction and alcohol precipitation steps. The tailed plasmid (3 μg) and the tailed ds cDNA (600 ng) were annealed and used to transform *E. coli* C600. Transformants were selected on LB plates containing Cm (3 μg/ml) and Tc (5 μg/ml).

Analysis of Hybrid Plasmids Carrying cDNA Inserts

From 19 transformants selected for their resistance to chloramphenicol and tetracycline, 16 were shown to be ampicillin sensitive. Analysis of polyacrylamide gel electrophoresis of clear lysates (13) of these clones showed that the cDNA inserts in the plasmids vary from 300-750 nucleotides. One of these plasmids, pGcDNA10, carrying the cDNA insert of 750 bp was purified by CsCl/ethidium bromide equilibrium gradient centrifugation and submitted to further analysis. Since the cDNA insert was introduced into the PstI site of the cloning vector by the poly(dA-dT) joining method, the site was lost; therefore to excise this insert it was necessary to treat the

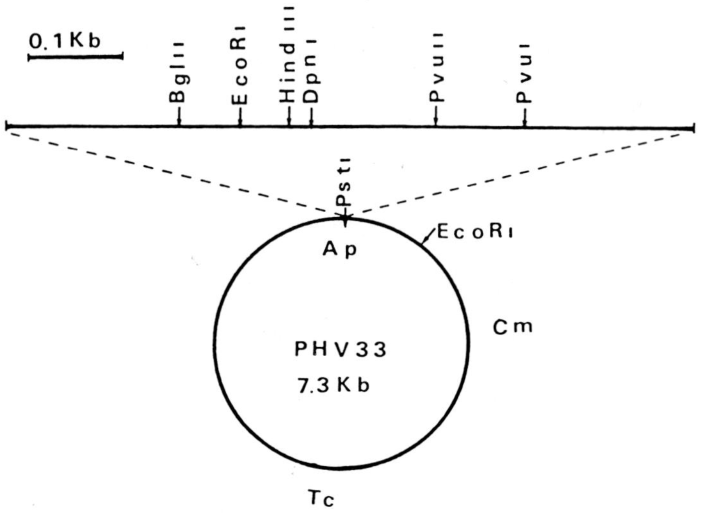

FIG. 3 *Endonuclease cleavage map of the cDNA insert (750 bp) of the pGcDNA10 hybrid plasmid.*

hybrid with exonuclease VII according to Goff and Berg (16). Fig. 3 shows the endonuclease cleavage map of the cDNA insert.

Is Poly(A)RNA Transcribed from DNA Segments Enriched in early Sporulation Genes?

The results of two different sets of experiments seem to favour this hypothesis.

Effect of netropsin. It was shown that this polypeptide antibiotic inhibits sporulation of *B. subtilis* without affecting vegetative growth (17). This is also true with *B. polymyxa* (Table 3). We have further shown that in both microorganisms netropsin strongly inhibits the formation of poly(A)RNA (Fig. 4 and Table 3). Based on the property of netropsin to inhibit DNA and RNA polymerase by binding to (dA-dT)-rich regions of template DNA and on the fact that netropsin treated cells are blocked at the O-I stage of sporulation, it has been suggested that early sporulation genes may be enriched in (AT) pairs (17,18). Experiments are in progress to further verify this hypothesis on the cloned spoOB gene of *B. subtilis* (19). On the other hand a detailed endonuclease restriction map of the cloned cDNA and of the cloned DNA fragments hybridizable to poly(A)RNA reported here should also provide information on the nature of this DNA.

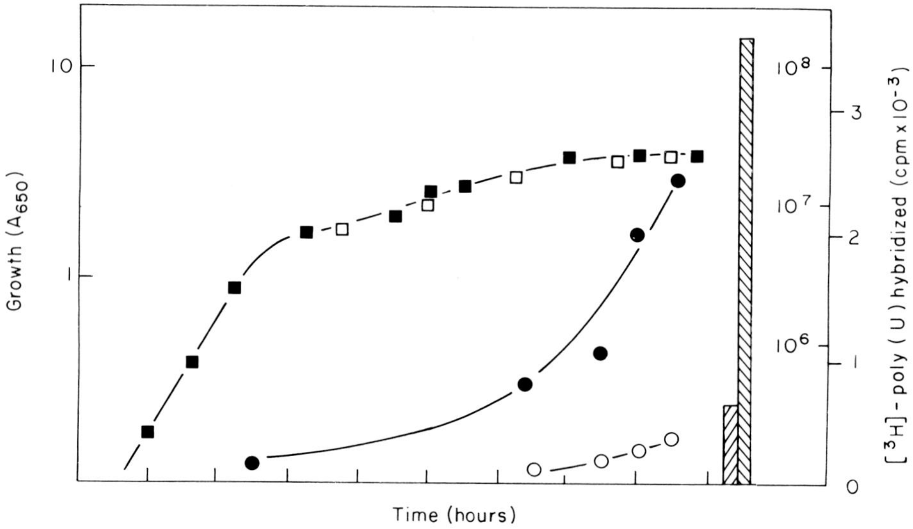

FIG. 4 *Effect of netropsin on sporulation and poly(A)RNA synthesis in* B. subtilis. *Growth in absence (■) and presence (□) of netropsin: Poly(A)RNA's synthesis in absence (●) and presence (○) of netropsin.*

TABLE 3 *Effect of netropsin on the synthesis and level of poly(A)RNA in* B. polymyxa.

	Poly(A)RNA formation as measured by:			
	Poly(U) filter binding*		(^{3}H)-poly(U) hybridization**	
Stage	-NET	+NET	-NET	+NET
Veg	1.4	0.65	800	250
t_2	10.6	1.10	12,070	400

*) Expressed as % labelled RNA bound to poly(U) filters.
**) Expressed as cpm of (^{3}H)-poly(U) hybridized/1 $O.D._{260nm}$ of unlabelled total RNA.

Transformation. When the hybrid plasmid DNA from about ten clones selected for their efficient hybridization to poly-(A)RNA was used for transformation of several early blocked sporulation mutants it was found that the DNAs from 3 of these clones were able to transform the spoOB, rec^- mutant to spo^+ with high efficiency (1). Although more clones need to be tested to draw a definite conclusion, these results suggest nevertheless that at least some poly(A)RNA appearing during sporulation might be transcribed from DNA segments enriched in early spore genes.

Mechanism and Signal for Polyadenylation

Analysis of the 3'-terminal sequence of a great number of eukaryotic mRNAs has revealed a common sequence of 5'-AAUAAA-3' in the 3'-untranslated regions between 11 and 30 nucleotides from the start of the poly(A) sequence (20). Furthermore, it was observed that deletions in the hexonucleotide sequence prevent polyadenylation of late SV40 mRNAs. It is also known that polyadenylation of a great number of mRNAs seems to involve two steps: an endonucleolytic cleavage of a primary transcript followed by poly(A) addition (20). Thus, polyadenylation seems to be a post-transcriptional event mediated by specific enzyme(s).

In bacteria the mechanism of polyadenylation remains unknown and there is also no evidence for the existance of a signal for polyadenylation. The availability of the cloned cDNA reported here offers the possibility to determine the nucleotide sequence adjacent to the poly(A)RNA sequence. This is presently being explored in our laboratory.

Function of Poly(A)RNA

In spite of the relatively large amount of information accumulated in the last 10 years on poly(A)RNA in eukaryotic mRNA very little is known about its regulatory function (for review, (21)).

Functional mRNAs exist which lack poly(A) sequences, therefore they have no essential role in gene expression, either at the level of transport from the nucleus to the cytoplasm or at the level of translation. Furthermore, vaccinia virus mRNA appears in polyadenylated and in non-adenylated forms, and both are translated *in vitro* with the same efficiency.

There are cases, however (for example encephalomyorditis and poliovirus) where mRNA also appears in two forms, but the presence of poly(A)RNA improves the infectivity (21).

Possible Role in Sporulation Process

There is now direct evidence in support of the assumption that the bacterial poly(A)RNA species act as mRNA, as they do in eukaryotic systems. Thus the isolation from *Bacillus brevis* of highly active protein synthesizing polysomes containing polyadenylated sequences has been recently reported by Kaufer *et al.* (22). These authors have also observed increasing amounts of poly(A)RNA at the end of exponential growth i.e. at the onset of sporulation. A concentration-dependent stimulation by poly(A)RNA of protein synthesis in an *in vitro B. subtilis* protein synthesizing system has been observed in our laboratory (unpublished results). Similar results have been previously reported by Graef *et al.* (23). Although the precise role of poly(A)RNA in sporulation still remains unknown, we would like to suggest a possible role based on our earlier finding that sporulating *B. subtilis* cells contain an exonuclease which degrades polynucleotides from the 3' end in the direction of the 5' terminus (24). It is therefore tempting to speculate that polyadenylation might be necesaary to protect the newly synthesized mRNA from degradation during sporulation. It is well known that prokaryotes as well as eukaryotes are able to control the synthesis of many proteins by keeping the amounts of their corresponding mRNAs at a certain level. This could be accomplished by controlling either the rate of degradation or protection of these mRNAs.

REFERENCES

1. Kerjan, P., Jayaraman, K. and Szulmajster, J. (1982) *Mol. Gen. Genet.* **185**, 448-453.
2. Kerjan, P. and Szulmajster, J. (1980) *Biochem. Biophys. Res. Commun.* **93**, 201-208.
3. Donis-Keller, H., Maxam, A.M. and Gilbert, W. (1977) *Nucl. Ac. Res.* **4**, 2527-2538.
4. Savage, M.J., Sala-Trepat, J.M. and Bonner, J. (1978) *Biochemistry* **17**, 462-467.
5. Southern, E.M. (1975) *J. Mol. Biol.* **98**, 503-517.
6. Kaur, S. and Jayaraman, K. (1979) *Biochem. Biophys. Res. Commun.* **86**, 331-339.
7. Jayaraman, K. and Murthy, S. (1982) *Mol. Gen. Genet.* **185**, 158-164.
8. Randerath, K. and Randerath, E. (1971) *Proc. Natl. Acad. Sci. USA* **69**, 1408-1412.
9. DiCioccio, R.A. and Strauss, N. (1973) *J. Mol. Biol.* **77**, 325-336.
10. Sumida-Yasumoto, C. and Doi, R.H. (1974) *J. Bacteriol.* **117**, 775-782.
11. Bonamy, C., Manca de Nadra, M.C. and Szulmajster, J. (1976) *Eur. J. Biochem.* **63**, 53-63.
12. Ehrlich, S.D. (1978) *Proc. Natl. Acad. Sci. USA* **75**, 1433-1436.
13. Klein, R.D., Selsing, E. and Wells, R.D. (1980) *Plasmid* **3**, 88-91.
14. Primrose, S.B. and Ehrlich, S.D. (1981) *Plasmid* **6**, 193-201.
15. Nelson, T. and Brutlag, D. (1979) *Methods Enzymol.* **68**, 41-50.
16. Goff, S.P. and Berg, P. (1978) *Biochemistry* **75**, 1763-1767.
17. Keilman, G.R., Tanimoto, B. and Doi, R.H. (1975) *Biochem. Biophys. Res. Commun.* **67**, 414-420.
18. Beaman, B.L., Burtis, K.C. and Doi, R.H. (1980) *Can. J. Microbiol.* **26**, 420-426.
19. Bonamy, C. and Szulmajster, J. (1982) *Mol. Gen. Genet.* (in press).
20. Fitzgerald, M. and Shenk, T. (1981) *Cell* **24**, 251-260.
21. Littauer, U.Z. and Soreq, H. (1982) *Prog. Nuc. Ac. Res.* **27**, 53-83.
22. Kaufer, N., Altman, M. and Dohren, H.W. (1981) *FEMS Microbiol. Lett.* **12**, 71-75.
23. Graef-Doods, E. and Chambliss, G.H. (1978) in *Spore VII* (Chambliss, G. and Vary, J.C., eds.), Am. Soc. Microbiol., Washington D.C., pp.237-241.
24. Kerjan, P. and Szulmajster, J. (1976) *Biochimie* **58**, 533-541.

THE ROLE OF HIGHLY PHOSPHORYLATED NUCLEOTIDES IN SPORULATION

†H.J. RHAESE, R. VETTER and U. KIRSCHNER

Institut für Mikrobiologie, Molekulare Genetik, Universität Frankfurt, Frankfurt, Federal Republic of Germany

SUMMARY

Adenosine-5',3'-bis triphosphate (p_3Ap_3) has been synthesized chemically and its structure determined by physico-chemical and enzymatic methods. Comparison with p_3Ap_3 from *B. subtilis* confirms the former tentative structure of this unusual nucleotide. Cloning and expression of the *spoOF* gene in *E. coli* and *B. subtilis* show that it codes for the enzyme adenosine-bis triphosphate (abt) synthetase. Antibodies against synthetic p_3Ap_3 specifically precipitate p_3Ap_3 in extracts of *E. coli* and *B. subtilis* strains containing cloned abt-synthetase (*spoOF*) gene. A functional abt-synthetase is clearly needed for sporulation.

INTRODUCTION

Initiation of sporulation in *B. subtilis* is caused by starvation of carbon, nitrogen, or phosphate sources. In searches for the molecular mechanism(s) of this phenomenon, several models have been proposed in recent years (1,2). None has yet been proven conclusively to be correct. The great complexity of the differentiation process in *B. subtilis* is likely to prevent the emergence of a complete picture of initiation within the immediate future.

Correlations between synthesis of signal molecules or certain changes in the concentration of micro- and macro-molecules and sporulation have been obtained, which support one or the other model.

In our efforts to continue to examine the sensor model of

initiation of sporulation proposed by us recently (3) we tried to ascertain the importance of the highly phosphorylated nucleotide p_3Ap_3 for sporulation. This presumed signal molecule has been shown to be synthesized upon carbon source starvation (4) and its appearance has been correlated with sporulation (5,6). Our finding, that a mutant (JH649) among a set of early blocked *spo0* mutants (5) cannot synthesize p_3Ap_3, allowed us to identify an Eco R1 restriction fragment of 1.3 megadalton (∿2 kb) size, which apparently carries the gene for the p_3Ap_3 synthesizing enzyme abt-synthetase. Transformation of the above mutants with this fragment resulted in sporogenous transformants which were also able to synthesize p_3Ap_3 (3).

We cloned this 1.3 megadalton fragment in mutant JH649 using the *B. subtilis* plasmid pBS161-1 (3) but obtained unstable clones, which always gave rise to sporogenous recombinants.

In order to investigate the effect of p_3Ap_3 on sporulation and to show that the cloned *spo0F* gene indeed codes for abt-synthetase, we recloned the 1.3 megadalton fragment in *E. coli* C600 and *B. subtilis* BD224 using the vectors pBR325 and pUB110, respectively. Furthermore, we intended to show that p_3Ap_3 synthesis is necessary for sporulation in our *B. subtilis* strains.

In this paper we report on the chemical synthesis of p_3Ap_3, present some data on its structure and then establish its identity with natural p_3Ap_3. Antibodies obtained with synthetic p_3Ap_3 will be shown to react with p_3Ap_3 synthesized in *E. coli* by the cloned *spo0F* gene protein. Finally, we describe the effect of increased p_3Ap_3 synthesis in a clone of *B. subtilis* BD224 carrying the *spo0F* gene on plasmid pUB110.

MATERIALS AND METHODS

Bacteria and Growth Conditions

B. subtilis strains 60015, JH649 and BD224 were grown in SYM (8). Spore numbers were determined as previously described (9).

Transformations

Transformations in *E. coli* (9) and *B. subtilis* (10,11) were performed as previously described.

Biochemical Methods

All methods concerning labelling, analysis, and measurements of p_3Ap_3 synthesis have been described previously (8).

Restriction enzymes and DNA ligase were obtained from New England Biolabs and used according to the protocol of the manufacturer.

Plasmids were isolated according to Birnboim (12).

Physico-chemical Methods

All methods used to elucidate the chemical structure of p_3Ap_3 and those used to compare synthetic with natural p_3Ap_3 will be described elsewhere (Rhaese, Kirschner, Zacharias; to be published).

Chemical Synthesis of Adenosine-bis triphosphate

A solution of 0.18 mmoles adenosine-5',3'-bis monophosphate (pAp) as tributylammonium salt dissolved in 3 ml 2-methyl-puridine and 0.39 ml tributylamine was treated with 7 moles diphenyl chloro-phosphate, dissolved in 1.63 ml dioxane for 2 h at 20°C. The reaction mixture was then treated with 2 ml of ether, the resulting precipitate washed once with 2 ml ether and the precipitate dried over phosphorus pentoxide in a desiccator. Tributylammonium pyrophosphate (0.55 mmoles) was dissolved in 0.6 ml dried dimethylformamide and added together with 0.5 ml of pyridine to the dried reaction mixture, which was then stirred for 45 min at room temperature. After flash evaporation to almost dryness, 5 ml of water were added, and the resulting solution extracted three times with 3.5 ml of ether. Isolation by column chromatography using DEAE-Sephadex A25 and a linear gradient of 0.2 - 0.6 M LiCl yields almost (90%) pure p_3Ap_3. This can be further purified using a step gradient of ammonium formate, pH 3.4 (13).

Preparation of Antiserum against Adenosine-bis triphosphate

Adenosine triphosphate (30 mg dissolved in 0.5 ml Freunds complete adjuvant and 0.5 ml of des.H_2O were injected s.c. into two rabbits. After 17 days 46.8 mg in 1 ml dest. H_2O were injected i.v. After 40 days the animals were sacrificed and serum prepared according to general procedures. The serum was stored at -80°C.

RESULTS

Chemical Synthesis of Adenosine-bis triphosphate

We have calculated that p_3Ap_3 may reach a concentration of maximally 1000 molecules within sporulating *B. subtilis* cells (Rhaese *et al.* to be published elsewhere). This amount is not sufficient to allow isolation of quantities large enough to induce antibody formation in a mammal. Therefore, we tried to synthesize p_3Ap_3 chemically.

The reaction scheme for synthesis of p_3Ap_3 is outlined in Fig. 1. Adenosine-5',3'-bis monophosphate (Serva) is activated according to Michelson (14) with diphenyl chloro-phosphate. The resulting triester is then treated with pyrophosphate. From the reaction mixture, p_3Ap_3 can be isolated in yields between 5 and 10% by column chromatography (see above).

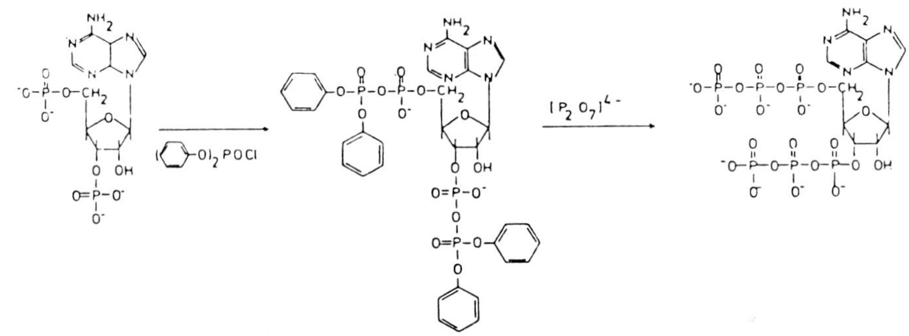

FIG. 1 *Reaction mechanism for the chemical synthesis of p_3Ap_3. Adenosine-5',3'-bis monophosphate (pAp) is treated with diphenyl chloro-phosphate. The resulting "activated" pAp is then pyrophosphorylated using inorganic pyrophosphate to yield p_3Ap_3.*

Properties of Chemical and Natural Adenosine-bis triphosphate

Isolation of p_3Ap_3 from *B. subtilis* (as ^{32}P-labelled material), *S. cerevisiae*, and *Tolipothrix sp* can be achieved by column chromatography as described (13). This material may be further purified by preparative high pressure liquid chromatography (HPLC) or column chromatography on DEAE sephadex (13).

Using both synthetic and natural p_3Ap_3 to investigate some of their physico-chemical properties, we obtained the following data summarized in Table 1. As can be seen, there is no difference between synthetic and natural p_3Ap_3. Furthermore, these data show that we have indeed synthesized p_3Ap_3 using pAp as basic material. Synthetic and natural p_3Ap_3 behave identically upon thin-layer chromatography (TLC) on PEI plates in different solvent systems. Likewise, the retention times on a SAX-10 or MCH-10 (reversed phase) column are identical. The UV-spectra of both substances are indistinguishable from that of ATP with a maximum at 258 nm. Field desorption mass spectroscopy yields a mass peak of 760 daltons, which corresponds to the dilithium

TABLE 1 *Some physico-chemical properties of adenosine-bis-triphosphate*

Physico-chemical method	Properties of Synthesised p_3Ap_3	Properties of natural p_3Ap_3
TLC on PEI, 1,5 M phosphate	R_F 0,08 - 0,09[1]	R_F 0,08 - 0,09[1])
HPLC, SAX-10 column	6,6 min[2])	6,6 min[2])
HPLC, reversed phase	33 min	32 - 33.5 min
UV maxima, wavelength	258 nm	258 nm
Mass spectroscopy, m.w. of di-lithium salt	760 d	760 d
Ratio of base/ribose/phosphate	1:1:6	1:1:6
^{31}P-NMR signals at pH 3.6	0,0 to 2ppm -7 to -12ppm -22 to -24ppm	0,0 to 2ppm -7 to -12ppm -22 to -24ppm

[1]relation to ATP
[2]retention time

salt of p_3Ap_3. Both substances have been shown to contain adenine, ribose, and phosphate in ratios of 1:1:6 and ^{31}P-NMR spectroscopy indicates that both molecules contain two α, two β, and two γ phosphates representing the 5' and the 3'triphosphates, respectively, of either side of p_3Ap_3.

Further evidence for the structure of both synthetic and natural substances described above is obtained from experiments summarized in Table 2. As shown, there is no difference between both substances concerning the hydrolysis by enzymes, acid or alkali nor in the periodate reaction. Bacterial alkaline phosphatase hydrolyzes fast giving rise to adenosine as the final product. Intermediate phosphorylated adenosine compounds have been observed, presumably because of stepwise hydrolysis. This reaction shows that there are monoesters in this molecule.

TABLE 2 *Enzymatic and chemical hydrolysis of adenosine-bis-tri-phosphate*

Enzymes and chemicals	Reaction and degradation products of synthetic p_3Ap_3 and natural p_3Ap_3
Bacterial alkaline phosphatase	fast hydrolysis, adenosine
Snake venom phosphodiesterase	slow hydrolysis, several phosphorylated products
Spleen phosphodiesterase	very slow hydrolysis, several phosphorylated products
ATPase	no reaction
RNase T2	no reaction
DNase	no reaction
Proteinase K	no reaction
Lipase	no reaction
Periodate oxidation	no reaction
1 N HCl, 1-0°C, 2 h	mostly adenine, some adenosine
0.3 N KOH	several unidentified degradation products

Slow and very slow hydrolysis by snake venom and spleen phosphodiesterases, respectively, is consistent with the observation that nucleoside triphosphates (like ATP) are hydrolyzed much slower than di-, oligo- or polynucleotides. Since p_3Ap_3 is a nucleoside-bis-triphosphate, its slow to very slow reaction, which has been described before (2) can be explained. Other enzymes like ATPase, RNase T2, DNase, proteinase K or lipase have no effect.

Hydrolysis by 1 N at 100°C for 2 h gives adenine as the only reaction product and 0.3 N KOH degrades p_3Ap_3 to give several phosphorylated substances, which have not been investigated further.

Details of the above described reactions have been (2) or will be published elsewhere.

The above described experiments show that we have indeed been able to synthesize adenosine-5',3'-bis triphosphate which is indistinguishable from a substance isolated from differentiating organisms like *B. subtilis*, *S. cerevisiae*, and the cyanobacterium *Tolipothrix sp.*

Antiserum against Adenosine-5',3'-bis triphosphate

In order to detect p_3Ap_3 in biological specimens, especially to verify cloning of the adenosine-bis triphosphate (abt) synthetase gene in *E. coli* (see below), we immunized two rabbits with synthetic p_3Ap_3 together with Freunds complete adjuvant as described in Materials and Methods. These sera

TABLE 3 *Binding of adenosine-bis-tri-phosphate to rabbit antiserum*

Serum		binding of $^3H-p_3Ap_3$ cpm	%	^3H-ATP cpm	%
7436	before[1]	32.500(100)		12.800(100)	
			75		12
	after[1]	8.400(25)		11.200(88)	
7437	before	32.900(100)		12.800(100)	
			61		12
	after	12.800(39)		11.300(88)	
control	before	34.200(100)		12.400(100)	
					0
	after	25.800(75)		12.400(100)	

1) before and after filtration through Amicon membranes

were tested for binding of $^3H-p_3Ap_3$ and ^3H-ATP.

As can be seen in Table 3, the test sera bind 75% and 61% of input p_3Ap_3, whereas the control serum binds 25%. Atp binds to all sera only weakly.

These data show that we have obtained antibodies against p_3Ap_3 and that these can be used to detect p_3Ap_3 in biological material. The specificity of the sera can certainly be improved since we have not yet coupled p_3Ap_3 to any strong immunogenic

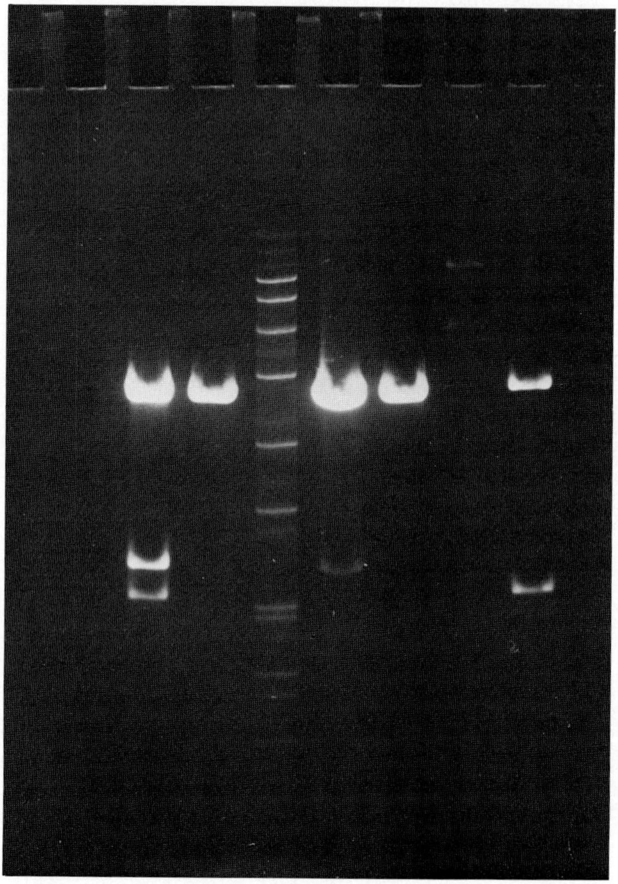

FIG. 2 *Agarose (1%) gel electrophoresis of Eco R1 treated lysates (12) of six clones from strain BD224 which was transformed by plasmid pUB110 with an insertion of a 1.3 megadalton fragment obtained from a λ clone (8). The 1.3 megadalton fragments present in clones 1 and 6 are visible as the fastest migrating bands in lanes 1 and 6.*

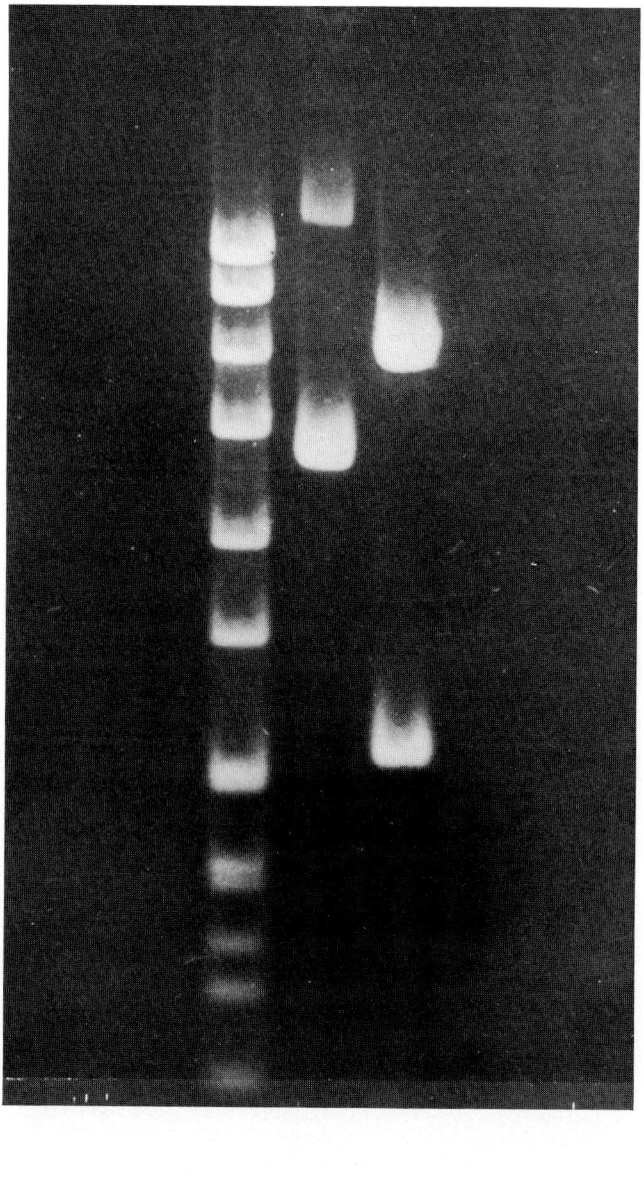

FIG. 3 *Agarose (1%) gel electrophoresis of the hybrid plasmid pBR325* spoOF *isolated from clone 10 of* E. coli *C600 (lane 2) and the same plasmid after Eco R1 restriction endonuclease treatment. The 1.3 megadalton fragment is the fastest moving band in lane 3. Lane 1 shows Eco R1 fragments of phage SPP1 as length standard.*

substance. But work is in progress to obtain monoclonal antibodies with high specificity.

Cloning of Adenosine-bis triphosphate Synthetase in *E. coli* and *B. subtilis*

As described previously (3,8) we have cloned a 1.3 megadalton fragment carrying the *spoOF* gene of *B. subtilis* in plasmid pBS161-1 and in phage λ. Since plasmid pBS161-1 is unstable in a recE4 strain, we tried to reclone the fragment in plasmid pUB110 in order to study in *B. subtilis* sporulation the role of the *spoOF* gene, presumed to code for abt synthetase.

For this purpose, the agarose-gel purified fragment was ligated into the Eco R1 site of plasmid pUB110 and strain BD224 (recE4) was transformed to kanamycin and neomycin resistance. Six antibiotic resistance clones were selected and plasmids isolated. As can be seen in Fig. 2, clone 1 (lane 1) and 6 (lane 6) contain fragments of approximately 1.3 megadalton. After transformation of strain JH649 (*spoOF*) with these fragments, only clone 6 gave 2.4 x 10^2 transformants/ml.

The 1.3 megadalton fragment of clone 6 was then recloned into the chloramphenicol resistance gene of plasmid pBR325 in *E. coli* C600. Out of 36 ampicillin and tetracycline resistant but chloramphenicol sensitive clones, clone 10 was used to isolate the recombinant plasmid. As shown in Fig. 3, the 1.3 megadalton fragment can be reisolated from this plasmid after Eco R1 cleavage (lane 3). Lane 2 shows plasmid pBR325/*spoOF*.

The reisolated fragment was used to transform mutant JH649 (*spoOF*). Between 5 x 10^2 and 10^3 transformants/ml were obtained in three different experiments.

Expression of the Cloned abt Synthetase Gene in *E. coli* C600 and *B. subtilis* BD224

To study the effect of p_3Ap_3 in *B. subtilis* via cloning and amplification of the abt synthetase gene, it was necessary to show that the 1.3 megadalton fragment carries the entire gene and to test, whether or not this gene is carried in both *E. coli* and *B. subtilis*.

For this purpose *E. coli* C600(pBR325), C600(pBR325/*spoOF*), *B. subtilis* BD224(pUB110) and BD224(pUB110/*spoOF*) were grown in SYM medium to late log phase in the presence of 0.5 mCi/ml of $^{32}P-H_3PO_4$. Starvation for nutrients and extraction of p_3Ap_3 was performed as described before (6,8). The extracts (2 ml each) were then applied on top of a DEAE-Sephadex A25 column (0.4 cm x 3.5 cm) and eluted with a linear gradient of 0.1 to 1 M LiCl (21 ml in each vessel). Fractions of 0.75 ml were collected. As shown in Fig. 4, an extract of strain C600 (pBR325/*spoOF*) contains p_3Ap_3 in remarkable amounts (lanes 4 and 5). An identical elution profile was obtained from an extract of BD224(pUB110/*spoOF*). In contrast, *E. coli* C600 (pBR325) containing the plasmid without *spoOF* gene does not synthesize any p_3Ap_3 as is shown by comparing several identical fractions from both *E. coli* strains (Fig. 5a and b). There is no doubt that *E. coli* containing the hybrid plasmid pBR325/*spoOF* produces a substance which seems to be identical with synthetic and natural p_3Ap_3 produced by sporulating cells of *B. subtilis*.

When p_3Ap_3 produced by strain DN224(pUB110) and BD224 (pUB110/*spoOF*) was measured quantitatively it was found that the strain carrying the *spoOF* gene on plasmid pUB110 produces

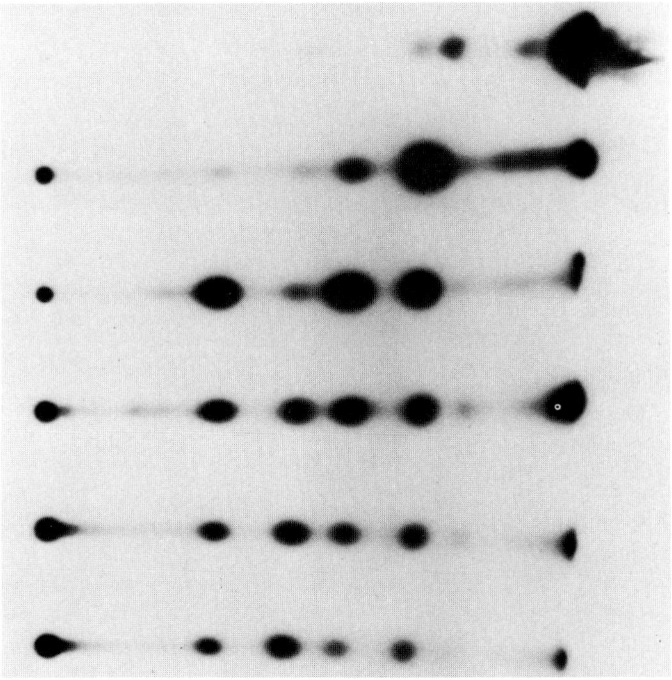

FIG. 4 *Autoradiogram of partially purified formic acid extracts from* E. coli *C600 (pBR325* spoOF*). Different fractions were spotted on PEI thin-layer plates and chromatographed with 1.5 M K/HPO*$_4$*, pH 3.4. Lanes 1 and 6 correspond to fractions 8, 18, 25, 34, 40, and 44, respectively. The spot near the origin, for example in fraction 34 (lane 4) corresponds to* p_3Ap_3.

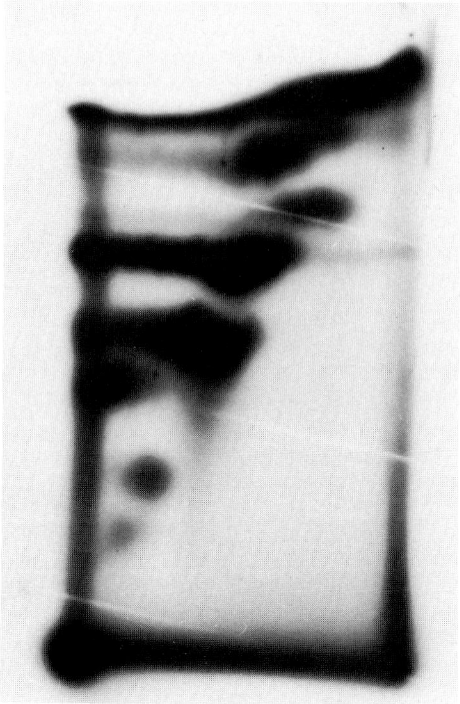

FIG. 5 *Autoradiogram of a two-dimensional thin-layer chromatogram of fraction 34 (Fig. 4, lane 4) from an extract of a)* E. coli *C600 (pBR325) and b)* E. coli *C600 (pBR325/*spoOF*). The right half (b) shows p_3Ap_3 near the origin. No trace of p_3Ap_3 is visible in an identical fraction from a strain C600 (pBR325) without the cloned* spoOF *gene.*

about 100% more p_3Ap_3 than the strain harboring plasmid pUB110 alone.

These experiments indicate that p_3Ap_3 is synthesized in *E. coli* by a *spoOF* gene coded protein. Likewise, the ca. 100% higher production of p_3Ap_3 in BD224 is obviously due to the pBR325 cloned *spoOF* gene.

Precipitation of Adenosine-bis triphosphate with Specific Antibodies

In order to finally prove that the *spoOF* gene of *B. subtilis* codes for abt synthetase and that it is expressed in both *B. subtilis* and *E. coli*, a 100 µl sample of fraction 34 (Fig. 4, lane 4) was incubated with 1 ml of serum 7436 for 1 h at room

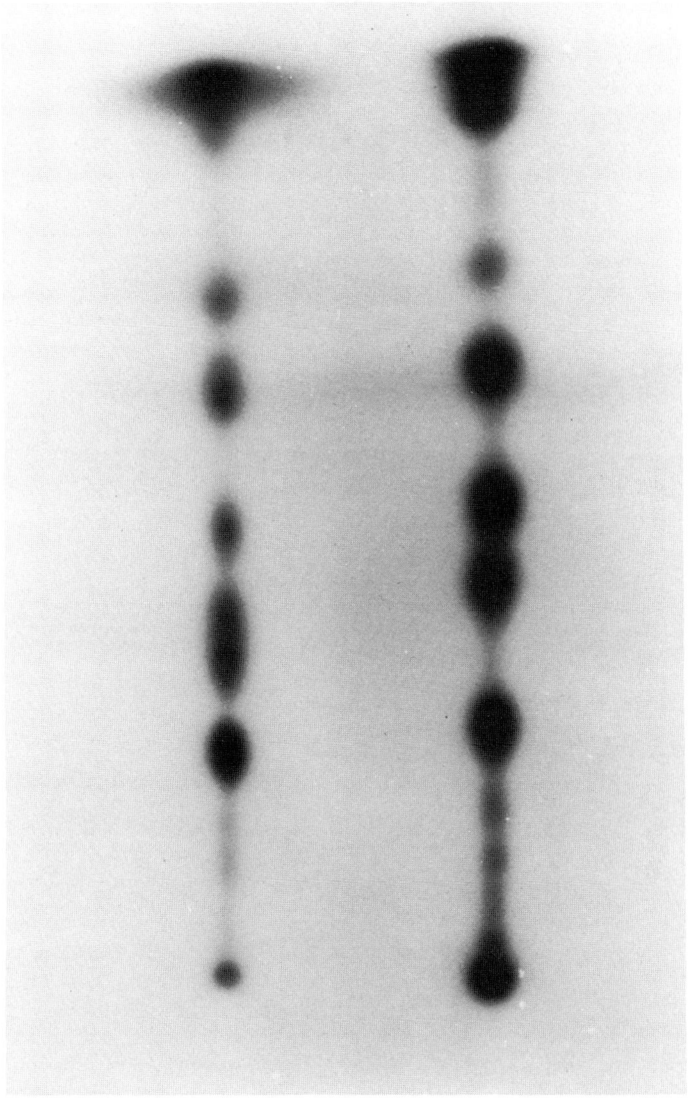

FIG. 6 *Autoradiogram of formic acid extracts from* E. coli *C600 (pBR325/*spoOF*) treated with antiserum against synthetic p_3Ap_3 (lane A) and a control serum (lane B). In lane B, p_3Ap_3 is visible near the origin, whereas it is absent in the antiserum treated sample (lane A). Chromatography is on PEI thin-layer plates using 1.5 M K/HPO_4, pH 3.4 as solvent.*

temperature. A 100 µl sample was then removed and counted in a scintillation counter. The rest was forced by centrifugation (2750 rpm, 40 min) through an Amicon filtration tube and washed with 1 ml of buffer. A 200 µl sample of the filtrate was counted as before.

An identical experiment was performed using the control serum (see Table 3). In four independent experiments, between 8 and 12% of total radioactivity was retained by serum 7436 as compared to control serum.

Fig. 6 (lane A) shows the specific removal of p_3Ap_3 from the extract of *E. coli* C600(pBR325/*spo0F*). A sample of the extract before addition of serum 7436 was chromatographed as shown in lane B. The lower radioactivity of the sample chromatographed in lane A is due to the washing procedure of the precipitate during filtration.

These experiments show that an antiserum against synthetic p_3Ap_3 reacts specifically with a substance produced by a protein of the cloned *spo0F* gene in *E. coli* and *B. subtilis*. There seems to be no doubt that abt synthetase is coded by the *spo0F* gene and is expressed even in an organism (*E. coli*) which has never been found to produce p_3Ap_3.

In order to obtain a more active antiserum against p_3Ap_3 we try at the moment to obtain monoclonal antibodies.

The Effect of *spo0F* Gene Amplification in *B. subtilis*

Amplification of the *spo0F* gene in *E. coli* has no detectable effect on this organism. In *B. subtilis*, however, one may expect an effect on sporulation, since the *spo0F* gene product is involved in regulation of sporulation (15). Therefore, sporulation and the fate of the cloned *spo0F* gene was investigated in six clones of BD224(pUB110/*spo0F*) in SYM medium. As a control five clones of strain BD224(pUB110) were also tested.

As can be seen in Table 4, strain BD224(pUB110) sporulates with a frequency of only 19 to 25% as compared to strain BD224 (without the plasmid) which sporulates to about 61%.

The six kanamycin, neomycin resistant clones of strain BD224 shown to contain the 1.3 megadalton fragment were cultivated for 24 h in 30 ml SYM to measure sporulation frequency (S/V x 100) and in TBAB to isolate plasmids. After the first cultivation period clone 1 was found to contain a rearranged plasmid and the sporulation frequency was similar to that of BD224 (pUB110). Clone 2 sporulated at a very low frequency (0.1%). Since there was no plasmid but still antibiotic resistance detectable, it was assumed that incorporation of the plasmid into the chromosome had occurred. Clones 3 to 6 did sporulate at a high frequency very well above strain BD224(pUB110). Their plasmids still harbored the 1.3 megadalton fragment.

TABLE 4 *Sporulation frequencies of BD 224(pUB 110), BD 224 (pUB 110/*spoOF*), and plasmid suitability in these strains*

Strain	Clone no.	1st cultivation period			2nd cultivation period			3rd cultivation period		
		s/v (%)	plasmid	Km^R,Neo^R,	s/v (%)	plasmid	Km^R,Neo^R,	s/v (%)	plasmid	Km^R,Neo^R,
BD 224	–	60–65		–	60–65	–	–	60–62	–	–
BD 224(pUB 110)	–	19–25	stable	+	19–25	stable	+	19–25	stable	+
BD 224(pUB 110/ *spoOF*)	1	24	rearr.	+	21	rearr.	+	20	rearr.	+
"	2	0.1	incorp.	+	0.05	incorp.	+	0.1	incorp.	+
"	3	50	stable	+	65	lost	–	64	lost	–
"	4	80	stable	+	88	stable	+	83	stable	+
"	5	73	stable	+	64	some-what rearr.	+	24	rearr.	+
"	6	7	stable	+	38	some-what rearr.	+	30	rearr.	+

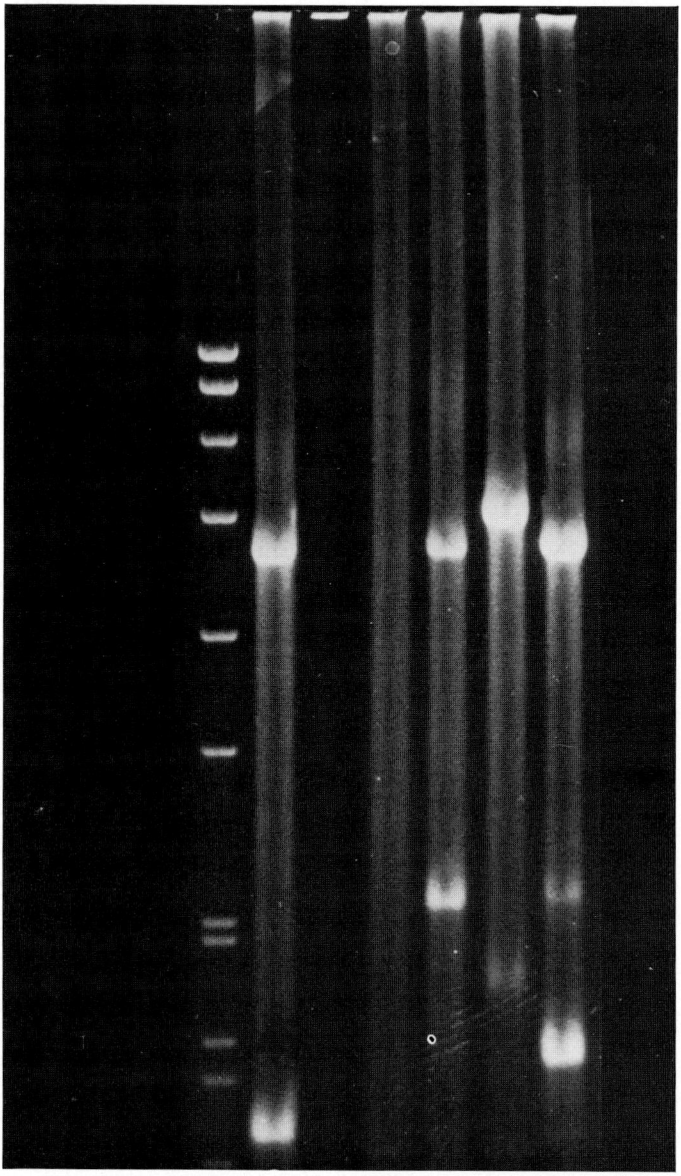

FIG. 7 *Agarose (1%) gel electrophoresis of hybrid plasmids pUB110/*spoOF *from six different clones of BD224 after Eco R1 restriction endonuclease treatment. Lanes 1 to 6 correspond to clones 1-6 of BD224 (pUB110/*spoOF*). For details see text.*

When spores of these cultures were again used to measure sporulation and plasmid composition it was found that except clone 4 no one had the original hybrid plasmid. Clone 4 which contained the correct 1.3 megadalton fragment sporulated at a frequency about 50% higher than BD224 (pUB110) and about 20% higher than BD224 (no plasmid). Clone 3 had lost the plasmid entirely (sensitive to antibiotics) and sporulated similar to strain BD224 (no plasmid). As before, clone 2 showed the same very low sporulation frequency and was kanamycin resistant. Clones 1, 5, and 6 still contained plasmid pUB110, were resistant to antibiotics, and sporulated to about 25%, however, the 1.3 megadalton fragments were altered.

Figure 7 shows an agarose gel of plasmid isolation according to Birnboim (12) of all six clones after prolonged cultivation in sporulation medium. This illustrates how a cloned *B. subtilis* gene, which is associated with sporulation can be altered in sporulating strains.

After many more growth and sporulation cycles clone 4 also lost its 1.3 megadalton fragment. The plasmid, however, is preserved in some of the spore preparations or can be regenerated by recloning it from the *E. coli* plasmid pBR325/*spoOF*.

DISCUSSION

There has been some uncertainty as to whether or not p_3Ap_3 found in sporulating cells of *B. subtilis* is indeed synthesized by the *spoOF* gene coded protein. If this can be shown unequivocally p_3Ap_3 must play an essential role in sporulation, because mutations in the *spoOF* gene cause asporogeny.

Three sets of experiments were used to show that the product of the *spoOF* gene synthesizes p_3Ap_3. First, p_3Ap_3 was synthesized chemically and its structure determined by physicochemical and enzymatic methods. Second, antiserum against p_3Ap_3 was obtained by injecting rabbits with synthetic p_3Ap_3 together with Freunds complete adjuvant. Third, the *spoOF* gene located on a 1.3 megadalton fragment was cloned in *E. coli* and *B. subtilis*.

In several experiments we could show by specific binding to antibodies that a substantial fraction of the nucleotide pool (6-10%) of an extract of *E. coli* (incapable of synthesizing p_3Ap_3 normally) which harbors the cloned *cpoOF* gene, consists of p_3Ap_3. Also, *B. subtilis* carrying the hybrid plasmid pUB110/*spoOF* produces between 10 and 100% more p_3Ap_3 than the corresponding strain without the cloned *spoOF* gene.

From these experiments we conclude that the *spoOF* gene indeed codes for abt synthetase, the enzyme capable of synthesizing p_3Ap_3.

Amplification of this gene in *B. subtilis* BD224(recE4) is

most interesting because of its instability in plasmid pUB110 and its effect on sporulation. When a medium (SYM) was used in which BD224 sporulates to about 60%, we found that the presence of plasmid pUB110 decreases the sporulation frequency to about 22%. Cloning of the *spoOF* gene in BD224 using pUB110 increases sporulation temporarily to approximately 88%. However, upon prolonged incubation different processes apparently can cause incorporation of the plasmid into the genome and a decrease in sporulation frequency to almost zero (0.05%). Alternatively, alterations or loss of the 1.3 megadalton fragment also decrease sporulation to the wild-type level of about 22% (Bd224 with pUB110).

Depending on which type of alteration finally predominates in a culture originating from one clone with a hybrid plasmid, sporulation frequencies can vary from 0.05 to 88%. There may be more types of changes detectable in strains with hybrid plasmids than we have observed by specifically screening for asporogenous clones (opaque colonies) on plates with media favouring sporulation. This is consistent with observations reported by Kawamura *et al.* (7). However, experiments reported here show that amplification of the *spoOF* gene in plasmid pUB110 resulting in increased synthesis of p_3Ap_3 upon nutrient starvation, increases sporulation frequencies as long as this hybrid plasmid is stable. But rearrangements occurring in a quite unpredictable manner may soon decrease or abolish spore formation.

We have not yet been able to grow any clone of BD224(pUB110/*spoOF*) without the above described phenomena. We have no satisfactory explanation as to why this instability exists, since other *spoOF* genes (*spoOB*) can be cloned in pUB110 (7, and experiments in our own laboratory) without the above described difficulties. It cannot be overlooked that p_3Ap_3 may play some role in this process.

ACKNOWLEDGEMENTS

We are grateful to Professor Siefert for his generous help in preparing the adenosine-bis triphosphate antiserum.

REFERENCES

1. Freese, E. (1981) in *Sporulation and Germination* (Levinson, H., Sonenshein, A.L. and Tipper, D.J., eds.), pp.1-12, American Society for Microbiology, Washington D.C.
2. Rhaese, H.J. and Groscurth, R. (1976) *Proc. Natl. Acad. Sci. USA* **73**, 331-335.
3. Rhaese, H.J., Groscurth, R., Vetter, R. and Gilbert, G. (1979) in *Regulation of Macromolecular Synthesis by Low Molecular Weight Mediators* (Koch, G. and Richter, D., eds.),

pp.145-149, Academic Press, New York.
4. Rhaese, H.J., Groscurth, R. and Rumpf, G. (1978) in *Spores VII* (Chambliss, G. and Vary, J.C., eds.), pp.286-292, American Society for Microbiology, Washington, D.C.
5. Rhaese, H.J., Hoch, J.A. and Groscurth, R. (1977) *Proc. Natl. Acad. Sci. USA* **74**, 1125-1129.
6. Rhaese, H.J. and Groscurth, R. (1979) *Proc. Natl. Acad. Sci. USA* **76**, 842-846.
7. Kawamura, F., Shimotsu, F., Saito, H., Hirochika, H. and Kobayashi, Y. (1981) *Sporulation and Germination* (Levinson, H., Sonenshein, A.L. and Tipper, D., eds.), pp.109-113, American Society for Microbiology, Washington D.C.
8. Rhaese, H.J., Groscurth, R., Amann, R., Kuhne, H. and Vetter, R. (1981) in *Sporulation and Germination* (Levinson, H., Sonenshein, A.L. and Tipper, D.J., eds.), pp.134-137, American Society for Microbiology, Washington D.C.
9. Cohen, S.N., Chang, A.C.Y. and Hsu, L. (1972) *Proc. Natl. Acad. Sci.* **69**, 2110-2114.
10. Chang, S. and Cohen, S.N. (1979) *Molec. Gen. Genet.* **168**, 111-115.
11. Anagnostoupolos, C. and Spizizen, J. (1961) *J. Bacteriol.* **81**, 741-746.
12. Birnboim, H.C. and Doly, J. (1979) *Nucleic Acid Res.* **7**, 1513-1523.
13. Shanmugasundaram, S. and Rhaese, H.J. (1982) *FEBS-Letters*, in print.
14. Michelson, A.M. (1964) *Biochim. Biophys. Acta* **91**, 1-18.
15. Hoch, J.A. (1976) *Adv. Genet.* **18**, 69-99.

GUANOSINE 5'-TRIPHOSPHATE BINDING PROTEINS IN *BACILLUS SUBTILIS* CELLS

L. VITKOVIĆ*, K.R. DHARIWAL*, E. FREESE* and D. GOLDMAN**

**Laboratory of Molecular Biology, National Institute of Neurological and Communicative Disorders and Stroke, National Institutes of Health*
and
***Laboratory of Clinical Science, National Institute of Mental Health, Bethesda, Maryland 20205, U.S.A.*

SUMMARY

All conditions examined so far which cause the initiation of massive sporulation also result in a decrease of GTP. Therefore, GTP in combination with a protein, may suppress sporulation. Proteins isolated from vegetative cells by affinity chromatography on GTP-agarose represented 3% of the total protein and included 218 polypeptides resolved in polyacrylamide gels by two-dimensional electrophoresis. The majority of these proteins had isoelectric points between 5 and 6 and molecular weights of between 30,000 and 90,000 daltons. We compared GTP-binding proteins of two standard strains, five different mutants blocked at stage 0, and one mutant blocked at stage 4 of sporulation.

Computer-assisted quantitative analysis of highly reproducible electropherograms revealed that meaningful comparisons can be made only between strictly isogenic strains. We found that *spo0A*, *spo0B*, and *spo0J* mutations did not affect the electrophoretic pattern of the 218 GTP-binding proteins despite their reported pleiotropic effects in vegetative cells.

INTRODUCTION

Sporulation of *B. subtilis* can be initiated in the presence of excess glucose, ammonium ions and phosphate by the partial depletion of guanine nucleotides (1). All conditions examined so far which cause initiation of massive sporulation also result in a decrease of GTP (2). Therefore, GTP in combination with a protein may suppress sporulation. Regulatory GTP-binding proteins have been discovered in eukaryotic cells. A guanine nucleotide binding protein (G/F or N) mediates hormonal activation of mammalian adenylate cyclase (3). Another purine nucleotide binding protein with highest affinity for GTP, regulates coupling of brown adipose tissue mitochondria (4).

Using two-dimensional electrophoresis, we therefore compared GTP-binding proteins of two standard *B. subtilis* strains, five different mutants blocked at stage 0 and one mutant blocked at stage 4 of sporulation. Our objective in this study was to assess the extent of pleiotropy of *spo0* mutations in vegetative cells by comparing the number and quantity of a limited, well-defined set of gene products.

MATERIALS AND METHODS

Bacterial Strains and Growth Conditions

The strains of *B. subtilis* used are listed in Table 1. Cells were grown overnight on plates containing tryptose blood agar base (33 g/l; Difco) plus 50 µg/ml tryptophan and phenylalanine as required. They were inoculated at an initial absorbency at 600 nm (OD_{600}) of 0.02 - 0.05 into synthetic medium (2) containing 100 mM glucose and 20 mM glutamate. When the cell

TABLE 1 *Origin and characteristics of the* B. subtilis *strains used*

Strain	Genetic symbol	Origin
61730	*trpC pheA1*	JH642, J.A. Hoch
61731	*trpC2 pheA1 spo0A12*	JH646, J.A. Hoch
61733	*trpC2 pheA1 spo0B136*	JH648, J.A. Hoch
61734	*trpC2 pheA1 spo0F221*	JH649, J.A. Hoch
62232	*trpC2 pheA1 spo0J87*	JH696, J.A. Hoch
60001	*trpC2*	168, J. Spizizen
62233	*trpC2 spo0K141*	Z31, J.G. Coote
61625	*trpC2 spo4E*	11T, P. Schaeffer

For information about the mutants see references 8, 9, and 10.

density reached an OD_{600} of 2, phenylmethylsulphonylfluoride (PMSF; 1.74 mM) and ice (100 g) were added; the culture was centrifuged at 10,000 x g for 5 min.

Isolation of GTP-binding Proteins

The following procedures were carried out at 4°C. Cells were washed four times with 100 mM 3-[N-morpholino]propanesulfonic acid (MOPS; pH 7.4), 1 M KCl, 5 mM EDTA, 1.74 mM PMSF (buffer A) followed by centrifugation at 10,000 x g for 10 min. Washed cells were frozen in ethanol-dry ice and stored at -70°C. Cells from 100 ml of culture were thawed in 3 ml of 50 mM MOPS (pH 7.4), 20 mM magnesium acetate, and 50 mM NH_4Cl (buffer B); and then broken by passage through a French pressure cell. Cell debris was removed by centrifugation at 35,000 x g for 20 min. The supernatant, containing 15 mg of protein, was equilibrated for 20 min with 5 ml of phosphocellulose (Cellex-P; BioRad) saturated with buffer B. The protein (5 mg) left in the supernatant after centrifugation for 5 min at 5,000 x g was equilibrated for 60 min with 5 ml of GTP-agarose (G-5257; Sigma) saturated with buffer B. The slurry was poured into a column (EconoColumn; BioRad), and the unadsorbed proteins were eluted with 15 ml of buffer B and discarded. The bound proteins were eluted with 5 ml of buffer B containing 1 M NH_4Cl. The eluent was dialyzed against one litre of H_2O overnight and then lyophilized. The GTP binding proteins were resuspended in sample buffer (2 mg/ml), boiled for 2 min, distributed into 20 µl fractions, and stored at -70°C. Sample buffer contained 1% (w/v) sodium dodecyl sulfate, 2.5% (v/v) mercaptoethanol, 10% (v/v) glycerol, 1% (v/v) ampholytes (pH 3-10; BioRad) and 1% (v/v) Nonidet-40 (Sigma).

Two-dimensional Electrophoresis

Two-dimensional electrophoresis was carried out according to O'Farrell (5) with use of isoelectric focusing (pH 5-7) in the first dimension and electrophoresis through a uniform 10% (v/v) acrylamide gel in the second dimension. Each gel was loaded with 40 µg of protein; 9-12 gels were electrophoresed simultaneously. Reproducibility was monitored in every experiment by measuring the pH gradient in an unloaded first dimension gel and by co-electrophoresis of protein molecular weight standards in the second dimension. Isoelectric focusing gels were equilibrated (5) and used immediately in the second dimension.

Staining and Drying of Gels for Transmission Densitometry

Polypeptides were detected by silver staining performed according to Morrissey (6). To obtain reproducible results, it was essential that all ingredients were prepared in deionized water. Silver staining can detect 0.02 ng protein/mm^2. The density is directly proportional to the protein concentration in the 0.05 - 2.0 ng/mm^2 range. By normalizing proteins on two-dimensional gels a ten-fold difference in protein loading can be tolerated (7). Stained gels were incubated in 5% glycerol + 7% acetic acid at room temperature for 20 min; sandwiched between two sheets of cellophane membrane backing (BioRad) and tightly clamped to a glass plate. Gels were dried in a hood at room temperature for 36-48 h to avoid physical distortion.

Quantitative Analysis

Gels were scanned at an optical density setting of 0-2 using a scanning microdensitometer (Optronics Corp., Model 1000HS). Semi-automated density measurements and gel comparisons were made using a PDP/11 computer (Digital Equipment Corp.) equipped with an image processor (DeAnza Systems Inc., Model IP5000) having 512 x 512 x 8 bit image arrays. All gel images were stored on a disk. An operator measured, on each gel in succession, the density of a particular polypeptide spot which was identified by its location relative to neighbouring spots (pattern recognition). This sequence was repeated for all polypeptides (7). Image densities were calculated as [(average density of a protein spot - background density in a protein free area) x area of the spot]. They were normalized to correct for differences in protein loading and staining. The normalization factor was the slope of a linear regression of all non-saturating densities (OD<2) in a gel matched to the corresponding values in an arbitrarily chosen reference gel. Image densities were converted to optical densities using a standard calibrated by the National Bureau of Standards. They are directly proportional to a standard quantity of protein. GTP-binding proteins from each strain were represented by three gels and 218 protein spots were measured in each gel.

RESULTS AND DISCUSSION

GTP-binding Proteins

Mutants blocked in the sporulation process and their parent strains (Table 1) were grown to the mid-exponential phase in synthetic medium. The mutants did not sporulate under shift-

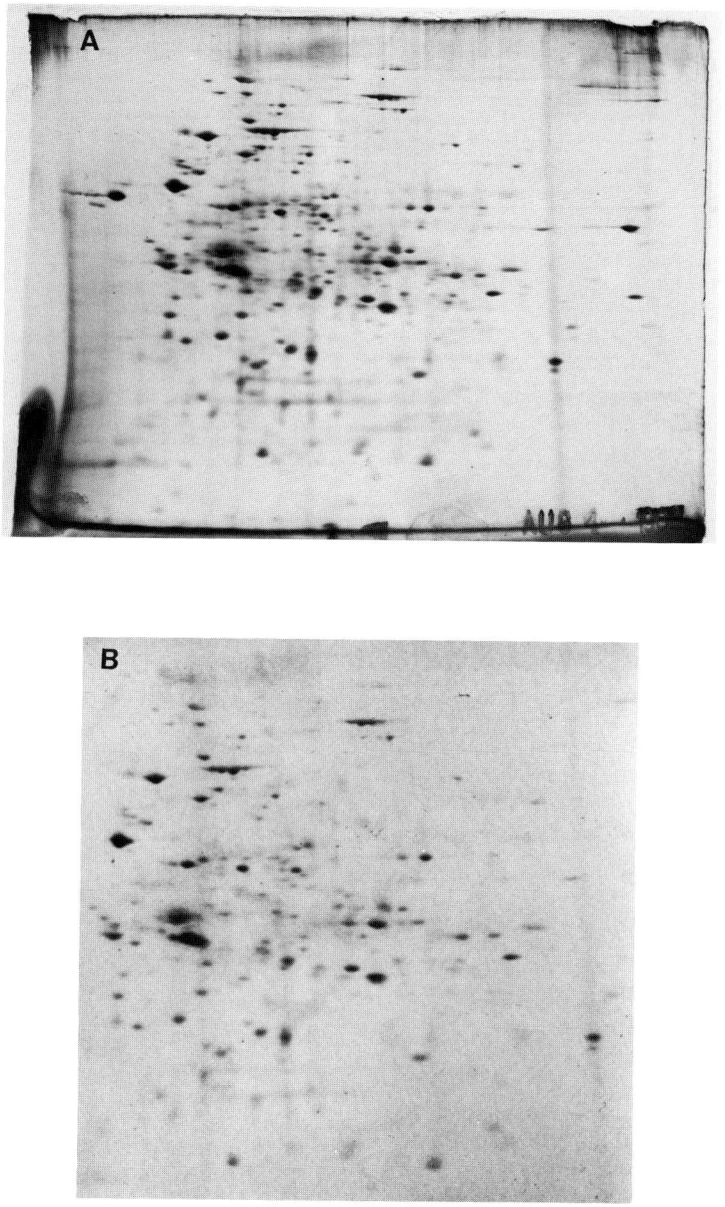

FIG. 1(A) *Reference gel used to normalize density measurements contained GTP-binding proteins from the strain 60001. Acidic, low molecular weight proteins are in the lower left corner of the gel.* (B) *Computer image of the reference gel.*

down conditions (9,10) nor with decoyinine addition (8). Proteins that did not bind to phosphocellulose but did bind to GTP-agarose were isolated. The specificity of binding was assessed with commercially purified enzymes. Under equilibrium conditions (1 mg protein per 1 ml GTP-agarose for 60 min at 4°C), approximately 60% of pyruvate kinase, which reacts with GDP, and 10% of lactate dehydrogenase, which has no affinity for guanine nucleotides, bound to GTP-agarose. GTP-binding proteins represented about 3% of the total protein. When the proteins were eluted with 10 mM GTP, 10 mM ATP or 1 M NH_4Cl and then separated on one-dimensional gels, essentially the same protein patterns were obtained. This indicates that the "GTP binding proteins" included proteins with affinity for ATP and presumably other nucleotides. Proteins eluted with 1 M NH_4Cl were separated in two-dimensional polyacrylamide gels by iso-electric point in the first dimension and by size in the second dimension. Highly sensitive silver staining of proteins in gels revealed 218 clearly visible and distinct spots (Fig. 1). The majority of GTP-binding proteins that were separated on pH 4.5 - 7.0 gels had isoelectric points between 5.0 and 6.0.

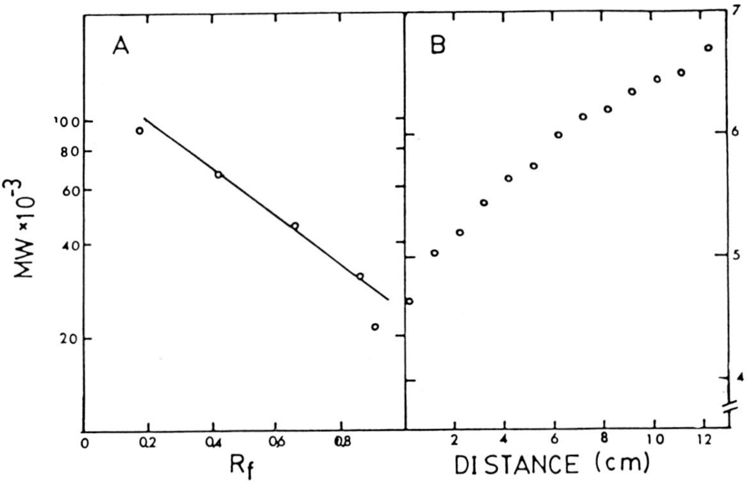

FIG. 2 *Molecular weight range (A) and isoelectric point range (B) of GTP-binding proteins.* **A.** *The molecular weights of protein standards were plotted on a logarithmic scale versus Rf (migration distance/dye front distance). Protein standards were: phosphorylase B (92,000), bovine serum albumin (66,000), ovalbumin (45,000), carbonic anhydrase (31,000), and soybean trypsin inhibitor (21,500).* **B.** *pH of 4 mm sections of an isoelectric focusing gel equilibrated in degassed H_2O were plotted versus gel length.*

Molecular weights of the majority of polypeptides were between 30,000 and 90,000 daltons (Fig.s 1 and 2). Protein patterns appeared to be very reproducible. For each strain, three two-dimensional electropherograms were quantified by computer-assisted density measurements described in Methods above.

Comparison of GTP-binding Proteins from Sporulating Strains and Non-sporulating Mutants.

We plotted, via the computer, the normalized quantities of each of the 218 GTP-binding proteins found in one sample against those in another sample and performed linear regression analysis. If the two samples were identical, all spots would be located on a straight line. This analysis revealed that the GTP-binding proteins from *spoOA*, *spoOB*, and *spoOJ* mutants were almost identical to the GTP-binding proteins from their isogenic sporulating parent (Fig. 3). In contrast, several GTP-binding proteins from mutants *spoOK* and *spo4E* were present in unequal amounts when compared to their sporulating parent (Fig.4).

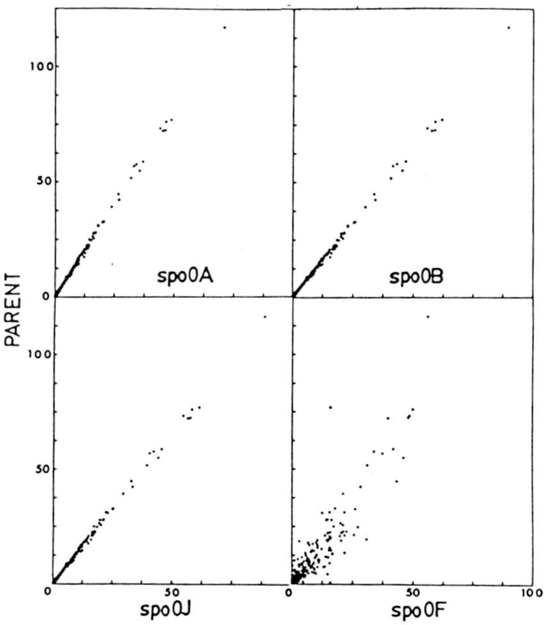

FIG. 3 *Comparison of GTP-binding proteins from strain 61730 with the GTP-binding proteins from strains 61731 (*spoOA*), 61733 (*spoOB*), 62232 (*spoOJ*), and 61734 (*spoOF*). The correlation coefficients were: 1.00 for 61731; 1.00 for 61733; 1.00 for 62232; and 0.90 for 61734.*

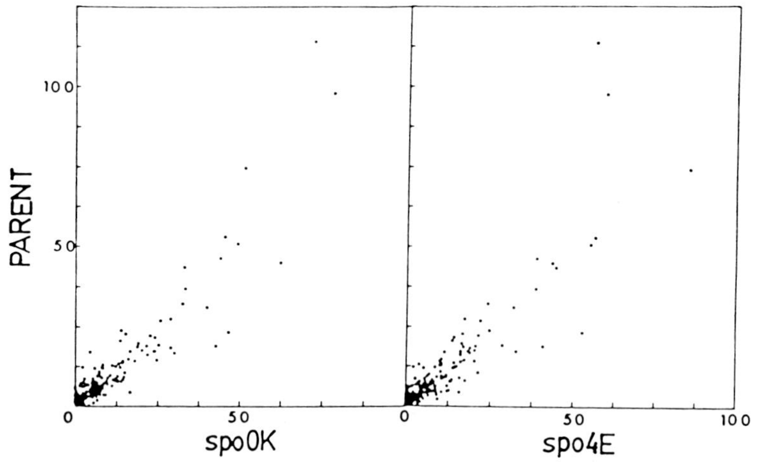

FIG. 4 *Mean densities of GTP-binding proteins from strain 60001 were plotted versus the mean densities of GTP-binding proteins from strains 62233 (spo0K) and 61625 (spo4E). The correlation coefficients were: 0,92 for 62233 and 0.89 for 61625.*

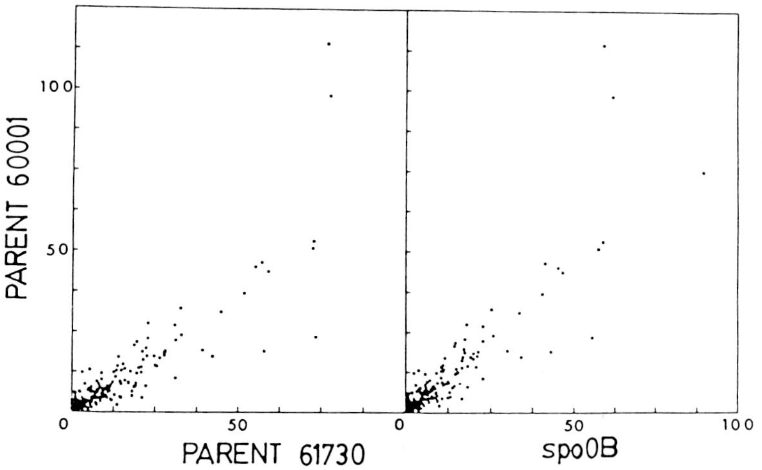

FIG. 5 *GTP-binding proteins from strain 60001 were compared with the GTP-binding proteins from the sporulating strain 61730 and its non-sporulating derivative 61733 (spo0B). The correlation coefficients were: 0.88 for 61730 and 0.88 for 61733.*

Because it is often assumed that mutants are essentially isogenic with their parents, it was tempting to attribute the differences in the quantities of GTP-binding proteins to the *spo* mutations and to further explore the mechanisms controlling the proteins present in greatly different amounts. Mutations in *spo0* genes are known to exert many pleiotropic effects (9). For example, mutations in the *spo0A*, *spo0B*, *spo0E*, *spo0F*, and *spo0H* genes caused an accumulation of proteins with an affinity for double-stranded DNA in vegetative cells (10). Mutant *spo0A*-5NA was reported to elaborate "*spo0A* specific polypeptides" during growth (11). Alternatively, the differences described here may reflect multiple mutations, resulting from a mutagenic treatment of the parent or from spontaneous mutations, accumulating over time, which modify proteins or influence control mechanisms. The latter possibility was strengthened by the comparison of two sporulating strains, 60001 and 61730, which have both been used for isolation of *spo* mutants. This comparison revealed that they also contained unequal amounts of most of the GTP-binding proteins (Fig. 5). Moreover, when the *spo0B* mutant, whose proteins correlated very well with the isogenic parent, was compared to standard strain 168, most of the protein amounts differed (Fig. 5). It appears that meaningful comparisons of protein patterns can be made only with strictly isogenic strains. Based on this criterion, we conclude that *spo0A*, *spo0B* and *spo0J* mutations have no effect on GTP-binding proteins of vegetative cells detectable by electrophoresis, whereas the other *spo* mutations have to be examined further. Two-dimensional electrophoresis of proteins combined with computer-assisted evaluation can now be used to investigate the induction of sporulation mutants deficient in cell division or septation.

ACKNOWLEDGEMENTS

We thank Peter Fortnagel for helping design the protocol for the isolation of GTP-binding proteins and Daniel Lewis for helping with the two-dimensional electrophoresis.

REFERENCES

1. Freese, E., Heinze, J.E. and Galliers, E. (1979) *J. Gen. Microbiol.* **115**, 193-205.
2. Lopez, J.M., Marks, C.L. and Freese, E. (1979) *Biochim. et Biophys. Acta* **587**, 238-252.
3. Neer, E.J. and Salter, R.S. (1981) *J. Biol. Chem.* **256**, 12102-12107.
4. Lin, C. and Klingenberg, M. (1982) *Biochemistry* **21**, 2950-2956.

5. O'Farrell, P.H. (1975) *J. Biol. Chem.* **250**, 4007-4021.
6. Morrissey, J.H. (1981) *Anal. Biochem.* **117**, 307-310.
7. Merril, C.R., Goldman, D. and Van Keuren, M.L. (1982) *Electrophoresis* **3**, 17-23.
8. Freese, E.B., Vasantha, N. and Freese, E. (1979) *Molec. Gen. Genet.* **170**, 67-74.
9. Piggot, P.J. and Coote, J.G. (1976) *Bacteriol. Rev.* **40**, 908-962.
10. Brehm, S.P., Hegarat, F. and Hoch, J.A. (1975) *J. Bacteriol.* **124**, 977-984.
11. Linn, T. and Losick, R. (1976) *Cell* **8**, 103-114.

ENHANCEMENT OF *BACILLUS SUBTILIS* MICROCYCLE SPORULATION BY S-ADENOSYLMETHIONINE

M.J. CLOUTIER, J.H. HANLIN, J.S. NOVAK and R.A. SLEPECKY

Department of Biology, Syracuse University, Syracuse, New York, U.S.A.

SUMMARY

Further elaboration of microcycle sporulation in *Bacillus subtilis* is presented. Although nutrient deprivation is essential for limited outgrowth to singlet primary cells, supplementation with specific metabolites (diaminopimelic acid, and adenine or S-adenosylmethionine) is necessary for the induction of microcycle sporulation. The role of S-adenosylmethionine as a factor is unknown but the involvement of methylation in sporulation is suggested. A requirement for DNA synthesis for microcycle sporulation to proceed has been demonstrated.

INTRODUCTION

We have previously described a microcycle sporulation system [the conversion of an outgrowing cell to a sporulating cell without an intervening cell division (1)] in *Bacillus subtilis* NCTC 3610 (2). By separating the germination and outgrowth phases and by systematic reduction of the concentrations of ingredients of a chemically defined outgrowth medium, the minimal concentrations allowing germination and limited outgrowth of germinated spores to primary singlet cells were determined. Sporulation ensued only with the addition of other metabolites. Additions at specific times of diaminopimelic acid (DAP) gave 30% microcycle sporulation.

This paper describes enhancement of sporulation beyond the 30% level. Sporulation was increased by the addition of adenine which is substitutable by S-adenosylmethionine. The requirement of DNA synthesis for microcycle sporulation was shown.

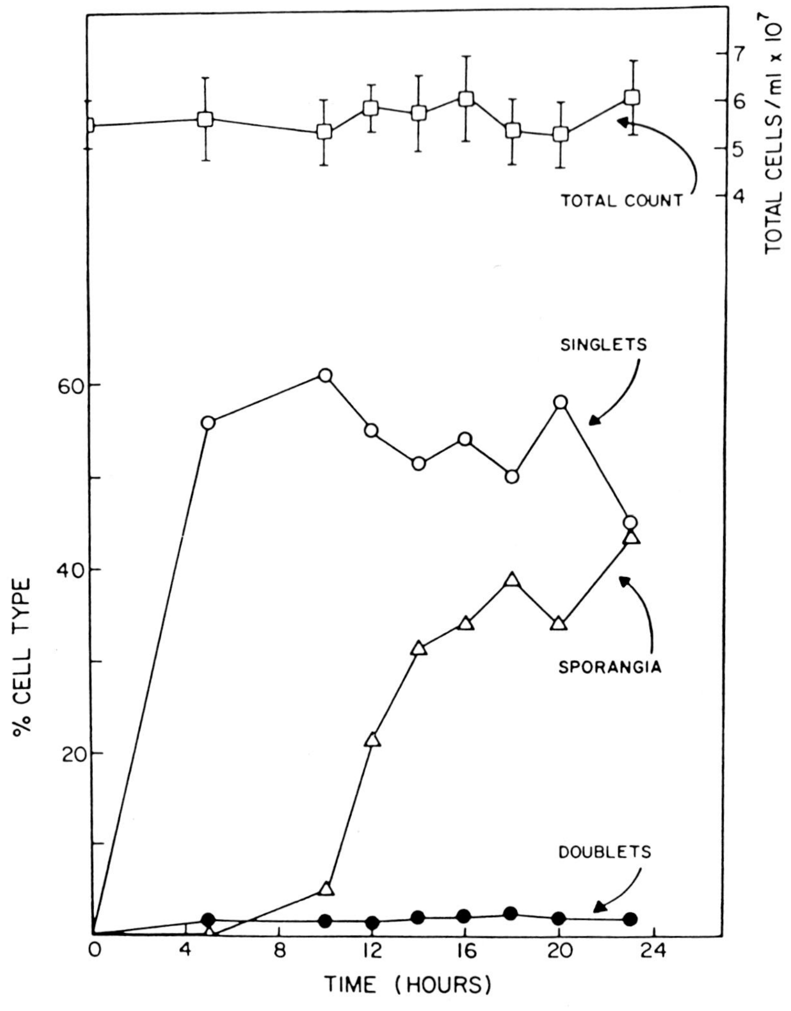

FIG. 1 *Total cell numbers and cell types percentages during spore to spore development as a function of time upon the addition of 30 μM adenine at 0 and 6 h, 10 μM doses of DAP added at 0, 1, 2, 3, 4, and 5 h. Symbols: □, total cell count, (bars indicate standard deviation of triplicate counts); o, singlets; Δ, sporangia; ●, doublets. Cell type percentages are the average of triplicate counts.*

METHODS

Frozen spores of *Bacillus subtilis* NCTC 3610 harvested from glucose salts medium (3) were thawed, vortexed and sonicated (to eliminate clumping). A 0.5 ml inoculum was pipetted into 15 ml sterile, plastic, screw capped centrifuge tubes. Spores were heat shocked at 85°C for 10 min. Then 4.5 ml of prewarmed germination medium (0.55 mM glucose, 0.56 mM L-alanine, 40 mM Tris, 134 mM KCl, pH 7.3) was added giving a final spore suspension density of about 5 x 10^7 spores per ml. Spores were incubated at 30°C for 150 min. After germination, tubes were centrifuged and the germinated spores were resuspended in 5.0 ml of prewarmed minimal outgrowth medium (1.0 mM glucose, 10 mM α-methylglucose, 40 mM Tris, pH 7.3, 7.5 mM K_2HPO_4, 3.25 mM KH_2PO_4, 0.2 mM M $SO_4.7H_2O$, 0.018 mM $FeSO_4.7H_2O$, 0.45 mM $CaCl_2$, 0.025 mM $MnCl_2.4H_2O$). The tightly sealed tubes were incubated at 30°C without shaking. Additions to the medium were made as indicated in the figure legends. Cell numbers were determined with a Petroff-Hausser counting chamber with a Zeiss microscope equipped with a 100x oil immersion lens. Delineation of cell types and counting procedures was as given previously (2).

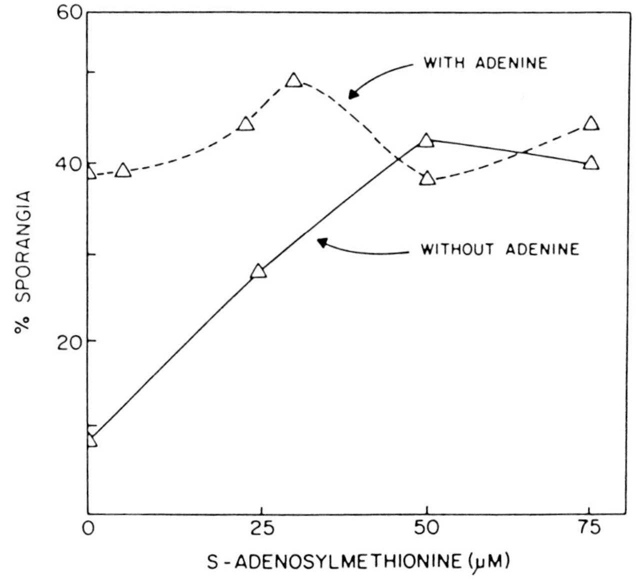

FIG. 2 *Percent sporangia after 24 h in outgrowth medium supplemented with S-adenosylmethionine (SAM), 10 μM DAP and, with and without 30 μM adenine all added at 0 h.*

RESULTS AND DISCUSSION

Upon further supplementation of the outgrowth medium with metabolite-rich mixtures (such as peptones, beef extract or yeast extract) or a vast array of individual "metabolites": general nutrition factors; protein; cell wall, membrane, nucleic acid precursors; various carbon and nitrogen sources; and compounds associated with spores; only the addition of adenine stimulated sporulation beyond that found with DAP supplementation. Addition of 30 μM adenine (Fig. 1) gave higher sporulation than DAP added alone (2) and sporulation was initiated three hours earlier. Adenine added without DAP was without effect. Additions of other purines, pyrimidines, adenosine, deoxyadenosine or inosine were ineffective.

Addition of S-adenosylmethionine (SAM) gave a response similar to that of adenine (Figs. 2 and 3). Lower levels of SAM enhanced sporulation slightly in the presence of adenine (Fig. 2). Higher concentrations of SAM were required in the

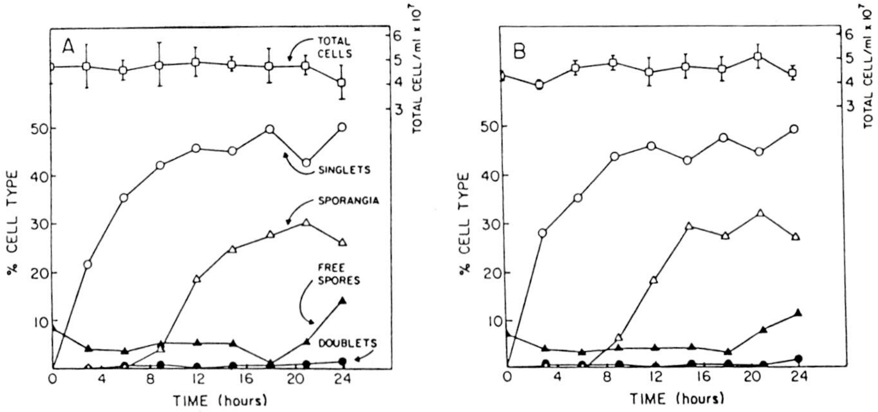

FIG. 3 *Total cell numbers and % cell types during spore to spore development as a function of time upon the addition of 10 μM DAP and 50 M SAM at 0 h. Symbols: □, total cell count (bars indicate standard deviation of triplicate counts); o, singlets; Δ, sporangia; ●, doublets; ▲, free spores. Cell type percentages are the average of triplicate counts. Panel A and B are separate cultures.*

absence of adenine and it was clear that SAM can substitute for adenine. Methionine addition or increased levels of sulphate with or without adenine did not enhance sporulation. The determination of cell numbers and progression of cell types as a function of time (Fig. 3) gave a similar pattern to that of adenine additions (Fig. 1) except that with SAM additional free spores were detected within the 24 h span. SAM was most effective when added between 0 and 2 h (Fig. 4).

Since SAM, an activated form of methionine, is associated with various forms of methylation in microorganisms and is involved in reactions in which methionine loses its methyl group (4), we sought ways to perturb methionine metabolism. L-Ethionine, a methionine analogue, was found to be a strong inhibitor of sporulation (Fig. 5). Very low concentrations allowed outgrowth to singlets but specifically inhibited the formation of sporangia. Since the methionine analogue L-ethionine is believed to be incorporated into proteins in place of methionine, perhaps proteins required for sporulation are made inactive; another possibility is that the L-ethionine may replace methionine in the synthesis of SAM.

L-Ethionine inhibition of sporulation was most effective when the compound was added at 2.5 h or earlier (Fig. 6). Inhibition of sporulation by L-ethionine was reduced in the presence of SAM (Fig. 7). That the inhibition by L-ethionine

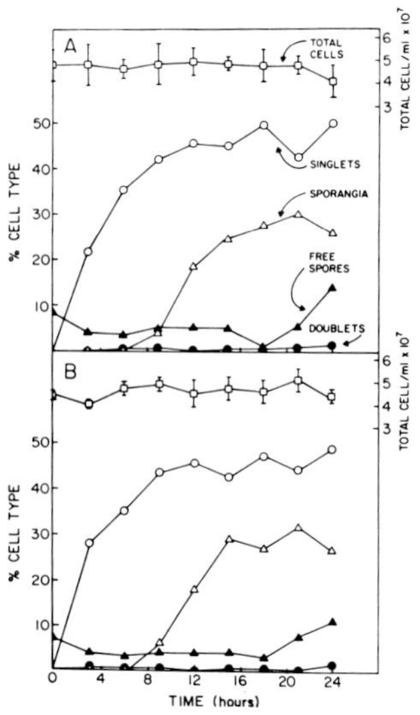

FIG. 4 *Percent sporangia after 24 h in outgrowth medium supplemented with 10 μM DAP at 0 h and 50 μM SAM added at the indicated times.*

occurred in the presence of adenine (Figs. 5 and 6) and can be reversed by the addition of SAM (Fig. 7) makes it highly unlikely that SAM is being used to provide adenine.

The time of inhibition by L-ethionine (Fig. 6) and the time of the effectiveness of SAM (Fig. 4) corresponded to the time of DNA synthesis as judged by the time of inhibition of sporulation by 6-(p-hydroxyphenylazo)-uracil (HPUra) (Fig. 8). HPUra will only inhibit new rounds of DNA synthesis (5), thus the experiment also demonstrates that a round of DNA synthesis

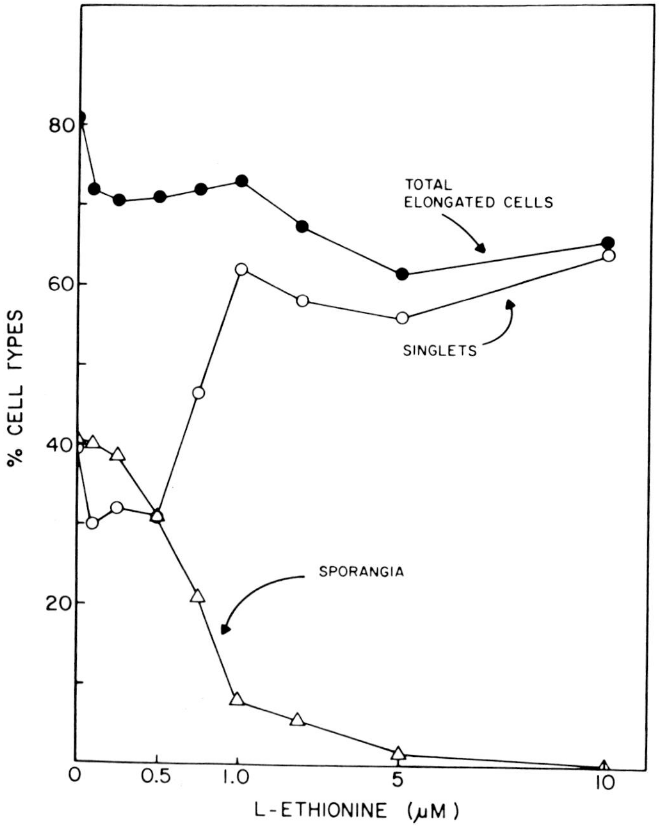

FIG. 5 *Percent cell types after 24 h in outgrowth medium supplemented with 10 μM DAP, 30 μM adenine and the indicated concentrations of L-ethionine all added at 0 h. Symbols: ●, elongated cells; o, singlets; Δ, sporangia.*

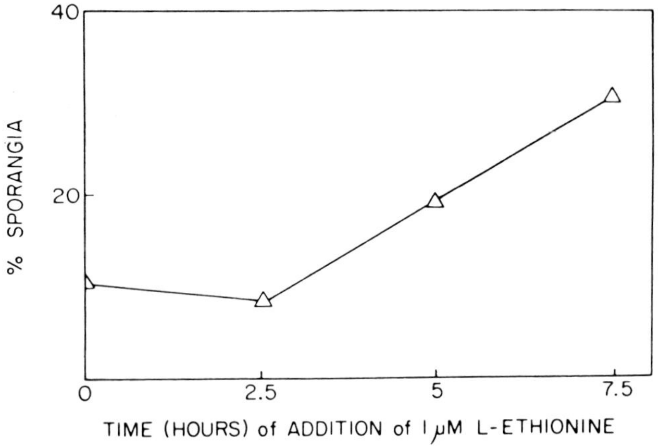

FIG. 6 *Percentage sporangia after 24 h in outgrowth medium supplemented with 10 μM DAP and 30 μM adenine added at 0 h and 1 μM L-ethionine added at the indicated times.*

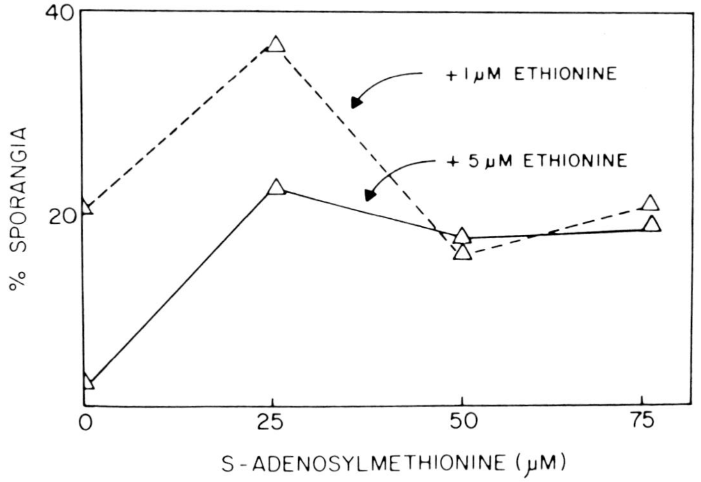

FIG. 7 *Percent sporangia after 24 h in outgrowth medium supplemented with 10 μM DAP, 1 μM or 5 μM L-ethionine and, the indicated concentrations of SAM all added at 0 h.*

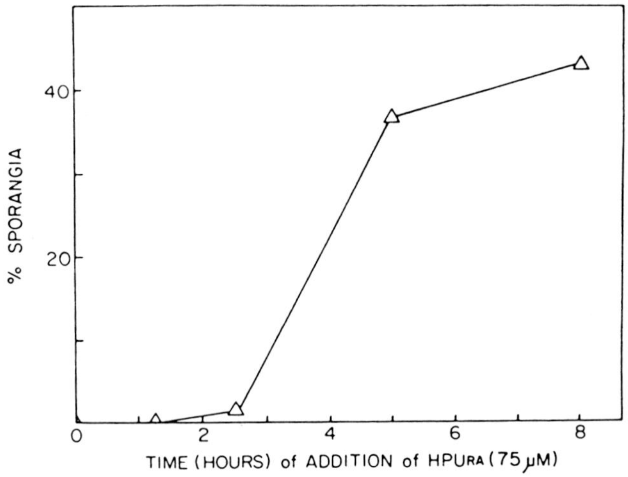

FIG. 8 *Percent sporangia after 24 h in outgrowth medium supplemented with 10 μM DAP and 50 μM SAM added at 0 h and, 75 μM 6-(p-hydroxyphenylazo)-uracil (HPUra) added at the indicated times.*

is required for microcycle sporulation in *B. subtilis* as was shown by us for *B. megaterium* (6).

Supplementation with specific metabolites (DAP and adenine or SAM) is required before singlet primary cells of *B. subtilis* can undergo microcycle sporulation. It has been proposed that DAP may be involved in the promotion of elongation of the primary cell or in septum formation (2). The role of SAM here is less known. The data allows one to suggest that SAM can substitute for adenine and the reversal of ethionine inhibition in the presence of adenine by SAM supports this view. Given the overwhelming evidence for SAM as an important biological methylating agent, it would appear that methylation may be involved in sporulation.

ACKNOWLEDGEMENT

This investigation was supported by grant PCM-7924785 from the National Science Foundation.

REFERENCES

1. Vinter, V. and Slepecky, R.A. (1965) *J. Bacteriol.* **90**, 803-807.
2. Cloutier, M.J., Hanlin, J.H. and Slepecky, R.A. (1981) in

Sporulation and Germination (Levinson, H.S., Sonenshein, A.L. and Tipper, D.J., eds.), pp.163-167, Am. Soc. Microbiol., Washington.
3. Schaeffer, P., Millet, J. and Aubert, J.P. (1965) *Proc. Natl. Acad. Sci. U.S.A.* **54**, 704-711.
4. Salvatore, F., Borek, E., Zappia, V., Williams-Ashman, H.G. and Schlenk, F. (1977) in *The Biochemistry of Adenosylmethionine*. Columbia Univ. Press, New York.
5. Brown, N.C. (1970) *Proc. Natl. Acad. Sci. U.S.A.* **67**, 1454-1461.
6. Mychajlonka, M. and Slepecky, R.A. (1974) *J. Bacteriol.* **120**, 1331-1338.

SPOROGENESIS IN *STREPTOMYCES*

C. HARDISSON, M.B. MANZANAL and A.F. BRAÑA

Departamento de Microbiología, Universidad de Oviedo, Oviedo, Spain

SUMMARY

Sporogenesis in *Streptomyces* involves the simultaneous formation of sporulation septa which divide the aerial hyphae into spore-sized compartments. In *Streptomyces antibioticus* the sporulation septa are formed by two thin layers which constitute the annuli, separated by a thick material which forms the deposits. Deposits, unlike the annuli, are only temporary assisting structures which are degraded during spore maturation. Sporulating hyphae of *Streptomyces* accumulate polysaccharide granules which are degraded during spore maturation. A possible role for this polymer is to supply a source of carbon or energy for the synthesis of spore components during maturation. Mature spores contain relatively high levels of trehalose. This sugar may be an energetic material to be used during germination or it may help the maturation by aiding in dehydrating the cytoplasm of the spore.

INTRODUCTION

The genus *Streptomyces* shows a relatively complex developmental cycle in which, from a branched vegetative mycelium submerged in the substrate, an aerial mycelium is formed. Through the sporulation process these aerial hyphae give rise to chains of spores which, when placed under favourable conditions, germinate and repeat the cycle.

Spores produced by *Streptomyces* differ from bacterial endospores both in the sporogenesis process and in their properties. They are exospores, and are not dormant, because they have detectable, although low, metabolic activities.

TABLE 1 *Some features of* Streptomyces antibioticus *spores*

Endogenous respiration	Q_{O_2} : 10.69 ± 0.62 1 O_2/h/mg dry weight
Adenine nucleotide levels	
AMP	0.20 - 0.30 nmoles/mg dry weight
ADP	0.10 - 0.15 nmoles/mg dry weight
ATP	0.20 - 0.25 nmoles/mg dry weight
Energy charge	0.47
Sugar uptake activity	Glucose, constitutive; galactose and fructose, inducible
Amino acid pool	20 amino acids
Nucleic acid precursors pool	5 nucleosides
Trehalose content	About 10% of the dry weight
Dipicolinic acid	Absent
Heat resistance	Slightly superior to vegetative cells

Table 1 shows some features of spores from *S. antibioticus*. As can be seen, in many aspects of *Streptomyces's* spores show many similarities with the spores of fungi.

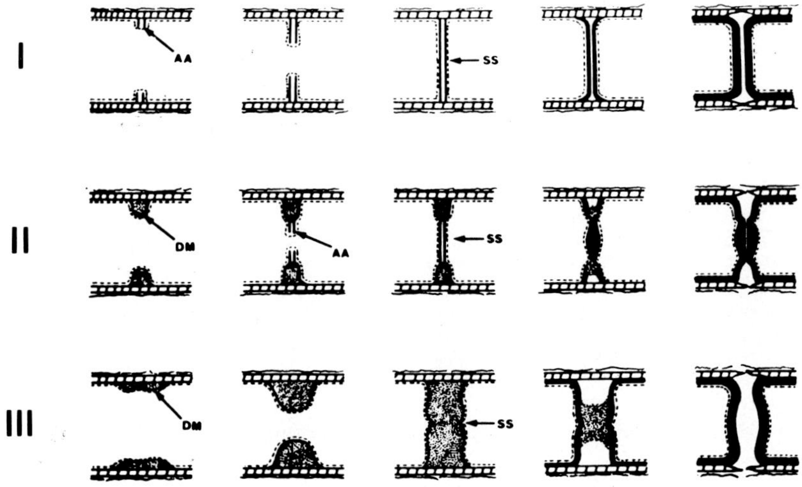

1

FIG. 1 *Diagrams showing the three basic types of sporulation septum formation in* Streptomyces. *AA, annuli; DM, deposits of septum material; SS, sporulation septum.*

Sporulation in *Streptomyces* is initiated with the simultaneous formation of specialized cross-walls, the sporulation septa (SS), which divide the pre-existing aerial hyphae into spore-sized compartments. The septa develop simultaneously and at regular intervals (1). These facts seem to indicate the existence of regulatory mechanisms which in answer to an appropriate signal or signals such as: nutrient depletion, unbalanced growth, reduction in the growth rate, etc., switch on the genes that control the sporogenesis process. Experimental observations suggest that cellular differentiation occurs in *Streptomyces* during secondary metabolism, together with the synthesis of antibiotics, pigments, reserve compounds, geosmine, etc. (2). It is possible that nutrient depletion is the signal for starting sporogenesis, and the effect of this starvation may be exerted through the reduction of the growth rate (2,3).

TYPES OF SPORE FORMATION

Electron microscopic studies (1,4) show that according to the structure of the SS and to the maturation process the genus *Streptomyces* can be divided into three groups (Fig. 1).

In Type I, the SS is formed from the beginning by two separate cross-walls. Once completed, each cross-wall thickens evenly during the maturation until it forms the lateral wall of the spore. Type II involves the formation of deposits of an amorphous material before the synthesis of the double annulus which completed the SS. The deposits will completely lyse during maturation, and only the margins are incorporated into the lateral wall of the spore. In Type III, the SS is formed by a single deposit of electron-dense material. Only in this Type is the material that constitutes the SS completely lysed, and the lateral wall of the spore is synthesized "de novo".

We must emphasise the singularity of these septa, whose structure is unique in the Kingdom Prokaryota and differs from the vegetative septa of the mycelium, which are similar to those of Gram positive bacteria. Another feature that must be underlined is the important role that controlled autolysis must play during spore maturation. The regulation of the enzymes involved in wall synthesis and lysis will probably play a fundamental role in the entire sporogenesis process.

Chemical Nature of the Sporulation Septum

In the previous observations, we had observed that in osmium fixed cells the material that constitutes the deposits is slightly less electron-dense than the outer layer of the SS (Fig. 2). In many cases (Fig. 3), the annuli are perforated by fine channels 8-10 nm in diameter, which may play an important role during sporogenesis, facilitating intercellular

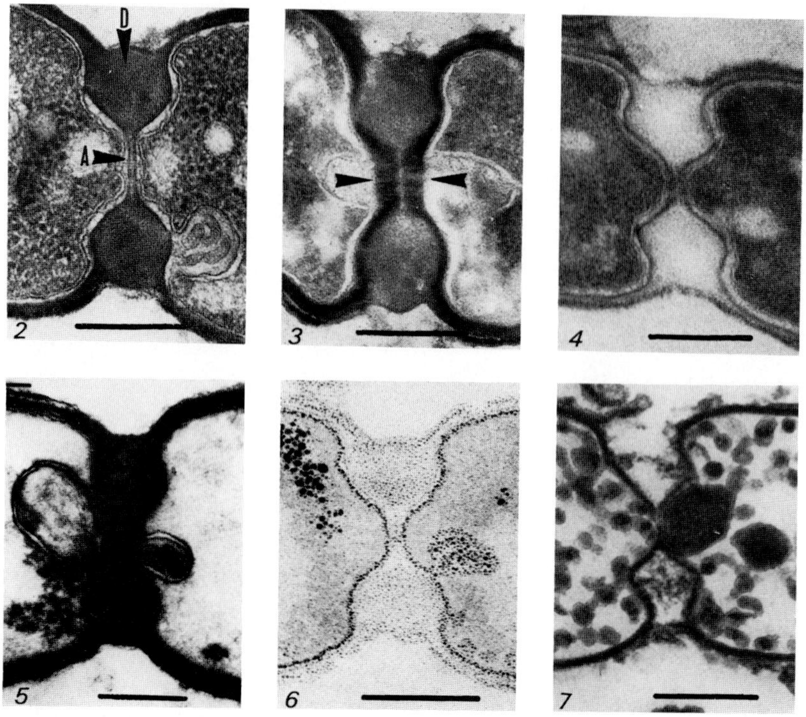

FIGS. 2-7 *Thin sections of sporulating hyphae of* S. antibioticus *showing the typical structure of the sporulation septum;* **(Figs. 2,3)** *Osmium fixed cells, the septum is formed by deposits (D) and annuli (A), note the presence of several perforations across the annuli (arrow-heads);* **(Fig. 4)** *Potassium permanganate fixed cell;* **(Fig. 5)** *Thin section stained by the phototungstic acid technique, all components of the septum are stained;* **(Fig. 6)** *Thin section stained by the silver proteinate procedure; the cell membrane, mesosomes and cytoplasmic granules are stained but the cell wall and the sporulation septum are not stained;* **(Fig. 7)** *Effect of lysozyme on the sporulation septum; after 5 min of incubation (100 μg/ml of lysozyme at 37°C) the deposits were completely degraded whereas the outer layer of the septum appeared unchanged. Bar represents 200 nm in all figures.*

connections between adjacent cells. When $KMnO_4$ fixation is used (Fig. 4), the differences in staining affinity are more evident and the SS consist of two thin electron-dense layers separated by two deposits of an electron-transparent material. except in the central portion, where only a fine gap is observed. These findings may be related to a different chemical composition between the structures that constitute the septum.

No data have been reported on the chemical nature of the material present in the SS. Obviously, a limitation in these studies is the difficulty of obtaining well isolated septa from the hyphae. For this reason, two different cytochemical techniques for revealing polysaccharides and enzymatic digestion with lysozyme were used to obtain more information about the chemical composition of the SS (5).

When thin sections of sporulating hyphae are stained with the phosphotungstic acid method of Rambourg (6), the cell membrane, cell wall, as well as the mesosomes and the SS, appeared heavily stained (Fig. 5). However, if thin sections are stained with the silver proteinate technique of Thiéry (7), the cell membrane and mesosomes are well stained, but the cell wall and the SS are not stained (Fig. 6). These data appear to indicate that the polysaccharide present in the SS may correspond to peptidoglycan, because this polymer is not revealed by silver proteinate, whereas it is intensely stained by the Rambourg technique, as previously reported in other bacteria (8). On the other hand, studies on enzymatic digestion showed that the deposits of the SS are highly sensitive to lysozyme whereas the margin of the deposits are resistant (Fig. 7).

All these results appear to confirm the existence of chemical differences between deposits and the margin of the deposits which includes the annuli. Deposits are formed by peptidoglycan, as confirmed by cytochemical staining and lysozyme sensitivity. The margin of the deposits is probably also formed by peptidoglycan, and its resistance to lysozyme could be due to localized chemical modifications of this polymer or to high degrees of peptide cross-linking, as has been suggested in different microorganisms (9,10). This interpretation is in agreement with the different susceptibility to autolytic breakdown shown by deposits and annuli, and may explain the fact that the deposits are lost during the sporogenesis, whereas the annuli are conserved. The deposits, unlike the annuli, are only assisting structures, functional until delimitation of the spore compartment is completed.

Occurrence of Glyogen and Trehalose during Sporogenesis

The study of the *Streptomyces* cell cycle presents some methodological difficulties. Thus, aerial mycelium formation and sporogenesis take place only on solid media, and although there are a few reports on sporogenesis in liquid media, these results have not been reproducible in other laboratories (11).

developmental cycle. After 24 h of incubation (Fig. 8) the colony is formed only by substrate hyphae. As can be observed, no polysaccharide granules are present in the cells. After 36 h of incubation, the colony consists of three different regions (Fig. 9). The lower region is occupied by lysed cells of substrate mycelium. In the central region most of the substrate hyphae are completely filled with polysaccharide granules. The upper region of the colony is occupied by a loose network of aerial hyphae. After 56 h of incubation (Fig. 10), sporogenesis has started in the upper region of the colony. This moment coincides with the appearance of polysaccharide granules in the sporulating hyphae. The sequential development of these glycogen granules during sporogenesis is shown in Figs. 11-13. Once the SS is completed, the number of polysaccharide granules increases, and the cytoplasm becomes filled with these granules. The point of maximum glycogen accummulation corresponds to an intermediate stage of spore maturation (Fig. 12). During later stages of spore maturation, the number of these granules progressively decreases, and in mature spores no granules are observed (Fig. 13).

It must be emphasized that glycogen is not detectable in mature spores either by cytochemical techniques or by chemical analyses. For this reason, the role of glycogen, if any, must be related to the sporogenesis process, but in any case it can not be considered as a reserve compound for spore germination. It is possible that the polymer plays the role of supplying a readily available source of carbon or energy or both for the synthesis of the spore components required for spore maturation. On the other hand, the initiation of synthesis of the SS may be correlated with the appearance of glycogen granules, because the depletion of free glucose may facilitate the derepression of sporulation genes, as has been suggested for endospore-forming bacteria (12,15).

As stated before, mature spores do not have any glycogen, however they have about 10-16% dry weight of trehalose, depending on the species of *Streptomyces*. This high content of trehalose may help the maturation of the spore by aiding the dehydration of the cytoplasm and the maintenance of dormancy, On the other hand, trehalose may be an energetic reserve material to be used during the early stages of germination, or it may serve both purposes.

CONCLUSION

A better understanding of the sporogenesis of *Streptomyces* requires several approaches: a morphological one, which improves our knowledge of the structural changes which take place during the process; a biochemical one, studying the changes in enzymatic activities playing key roles during the morphological changes and their regulation; and finally a genetic one, characterizing mutants blocked in the different stages. The genetic approach will probably be the most useful. Although the isolation of mutants blocked at the stage of aerial mycellium and unable to sporulate is not a very difficult task, their characterization has been proven to be very difficult. In no case has it been possible to relate a mutant to an altered function. As has been suggested (16), the application of DNA recombinant technology may help in the advancement of this knowledge, by cloning sporulation genes and expressing them in appropriate hosts in order to identify their products.

ACKNOWLEDGEMENTS

The work has been partially supported by a grant from the Fondo Nacional para el Desarrollo de la Investigación Científica y Técnica.

REFERENCES

1. Hardisson, C. and Manzanal, M.B. (1976) *J. Bacteriol.* **127**, 1443-1454.
2. Kalakoutskii, L.V. and Agre, N.S. (1976) *J. Bacteriol. Rev.* **40**, 469-524.
3. Dawes, I.W. and Mandelstam, J. (1970) *J. Bacteriol.* **103**, 529-539.
4. Manzanal, M.B. and Hardisson, C. (1978) *J. Bacteriol.* **133**, 293-297.
5. Braña, A.F., Manzanal, M.B. and Hardisson, C. (1981) *Can. J. Microbiol.* **27**, 1060-1065.
6. Rambourg, A. (1967) *C.R. Hebd. Seances Acad. Sci.* **265**, 1426-1428.
7. Thiéry, J.P. (1976) *J. Microsc. (Paris)* **6**, 987-1018.
8. Rousseau, M. and Hermier, J. (1975) *J. Microsc. Biol. Cell.* **23**, 237-248.
9. Ingram, L.O. and Aldrich, H.C. (1974) *J. Bacteriol.* **118**, 708-716.
10. Trentini, W.C. and Murray, R.G.E. (1975) *Can. J. Microbiol.* **21**, 164-172.
11. Ensign, J.C. (1978) *Ann. Rev. Microbiol.* **32**, 185-219.
12. Bergere, J.L., Rousseau, M. and Mercier, C. (1975) *Ann. Microbiol. Inst. Pasteur* **126A** 295-314.
13. Strasdine, G.A. (1972) *Can. J. Microbiol.* **18**, 211-217.
14. Braña, A.F., Manzanal, M.B. and Hardisson, C. (1982) *Can. J. Microbiol.* **28**, 1320-1323.
15. Morris, J.G. and Robson, R.L. (1973) in *Regulation de la*

Sporulation Microbienne pp.87-89, Centre National de la Reserche Scientifique, Paris.
16. Chater, K.F. and Merrick, M.J. (1979) in *Developmental Biology of Prokaryotes* (Parish, J.H. ed.), Vol. 1, pp.93-114, Blackwell Scientific Publications, Oxford.

Spore Resistance & Dormancy

MECHANISMS OF HEAT RESISTANCE

A.D. WARTH

CSIRO Division of Food Research, PO Box 52,
North Ryde, New South Wales, 2113 Australia

SUMMARY

Partial dehydration is a tenable general mechanism for stabilization of spore proteins to heat. A water activity of 0.73 corresponding to a water content of 20%, was required to stabilize *in vitro*, a spore enzyme to the same extent as found *in vivo*. Removal of Ca dipicolinate and other low molecular weight spore substances did not have a major effect, indicating that these substances do not have unique general stabilizing properties. Dehydration of the core is maintained by tension in the cortical layers. Swelling pressure is only in the radial direction, between layers.

INTRODUCTION

In order for a spore to survive high temperatures, all vital, potentially labile components must be stabilized. Some general mechanism is therefore probably required, although a number of specific adaptations of individual components will probably be found. Proteins are the most important class of heat-labile compounds to be considered. It is not obvious at present if non-protein cell components, such as lipid bilayers or DNA, would require special stabilization. This discussion will be concerned only with proteins.

At the molecular level, three types of mechanism may contribute to protein stability in the spore: spore proteins may be intrinsically stable, stabilizing substances may be present in the spore, or the solvent water may be partly removed. Spore cytoplasmic enzymes in general are not heat stable but, depending on the species, are stabilized within the spore by about 45°C (1). At present there is no strong evidence for the existence of stabilizing factors in the spore, but the unique and ubiquitous presence of calcium dipicolinate (CaDPA) and its close physiological association with heat resistance does suggest such a role. On the other hand it is well known that many proteins can be greatly stabilized to heat by drying, and that the spore interior has a low water content. Consequently it has been commonly assumed and accepted that dehydration could provide a general explanation of spore heat resistance. The adequacy of dehydration as a general mechanism for stabilization will first be considered. From the dehydration necessary to stabilize glucose 6-phosphate dehydrogenase (G6PDH) *in vitro* to the same degree as found in spores, an estimate can be made of the water activity (a_w) which might exist in spores and the pressure needed to maintain it. Evidence for stabilizing factors in the spore can be sought by studying the effect of purification of the enzyme on heat stability at reduced a_w. In the second section, models for the disposition of forces in the cortex and coats are discussed. Finally, the physiological mechanisms that could generate these forces are suggested.

RESULTS AND DISCUSSION

Effect of Dehydration on Heat Stability

Stabilization of organisms and proteins. Despite the general acceptance of dehydration as a stabilization mechanism, most of the evidence relating to survival of organisms at temperatures comparable to bacterial spores is anecdotal or qualitative. Most organisms in fact survive drying poorly and are often less heat stable at reduced a_w (2). Nevertheless it is well documented that some organisms can be very heat resistant when dry. Mosses withstood temperatures up to 110°C (3), bean embryos survived 45 min at 100°C (4) and *Salmonellae* dried in the presence of glutamate survived 7 h at 100°C (5). Many resting forms, e.g. seeds, pollen, and non-bacterial spores, are more heat resistant than vegetative forms and generally have low water contents, but this is not well quantitated. Some bacteria can be stabilized by reduction in the water activity (a_w) either by equilibration at defined humidities, or by addition of high concentrations of solutes to the media (6,7). The increases, although considerable and of practical importance, do not give the order of heat resistance found in spores and suggest that a large reduction in a_w would be necessary in spores. Many reasons may exist for the sensitivity of some organisms to drying (8) and solutes (9). It appears though, that dehydration is a tenable general stabilization

TABLE 1 *Dehydration required to stabilize proteins to 100 C*

Protein	Stabilization[a] (°C)	Water content (%)	Relative humidity (%)
Myoglobin (42)	23	8	
Haemoglobin (43)	48	11	
Ovalbumin (43)	30	33	0.7 at 90°
Ovalbumin (44)		11	
Chymotrypsinogen (45)	32	11	
Lysozyme (46)	30	15	
β-Lactoglobulin (47)	20	22	0.86 at 25°
Tropocollagen (48)	34	24	
Collagen (43)	45	28	0.78 at 25°

[a]Difference between 100°C and the denaturation temperature at high a_w. Most denaturation temperatures were measured by differential scanning calorimetry.

mechanism, but that other specific factors will also be involved.

For those proteins that have been well studied, drying causes very large and consistent increases in heat stability (Table 1). To achieve a denaturation temperature of 100°C, generally requires drying to water contents about 11%. For ovalbumin, this corresponds to an a_w at 90°C of 0.7. Water absorption isotherms at the heating temperatures are not available for other proteins, and water binding at 25°C is generally significantly (20-30%) higher, which may correspond to large differences in a_w. Using data for 25°C, the water contents imply removal of half to two thirds of the water bound to the proteins (10), and a_w's of 0.7 - 0.85. Generally insignificant stabilizations are found with initial large reductions in water content and only when bound water starts to be removed, at $a_w<0.93$, does stability greatly increase. The data suggest that dehydration could be general in its stabilizing effects on proteins, and that a_ws around 0.75 would be necessary to achieve heat stabilities comparable with those of spore proteins *in vivo*.

Stabilization of Spore Enzymes

In order to evaluate the contribution that dehydration and protective factors may make to the heat resistance of spore cytoplasmic proteins *in vivo*, data are required on the heat stability of spore enzymes as a function of water activity and content, in both the presence and absence of potential stabilizing substances. A preliminary report of this work has been published (11).

Spore cytoplasm was extracted from disrupted spores of *Bacillus cereus*. This should contain all the components of the spore core including any potential stabilizing factors. In the intact spore, water content is regulated by osmotic pressure but, due to limitations of available membranes, it is not possible to achieve adequate concentration by reverse osmosis *in vitro*. Water content was therefore controlled by equilibration of the freeze-dried extract in a bithermal heating apparatus (11). In attempting to recreate the conditions within the spore core, some possible differences should be kept in mind. Proteins suffer denaturation on surfaces and on drying and rehydration, proteolysis of proteins is difficult to prevent entirely (12), and possibly there is special packing of some spore constituents *in vivo*. G6PDH was chosen as the principal test enzyme because in *B. cereus* heat stability was similar to that of the cell and it did not have unusual properties (1). For comparison, glucose dehydrogenase (GDN) and an intrinsically heat stable enzyme, leucine dehydrogenase (LDH) were also studied.

Heat stability of G6PDH in solution and at high a_w. In solution, there was no difference between the heat stability of G6PDH in the total extract, and after partial (10x) purification, indicating that substances removed during the purification were not effective at stabilization in dilute solution. Addition of Ca^{++}, Mg^{++}, DPA, and EDTA (1 mM) also had insignificant effects. It is not surprising of course that stabilization is not seen under these conditions, as in the spore, potential stabilants would be present in much greater concentrations. Under much more concentrated conditions, in the freeze-dried material equilibrated to high a_w, the inactivation rate did decrease (11), but by a trivial amount, equivalent to an increase in inactivation temperature of 1-2°C. This can be compared with the increase of 30°C that this enzyme shows within the spore (12). Therefore under conditions of relatively low water content, but high a_w, direct stabilization by components of spore cytoplasm was not important.

Effect of water activity on enzyme stability. Reduction in a_w dramatically increased enzyme stability. At $0.90a_w$, G6PDH was inactivated more slowly at 65°C than it was at 45°C and $0.99a_w$. This is equivalent to a decrease in inactivation rate of more than 1,000-fold. Lower a_ws gave additional large increases in stability. Enzyme inactivation temperature (temperature at which $k=0.010\ min^{-1}$) showed a steady increase with reduction in a_w, reaching nearly 140°C at low a_w (Fig. 1). GDH and LDH,

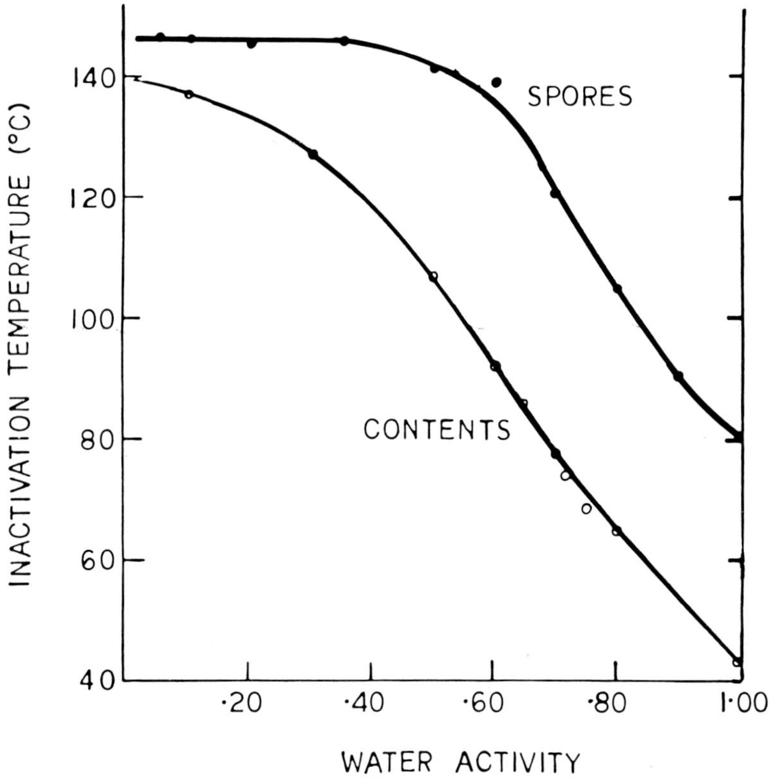

FIG. 1 *Effect of water activity on the heat stability of glucose 6-phosphate dehydrogenase* in vitro *and* in vivo. o, *Spore contents;* ●, *Spores. Inactivation temperature was the temperature at which the inactivation rate constant* $k = 0.01\ min^{-1}$.

likewise were stabilized. At low a_w all three enzymes had very similar stabilities despite their great differences in stability in solution (11). The similar inactivation rates of dry proteins provides an explanation for the greatly reduced variation in heat resistance of partly dry spores compared with their wet heat resistance (13).

These results demonstrate that it is possible to stabilize spore enzymes to at least the degree found *in vivo* by partial dehydration. The stabilization found was comparable to that reported for purified proteins (Table 1). It is very likely that partial dehydration can provide a feasible general mechanism for stabilization of heat labile proteins in the spore.

Heat inactivation of enzymes in spores. Reduction in the a_w at which spores were heated increased the heat stability of the three enzymes within the spore. The plot of inactivation temperature against a_w for G6PDH paralleled that of the extract with a displacement of about $0.3a_w$ (Fig. 1). This suggests that the spore structures that were damaged by disruption had an effect equivalent to a reduction in a_w of about 0.3. The progressive increase in enzyme stability *in vivo* in response to changes in the external a_w has important implications for different models of how the spore controls the water content of the core, and will be discussed later.

At 0.4 a_w and below, enzyme stability in the spore was constant at a level slightly greater than in the extract at low a_w. In this regard enzyme stability differed from viability, which showed a marked reduction in heat resistance at low a_ws (13,14). Therefore heat killing of dry spores may not be caused by enzyme inactivation.

Estimation of the water activity and content giving the same stabilization as found in vivo. The inactivation rate constant for G6PDH in spores heated in water at 85°C was 0.052 min^{-1}. With freshly prepared spore extracts, equilibrated directly in the heating tubes, an a_w of 0.73 was required to obtain the same inactivation rate at 85°C. Some factors could effect the accuracy of this result. Although the spores were disrupted and dried quickly in the presence of protease inhibitor, some changes in the composition and physical structure of the cytoplasm, due to dilution, proteolysis, and surface stresses during drying would seem inevitable. It is possible that spore cytoplasm *in vivo* is physically structured in a way unattainable *in vitro*. Hydration and redrying reduced G6PDH stability at $0.7a_w$, suggesting that less reduction in a_w would have been needed if the drying step had not been necessary.

At an a_w of 0.73, contents contained 20% w:w water at 85°C (Fig. 2). At 25°C, 28% water was bound. This well known effect makes it difficult to apply data obtained at laboratory temperatures to processes occurring at spore death temperatures. However both values are well within the range of spore core water contents which are implied by refractive index (15,16,17) and buoyant density data. Species differ in the effectiveness of the heat protective mechanisms of their spores (18). Probably much of this variation is in the degree of dehydration of the core, and Beaman *et al.* (19) have shown a correlation between refractive index and heat resistance.

Existence of stabilizing factors in the core. The above results were obtained in the presence of any soluble compounds which might stabilize enzymes. It has already been shown that

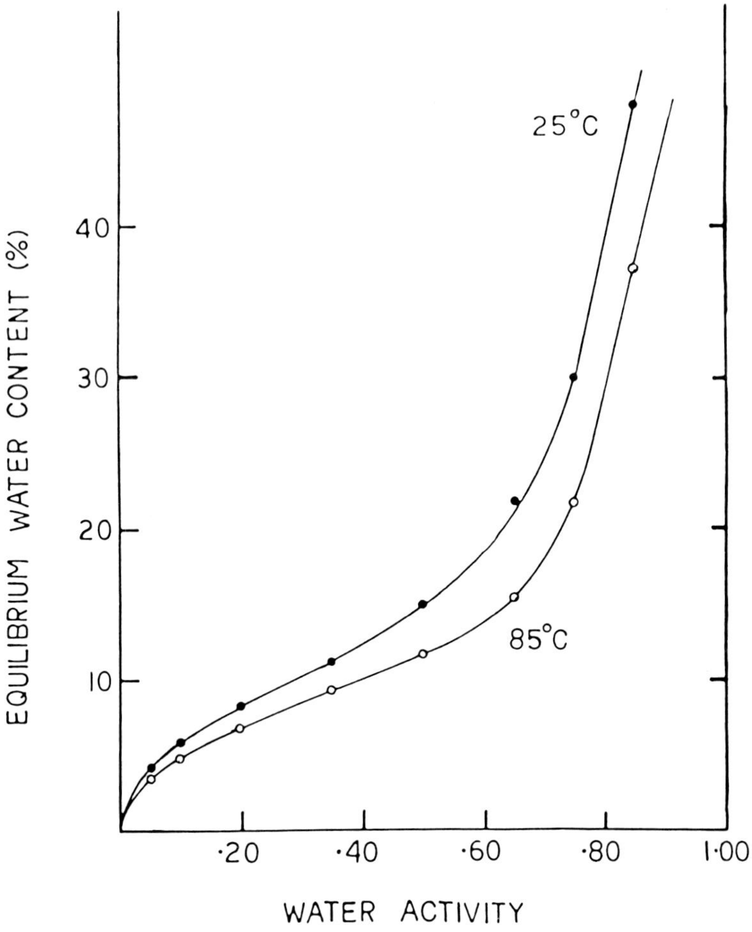

FIG. 2 *Water adsorption isotherms at 25 C and 85 C for spore contents.*

such factors were not important for G6PDH at high a_w, however they might be effective only at lower a_ws. In particular the effect of the low molecular weight solutes was studied by removing them by gel filtration. This had only a slight effect on heat stability at 85°C at 0.7 a_w (11), indicating that Ca^{++}, DPA, phosphoglycerate, glutamate, etc. do not have a major direct general stabilizing action. Partial purification of the enzyme resulted in some loss of stability, suggesting the existence of high molecular weight stabilants. This interpretation should be treated with caution as small samples (<50 μg) which may suffer from surface and transitory effects were used. Indeed it was found that common proteins, albumin and ovalbumin,

4) pressure exists between the system and one at high a_w. Each of these mechanisms are used by other dry life forms such as seeds. For spores, Lewis *et al.* (26) have demonstrated the implausibility of a water impermeable barrier. Water adsorption isotherms (Fig. 1; 27) and the ready solubility of spore contents show that spore cytoplasm does not have an unusually low affinity for water, and, since spores are formed in aqueous media of high a_w, mechanism 3 is not applicable. We are left with the inescapable conclusion that, for a spore in water, a partly dry interior requires the existence of a pressure which may be called the "turgor" or "osmotic" pressure of the spore. The need for a pressure has been widely accepted, and several models for the generation and maintenance of it have been proposed (23,25,26).

Consider the possible disposition of forces in a dormant spore, consisting of a core, a thick cortex and coats (Fig. 3). Pressure applied to the core is due to tension in the surrounding layers. The layer adjacent to the core can apply pressure directly, but each successive layer must transmit its pressure by compression of those beneath it. Intermediate layers then are in tension parallel to the surface, but are under compression in the radial direction. Which layers in the spore are under tension? Hydrolysis of the cortex with lysozyme and other agents causes loss of refractility and heat resistance, and blocking cortex synthesis prevents heat resistance (28). On the other hand, the lack of any major effect on heat resistance when much of the coat is extracted chemically, together with the existence of heat resistant coatless and coat defective spores, strongly suggests that the cortex, but not the coats, carries the tension responsible for the pressure on the core (23). This is not consistent with an osmoregulatory model in which the cortex exerts a simple swelling pressure (29). Swelling of a uniform (isotropic) cortex structure gives an increase in the enclosed (core) volume, and hence pressure cannot be applied to the core in this way. If the cortex is treated as a fluid, swelling will only produce a pressure when it is constrained by the coats, which are therefore essential to this model. Hydrolysis of the cortex will increase the osmotic swelling pressure, and heat resistance should be very dependent on the strength of the coats, quite the reverse of what is observed. On the other hand, an anisotropic cortex with swelling pressure confined to the radial direction but under tension in the plane of the surface (23), is consistent with the above and with the structure of the cortex seen in electron micrographs of disrupted spores. Typically these show an open multilayered network with extensive folding of the inner layers (Fig. 4). In the radial direction, the cortex appears very extensible and has expanded inwards when the radial

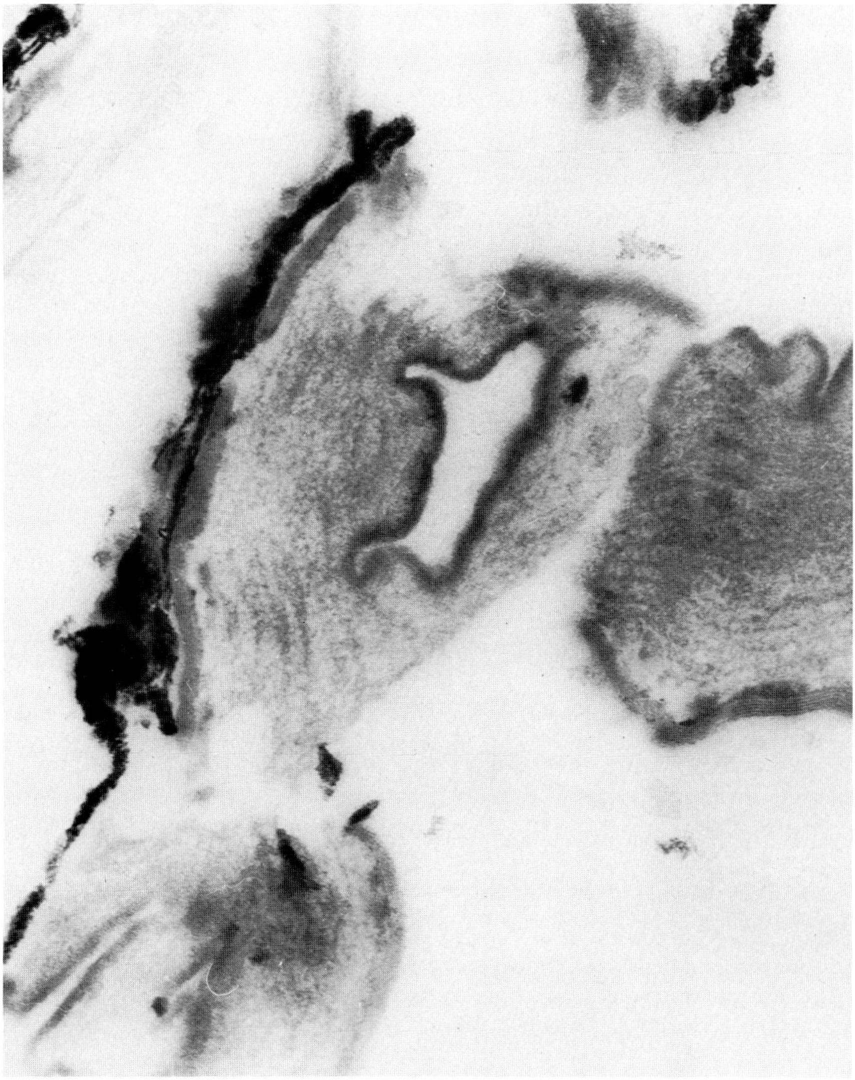

FIG. 4 *Electron micrograph of disrupted spores of* Bacillus coagulans, *showing the cortex expanded in the radial direction inwards, but not laterally, with consequent folding of the inner layers.*

pressure previously applied to the core was relieved by disruption. Signs of outward or lateral expansion, required for a fluid or isotropic cortex are not observed. In particular there is no evidence of cortex extruding through breaks in the coats of damaged spores. In the plane of the surface, relief of the tension, does not give a large contraction. These properties suggested that the glycan chains of the peptidoglycan were orientated parallel to the surface (23) as is thought to be the case for vegetative cell walls (30).

Estimation of the tension in the cortex. Maintenance of the internal a_w at 0.73 would require a pressure of 52 MPa (513 atmospheres) in an ideal system with no opposing forces due to deformation of structures and a constant partial molar volume for water ($P = RT/V . \ln a_w$). The apparent partial molar volume of water did not change much in ovalbumin at this a_w (31), and the presence of CaDPA may reduce deformations of macromolecules (see above). It is possible that the real pressure in the spore may not be much greater than 50 MPa, and may be less if the a_w estimate is low. This value may be compared with typical turgor pressures in gram positive bacteria of 2-3 MPa, and the hydrostatic pressures causing germination of 80 MPa (32). For a core diameter of 650 nm and a cortex thickness of 100 nm, this requires a stress in the cortex of 91 MPa. Assuming, for *E. coli*, a turgor pressure of 0.5 MPa, a peptidoglycan layer 4.2 nm thick and a radius of 350 nm, the stress in the peptidoglycan would be 21 MPa. A stress of 25 MPa was calculated for *B. stearothermophilus* cell walls (33). Since a growing cell has to cope with growth and variations in turgor pressure, and the peptidoglycan network may not be as tightly packed as assumed, a specialized peptidoglycan such as the cortex could probably readily sustain a stress of the magnitude calculated above.

For the outer layers of the cortex to contribute to the pressure on the core, the pressure from these layers has to be transmitted through the layers beneath. This cannot occur if the cortex is very expanded, because expanded polymers have low swelling pressures. In the intact spore, therefore, the cortex is probably more tightly packed than previously considered (34).

The pressure required for dehydration to $0.73 a_w$ is very much greater than the 3 MPa envisaged in the osmoregulatory model (29). Water absorption isotherms (27), show that spore contents are still very hydrated at the equivalent a_w (0.98), and this a_w gives sufficient stabilization to proteins. Clearly a simple swelling due to the charged groups in the cortex is an inadequate explanation. This model is also untenable because in such ion-exchangers, swelling is largely suppressed at much lower ionic strengths than the 2M $CaCl_2$ found necessary to heat-sensitize spores (29). Although carboxyl groups may be important for cortex function their role as cation exchange sites essential for swelling the cortex may have been overemphasized. Swelling pressure of hydrophilic polymers does

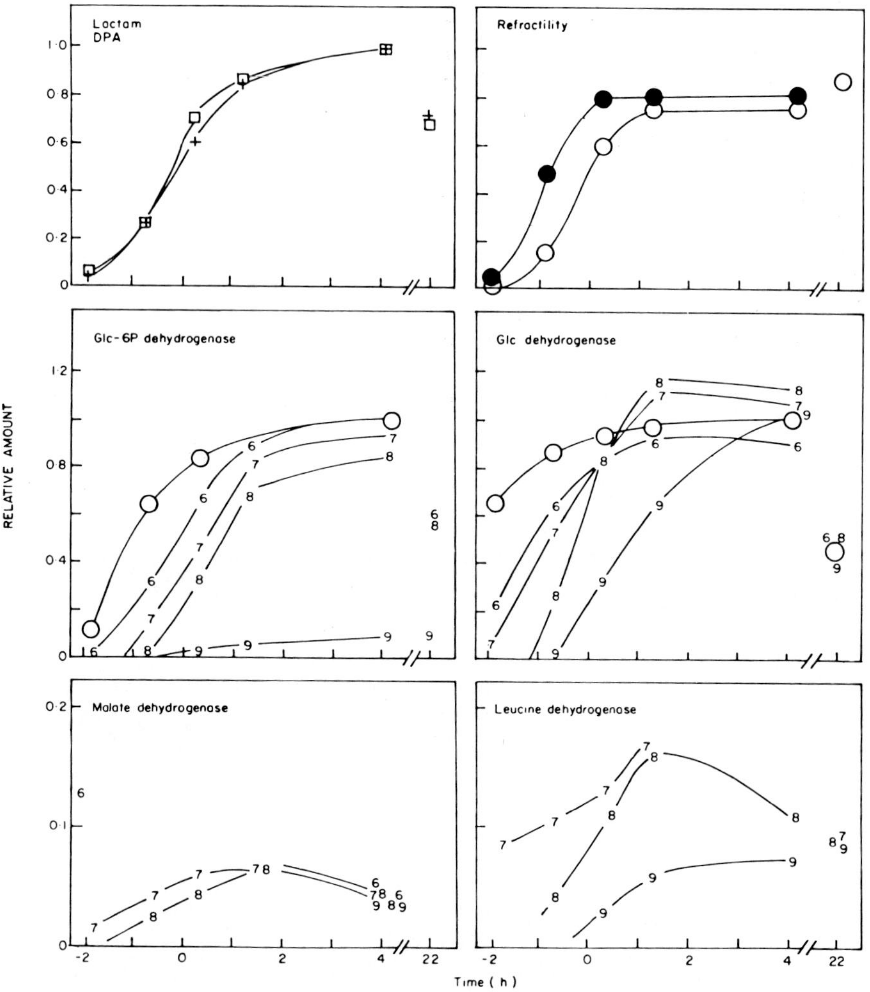

FIG. 5 *Development of heat stabilization of spore enzymes in relation to cortex and DPA synthesis during spore formation in* Bacillus cereus. *Symbols:* ■, *muramic lactam;* +, *DPA;* ●, *phase-grey + refractile spores;* o, *phase-white + refractile spores; 0.6.7.8.9. heated for 10 min at 25°, 60°, 70°, 80° and 90°C respectively.*

not depend only on ionized groups. If dehydration were maintained only by the ionized groups, then, near the isoelectric point around pH 3.5, spores should be extremely unstable. Electron microscopy of disrupted- and acid-popped spores shows the cortex remaining highly expanded over a wide range of pHs, and at high cation concentrations (Ohye and Murrell, unpublished results). This observation suggests that the inward expansion following relief of pressure from the core is a property of the whole cortex structure, not just the charged groups.

A major difficulty with the osmoregulation concept is the effect of reduction of R.H. on spore (13) and enzyme (Fig. 1) heat resistance. These both increased steadily as the R.H. was reduced, and enzyme stability *in vivo* paralleled that *in vitro*. Yet if dehydration depended greatly on coupling to an osmotic pressure, a much reduced response would be expected. Reimposition of heat resistance on germinated (35) or DPA^- spores (36) by sucrose does not necessarily give a measure of the osmotic pressure needed to dehydrate the core. These spores lack DPA but may have a functional cortex. Possibly sucrose functions by way of compensating for the loss of DPA.

Mechanisms for Generation of Pressure in the Cortex

How might the requisite tension and pressure be generated in the cortex? Unfortunately this involves the molecular details of stereo-specific and strain-sensitive control of peptidoglycan biosynthesis, of which nothing is known. It is very likely however that peptidoglycan is initially polymerized under strain-free conditions (37), in an ordered tightly packed conformation, with the typical complement of peptide side chains. Under these conditions very high (>50 MPa) theoretical swelling pressures exist and could be directed to compression of the interior, and fixed by cross-linking of the glycan chains so as to bear the tension. Free energy is also available from hydrolysis and release of peptide fragments, and there is evidence that diaminopimelate is in fact recycled during cortex formation (38). The replacement in spore peptidoglycan of peptide substituents or muramic acid with muramic lactam units and alanine side chains is the end result of vectorially directed biosynthesis. The initial swelling pressure is not due simply to ionized groups but results largely from the entropy inherent in an ordered hydrophilic structure.

Large volume reductions are not required during cortex synthesis, as the core has already been reduced to 50% of its previous volume and is reduced by only a further 8-12% on maturation (23,39). Cortex and coats are formed over a period of 3 h, at the same time as DPA synthesis and the appearance of refractility. Muramic lactam is also synthesized during this period in close parallel to DPA synthesis (40). During this time, heat stabilization of spore enzymes (Fig. 5) and spore heat resistance (41) increased in progressive increments. This indicates that the cortex is not synthesized as a whole, and then matured, but that successive layers are assembled and tensioned, building up the pressure, which in conjunction with CaDPA uptake to the core, produces the dessicated heat resistant state.

REFERENCES

1. Warth, A.D. (1980) *J. Bacteriol.* **143**, 27-34.
2. Mazur, P. (1980) *Origins Life* **10**, 137-159.
3. Lange, O.L. (1955) *Flora* **142**, 381-389.
4. Siegel, S.M. (1953) *Bot. Gaz.* **14**, 297-312.
5. Annear, D.I. (1964) *Aust. J. Exp. Biol. Med. Sci.* **42**, 717-722.
6. Corry, J.E.L. (1973) *Prog. Ind. Microbiol.* **12**, 73-108.
7. Corry, J.E.L. (1976) *J. Appl. Bacteriol.* **40**, 277-284.
8. Strange, R.E. and Cox, C.S. (1976) in *The Survival of Vegetative Microbes* (Gray, T.R.G. and Postgate, J.R., eds.), pp.111-154, Society for General Microbiology, Cambridge.
9. Rose, A.H. (1976) in *The Survival of Vegetative Microbes* (Gray, T.R.G. and Postgate, J.R. eds.), pp.155-182, Society for General Microbiology, Cambridge.
10. Kuntz, I.D. and Kauzmann, W. (1974) *Adv. Protein Chem.* **28**, 239-345.
11. Warth, A.D. (1981) in *Sporulation and Germination* (Levinson, H.S., Sonenshein, A.L. and Tipper, D.J., eds.), pp.249-252, American Society for Microbiology, Washington, D.C.
12. Setlow, P. (1975) *J. Bacteriol.* **122**, 642-649.
13. Murrell, W.G. and Scott, W.J. (1966) *J. Gen. Microbiol.* **43**, 411-425.
14. Alderton, G. and Snell, N. (1970) *Appl. Microbiol.* **17**, 745-749.
15. Ross, K.F.A. and Billing, E. (1957) *J. Gen. Microbiol.* **16**, 118-225.
16. Leman, A. (1973) *Jena Rev.* 263-270.
17. Gerhardt, P., Beaman, T.C., Corner, T.R., Greenamyre, J.T. and Tisa, L.S. (1982) *J. Bacteriol.* **150**, 643-648.
18. Warth, A.D. (1978) *J. Bacteriol.* **134**, 699-705.
19. Beaman, T.C., Greenamyre, J.T., Corner, T.C., Pankratz, H.S. and Gerhardt, P. (1982) *J. Bacteriol.* **150**, 870-877.
20. Lindsay, J.A. and Murrell, W.G. (1980) *Spore Newsletter* **7**, 46.
21. Williamson, F.B. (1977) in *Spore Research 1976* (Baker, A.N., Wolf, J., Ellar, D.J., Dring, J.G. and Gould, G.W., eds.),

pp.55-67, Academic Press, London.
22. Rajan, K.S., Jaw, R. and Grecz, N. (1978) *Bioinorg. Chem.* **8**,
23. Warth, A.D. (1978) *Adv. Microb. Physiol.* **17**, 1-45.
24. Klibanov, A.M. (1979) *Anal. Biochem.* **93**, 1-25.
25. Gould, G.W. (1977) *J. Appl. Bacteriol.* **42**, 297-309.
26. Lewis, J.C., Snell, N.S. and Burr, H.K. (1960) *Science* **132**, 544-545.
27. Marshall, B.J. and Murrell, W.G. (1970) *J. Appl. Bacteriol.* **33**, 103-129.
28. Imae, Y. and Strominger, J.L. (1976) *J. Bacteriol.* **126**, 907-913.
29. Gould, G.W. and Dring, G.J. (1975) *Nature (London)* **258**, 402-405.
30. Labischinski, H., Barnickel, G., Bradaczek, H. and Giesbrecht, P. (1979) *Eur. J. Biochem.* **95**, 147-155.
31. Bull, H.B. and Breese, K. (1968) *Arch. Biochem. Biophys.* **128**, 497-502.
32. Clouston, J.G. and Wills, P.A. (1970) *J. Bacteriol.* **103**, 140-143.
33. Algie, J.E. (1980) *Curr. Microbiol.* **3**, 287-290.
34. Gould, G.W. and Dring, G.J. (1974) *Adv. Microbial. Phys.* **11**, 137-164.
35. Dring, G.J. and Gould, G.W. (1975) *Biochem. Biophys. Res. Commun.* **66**, 202-208.
36. Bhothipaksa, K. and Busta, F.F. (1978) *Appl. Environ. Microbiol.* **35**, 800-808.
37. Koch, A.L., Higgins, M.L. and Doyle, R.J. (1981) *J. Gen. Microbiol.* **123**, 151-161.
38. Frehel, C., DeChastellier, C. and Ryter, A. (1980) *Can. J. Microbiol.* **26**, 308-317.
39. Murrell, W.G. (1981) in *Sporulation and Germination* (Levinson, H.S., Sonenshein, A.L. and Tipper, D.J., eds.), pp. 64-77.
40. Wickus, G.G., Warth, A.D. and Strominger, J.L. (1972) *J. Bacteriol.* **111**, 625-627.
41. Balassa, G., Milhaud, P., Raulet, E., Silva, M.T. and Sousa, J.C.F. (1979) *J. Gen. Microbiol.* **110**, 365-379.
42. Hagerdal, B. and Martens, H. (1976) *J. Food Sci.* **41**, 933.
43. Takahashi, K.G.I. and Shirai, F.L. (1980) *Nippon Nogei Kagaku Kaishi* **54**, 357.
44. Altman, R.L. and Benson, S.W. (1960) *J. Am. Chem. Soc.* **82**, 3852-3857.
45. Fujita, Y. and Noda, Y. (1981) *Int. J. Pep. Protein Res.* **18**, 12-17.
46. Fujita, Y. and Noda, Y. (1978) *Bull. Chem. Soc. Japan* **51**, 1567-1568.
47. Ruegg, M., Moor, U. and Blanc, B. (1975) *Biochim. Biophys.*

Acta **400**, 334-342.
48. Luescher, M., Ruegg, M. and Schindler, P. (1974) *Biopolymers* **13**, 2489-2503.

PHYSIOLOGICAL BIOPHYSICS OF SPORES

R.E. MARQUIS, E.L. CARSTENSEN, G.R. BENDER and S.Z. CHILD

Departments of Microbiology and Electrical Engineering, The University of Rochester, Rochester, NY 14642, U.S.A.

SUMMARY

Information obtained by use of simple, nondestructive, physical techniques is surveyed to obtain a clearer picture of the nature of dehydration during sporogenesis, the physical states of spore electrolytes and the relationship of heat resistance to mineralization. Major dehydration during stage III of sporogenesis seems to depend on osmotic-metabolic mechanisms involving the inverted, outer, forespore membrane. Subsequent dehydration and the maintenance of the dehydrated state appear to depend on cortical peptidoglycan elasticity. Minerals in the spore core are immobilized, but those in enveloping structures of many spores are mobile. Heat resistance appears to be acquired incrementally in association with dehydration and specific mineralization.

INTRODUCTION

Living organisms faced with harsh environmental conditions must adapt or die. The bacterial endospore is a prime example of a cryptobiotic cell designed for survival during difficult times. Its survival mechanisms are primarily cellular rather than molecular. For example, heat resistance is not the result of synthesis of heat resistant molecules but instead depends on changes in cell structure and composition that provide a stabilizing environment. The changes in the spores that appear to be most important for heat resistance are dehydration and mineralization. This paper will focus on these changes and on information obtained with simple, nondestructive, physical

TABLE 1 *Dextran-impermeable volumes per g, dry weight, for various cells*

Organism	Cell type	Average[1] dextran impermeable volume (ml/g)	Calculated[2] water content (ml H_2O/g)
Bacillus megaterium ATCC 19213	vegetative	7.3	6.6
	protoplast[3]	8.0	7.3
	sporulating cell (stage III)	6.0	5.3
	isolated forespores[3] (stage III)	3.1	2.4
	mature spores[4]	2.5	1.8
	decoated spores	1.8	1.1
	germinated spores	5.4	4.7
	outgrowing cell	7.4	6.7
Bacillus cereus terminalis	vegetative	4.4	3.7
	mature spores	2.4	1.7
Bacillus subtilis niger	vegetative	4.1	3.4
	mature spores	2.3	1.6

Bacillus sphaericus	vegetative	3.9	3.2
9602	mature spores	2.3	1.6
Neurospora crassa	conidia	3.0	1.3
Streptococcus faecalis	vegetative	4.1	3.4

[1]Averages of at least duplicate determinations with thick suspension.
[2]The dry matter of all cells was assumed to have a volume of 0.7 ml/g (8).
[3]The forespores and protoplasts were suspended in dextran solutions with 0.75 osmolal sucrose as osmotic stabilizer.
[4]Various salt forms (Na, K, H, Ca and Mn) obtained by complete ion exchange had essentially the same dextran-impermeable volume per g as the native spores (28).

techniques which can be applied to the physiological interpretation of spore resistance.

Dehydration

There is now little doubt that bacterial spores in aqueous suspension are less hydrated than the vegetative cells from which they are derived. The refractile appearance of the spores in the phase microscope clearly supports this view. That at least some region of the spore has a high-solids and low-water content is indicated unequivocally by data obtained with interference microscopy (1) and quantitative immersion refractometry (2,3,4). The recent careful determinations of water contents for a variety of spores in aqueous suspension by Beaman *et al.* (5) indicate that "water content varies from about 0.45 to 0.65 g of water per g of wet spore". Most vegetative cells contain between 0.75 and 0.90 g of water per g of wet weight and so are clearly more hydrated than spores.

A parameter we have found useful for comparison of hydration states among various cells is the dextra-impermeable volume per g, dry weight. This parameter is easily determined by use of the thick-suspension or space technique described in detail previously (6) and high-molecular-weight dextrans which cannot penetrate cell walls or membranes. Experimental values are readily interpreted as total cell volumes per unit of dry weight, including the volume of cell water and the volumes of the nonaqueous components of the cell. Generally, average volumes for the latter are about 0.7 ml/g (7), and so the dextran-impermeable volume per g minus 0.7 ml is about equal to the volume of cell water per g, dry weight.

The data presented in Table 1 indicate that the spores tested consistently had lower values for dextran-impermeable volume per g than did vegetative cells of the same strains. Average values for mature spores ranged only from 2.3 to 2.5 ml/g. The calculated water contents for spores were 1.6 to 1.8 ml H_2O/g, or about 0.61 to 0.64 ml H_2O per g, wet weight. Average values for vegetative cells ranged more widely from 3.9 to 7.3 ml/g. Calculated water contents for vegetative cells were 3.2 to 6.6 ml H_2O/g, or about 0.76 to 0.87 ml H_2O per g, wet weight. Fungal conidia isolated from *Neurospora crassa* were found to be more hydrated than bacterial spores but less hydrated than vegetative cells.

Enveloping structures play major mechanical roles in maintaining the states of hydration of vegetative cells. Growing vegetative cells are turgid. They concentrate solutes, which lower the water activity of the cytoplasm. Since the cells are highly permeable to water, environmental water flows in across the cell membrane, and the cells begin to swell. However, the cell wall is an elastic net of peptidoglycan which develops tension when stretched. The cell wall elasticity then resists further swelling, and in the final turgid state the cytoplasmic water is under hydrostatic pressure, which increases the internal water potential until it is equal to the potential of the water outside the cell.

Even though spores contain much less water than vegetative cells, they still are fully permeable to water (7,8). The physiologic mechanisms for maintaining low water levels in spores in aqueous suspensions are not completely defined at present. The data presented in Table 1 for spores of *B. megaterium* ATCC 19213 suggest that coat layers do not play a major mechanical restraining role. Their removal did not result in spore swelling. In fact, the value for dextran-impermeable volume per g decreased, and electron micrographs of sections of these coatless spores prepared by Dr. H.S. Pankratz of Michigan State university showed no evidence of cortex swelling. Data obtained with spores of the QMB 1551 strain of this organism presented at this meeting by Gerhardt *et al.*(9) suggest some swelling of the cells associated with stripping of coat layers. At this time, it seems best to conclude that coat layers could play auxiliary mechanical restraining roles but that the state of dehydration required for refractility is maintained primarily through mechanical restraint caused by the elasticity of the peptidoglycan network of the cortex.

Dehydration begins early in sporulation shortly after engulfment. Again, phase microscopy provides very direct evidence for dehydration in terms of the image of a phase dark forespore or prespore in a less dark sporangial cytoplasm. Unfortunately, this system of a cell within a cell is not amenable to quantitative optical techniques such as immersion refractometry or interference microscopy. However, Murrell (10) has estimated dehydration in terms of decreased volumes of forespores seen in electron micrographs of sectioned sporulating cells. The cells had to be fixed and dehydrated for electron microscopy, and one would expect alterations in structural relationships associated with these processes. Also there are problems in estimating cell volumes from cross sectional diameters and in assessing stages of sporulation in populations. Despite these difficulties, it was concluded that contraction of the forespore, presumably due to dehydration, occurs early in stage III prior to cortex formation or mineralization.

As indicated by the data in Table 1, the average dextra-impermeable volume of isolated *B. megaterium* forespores in stabilizing 0.75 osmolal sucrose solutions was found to be about 3.0 ml/g, compared with values of 2.5 ml/g for mature spores and 7.3 ml/g for vegetative cells. Determinations of the osmolality for incipient plasmolysis for sporulating cells

TABLE 2 *Dielectric properties of bacterial spores and vegetative cells.*

Cell	Environmental conductivity (mho/m)	Low frequency (1 MHz) cell conductivity (mho/m)	High frequency (50 MHz) cell conductivity (mho/m)	Frequency for cell dielectric constant of 200 (MHz)
Bacillus megaterium[1]	0.07	0.11	0.25	8
vegetative	0.70	0.46	0.62	10
B. megaterium mature	0.05	0.06	0.11	3
spores	0.50	0.17	0.27	4
B. megaterium decoated	0.05	0.06	0.09	< 1
spores	0.50	0.28	0.31	< 1
B. megaterium, stage III	0.03	0.01	0.03	< 1
forespores	0.36	0.19	0.29	4
B. cereus T mature	0.01	0.02	0.03	< 1
spores	0.40	0.14	0.24	3

[1]ATCC 19213 strain

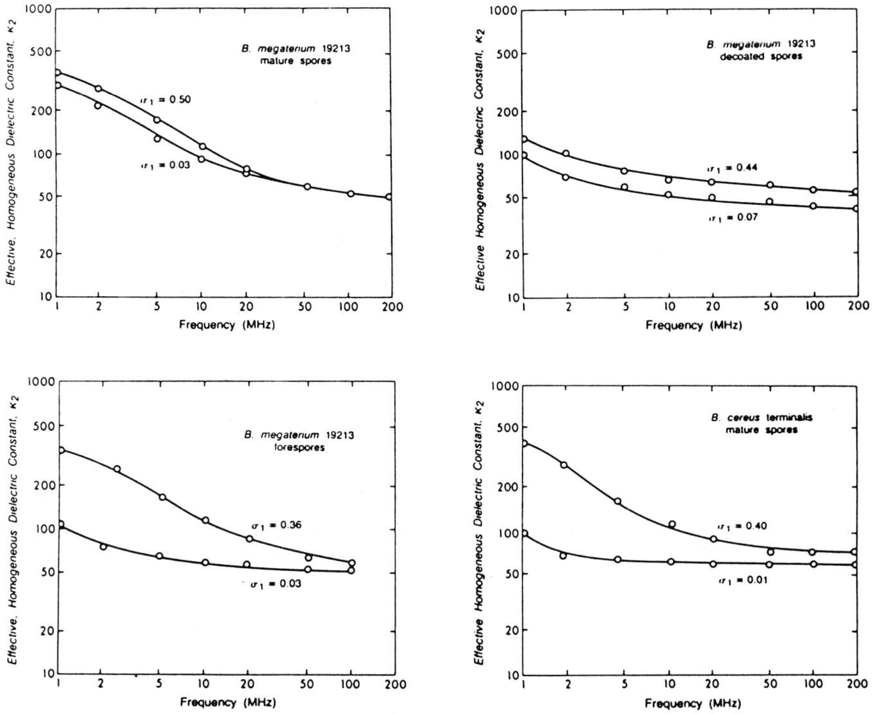

FIG. 1 *Effective homogeneous dielectric constants as a function of current frequency for mature spores, decoated spores of* B. megaterium *ATCC 19213 and for mature spores of* B. cereus terminalis. *Values for* σ_1 *the environmental conductivity are in units of mho/m.*

Cell membranes act as effective barriers or insulators for the cell interior and the cell dielectric constant is large. However when high-frequency current of 50 MHz is used, the cell membranes are capacitatively short-circuited, and mobile ions anywhere in the cell, including those in the core, can take part in current conduction.

Typical dielectric data are presented in Table 2 and Fig. 1 for spores and vegetative cells of *B. megaterium* ATCC 19213 and for spores of *B. cereus* T. The curves for effective, homogeneous dielectric constant κ_2 for mature *B. megaterium* spores plotted against frequency of the probing current show the so-called β or Maxwell-Wagner dispersion. At low frequency, the membranes act as effective insulators, and κ_2 values are high. As the frequency is increased, the cell membranes disappear dielectrically until finally the dielectric constant values decrease to slightly less than the dielectric constant of 78.3 for water at 25°C. The characteristic frequency (f_r) for the dispersion, at the inflection point of the curve, can be related to internal conductivity by use of the equation,

$$f_r = \sigma_i / 2\pi a C_m$$

where σ_i is the internal conductivity, a is the cell radius and C_m is the membrane capacitance, about 0.01 F/m^2 for biological membranes. [A detailed model for use in interpreting the dielectric properties of spores has been presented previously (25)]. Since spore radius would be expected to change very little in response to changes in environmental salt levels, and C_m also should remain constant for the spores used for dielectric studies, f_r can be related directly to σ_i with a simple constant.

The characteristic frequency for *B. megaterium* spores in media of low ionic strength is much lower than that for vegetative cells (Table 2), and spore conductivities at 1 or 50 MHz are lower. When the ionic strength of the suspending medium was increased with added NaCl, the characteristic frequency shifted somewhat to the right to higher values indicating that the spores were to an extent invaded by environmental ions which increased σ_1. Values in the last column of Table 2 indicate frequencies at which κ_2 = 200, and changes in these values reflect changes in f_r.

It is clear that the electrolytes in *B. megaterium* spores are, on average, much less mobile than are those in vegetative cells. Moreover, the mobile ions in *B. megaterium* spores appear to be those in the cortex. When the coat-outer-membrane complex was removed by means of the procedure described by Aronson and Horn (26), the f_r value for the spores dropped to below 1 MHz. Thus, removal of the coat-outer-membrane removed a dielectrically effective membrane so that now the only effective membrane was the inner membrane surrounding the core. Environmental ions could invade the decoated spores, as indicated by changes in conductivity when the cells were transferred to media of high ionic strength. However, the conductivity change was apparent even at low frequency, and f_r was not shifted greatly. In essence, it appeared that the ions penetrated the cortical peptidoglycan but not the core. Thus, these *B. megaterium* spores showed extreme electrostasis of the core but contained mobile ions in the cortex.

In contrast, intact *B. cereus* T spores (Table 2, Fig. 1) in media of low ionic strength had few mobile ions anywhere in the cell, and f_r values were well below 1 MHz. However, when the spores were transferred to media of high ionic strength, environmental ions invaded, conductivity at high and low frequency increased, and the characteristic frequency shifted dramatically to higher values. We initially thought (27) that the ions invaded the outer-membrane-coat complex to gain access to the cortex. However, we now feel that the ions invade only the exosporium, which must be able to function as a dielectrically effective membrane. Unfortunately, the exosporium is not easily removed from *B. cereus* T spores without other damage to the cells, and so it is difficult to test our hypothesis.

Forespores of *B. megaterium* isolated at stage III of sporulation behaved dielectrically much as did mature *B. cereus* T spores (Table 2 and Fig. 1) in that they had very low inherent conductivity but could be invaded by environmental ions. The shift in f_r in Fig. 1 is major. However, the basis for low forespore conductivity and low f_r values in media of low ionic strength was a very low content of mineral ions - at a level approximately equal to that required simply for neutralization of excess negative charges on cell polymers (11). Forespores at this stage did not contain detectable dipicolinate and had average calcium levels of only 0.026 μmole/mg cell dry weight. They also contained very little potassium, an average of 0.12 μmole/mg, or about one fifth of the amount in sporulating vegetative cells. Thus, the dielectric behaviour of forespores reflected their low level of mineralization, while the superficially similar behavior of mature spores of *B. cereus* reflected electrolyte immobilization in highly mineralized cells.

We have recently been able to investigate in a definite way the effects of specific mineralization on spores of *B. megaterium* ATCC 19213. These spores have remarkable acid resistance, and suspensions of the cells can be titrated to pH values as low as 1.8 to 2.0 without popping or otherwise dying. This acid resistance allows for complete exchange of minerals, including those in the core. Thus, we could prepare viable H forms of the spores lacking entirely Ca, Mg, Mn and K by titrating the suspensions to a pH value of 2 and heating them at 60°C. Then the H spores could be converted to their salt forms by back titration with appropriate base solutions and heating at 60°C. Presumably the heating was necessary to increase the permeability of the cells to the exchanging minerals. The exchange did not significantly affect the state of hydration. For example, the K form had essentially the same dextran-impermeable volume per g as the Ca form, or the native form listed in Table 1. Therefore, the dehydrated state of the spore does not depend on electrostatic interactions involving divalent cations in the core or the cortex. Since the dehydrated state of the spore does not appear to depend on these interactions, it is reasonable to consider that the elasticity of the cortex may serve to maintain the dehydrated state of the core, and in some cases, such as in the QMB 1551 strain of *B. megaterium*, the coat layers may also serve a mechanical restraining function. Certainly, any damage to the cortex of decoated spores results in rapid hydration and germination.

It is clear from studies of ion-exchanged spores that heat resistance depends very much on specific mineralization (28), as does pressure resistance (29). The Mn form of *B. megaterium* ATCC 19213 spores was the most heat resistant of the various forms tested, followed in order by the native, calcium, magnesium, potassium, sodium and hydrogen forms. The ammonium form was nearly as heat sensitive as the H form. In fact, it is the heat sensitivity of the H form that presents difficulty in obtaining complete ion exchange. Unless the heating of the spores at 60°C is kept to only a few hours, heat-sensitive H forms can be killed. However, the H-form spores were much more heat resistant than were vegetative cells. For example, D values for H spores at pH 2 or 6.5 and 60°C were greater than 480 min, while D values for vegetative cells at 60°C and pH 7.5 averaged 1.25 min. Therefore, heat resistance in spores appears to be incremental rather than all-or-none. As documented by Warth (29), the basic heat resistance of vegetative cell polymers, especially proteins, is reflected in the heat resistance of spores formed by various species. Thermophiles and thermotolerant bacteria generally produce the most heat-resistant spores. In fact, it appears that the spore state increases heat resistance by some 38 to 46°C above the vegetative state for specific enzymes (30). However, the actual increase depends on many factors. Part of the increase seems to be independent of mineralization, as indicated by our finding that even H forms or K forms of spores are more heat resistant than vegetative cells. Possibly this resistance is related to the dehydrated state of the cell and it can be increased by further dehydration when spores are dried (31). The results of studies with ion-exchanged spores indicate clearly that this basic resistance can be increased markedly by mineralization, especially with Mn or Ca. In other words, there seems to be a component of heat resistance associated with spore structure and dehydration and another component associated with mineralization.

ACKNOWLEDGEMENTS

The investigations of the authors were supported by award number DAAG29-80-C-0051 from the U.S. Army Research Office with Philipp Gerhardt as principal investigator.

REFERENCES

1. Leman, A. (1973) *Jena Rev.* **5**,
2. Ross, K.F.A. and Billing, E. (1957) *J. Gen. Microbiol.* **16**, 418-425.
3. Ross, K.F.A. (1967) in *Phase Contrast and Interference Microscopy for Cell Biologists*, pp.130-134, St. Martin's Press, New York.
4. Gerhardt, P., Beaman, T.C., Corner, T.R., Greenamyre, J.T. and Tisa, L.S. (1982) *J. Bacteriol.* **150**, 643-648.
5. Beaman, R.C., Greenamyre, J.T., Corner, T.R., Pankratz, H.S. and Gerhardt, P. (1982) *J. Bacteriol.* **150**, 870-877.
6. Marquis, R.E. (1981) in *Manual of Methods for General Bacteriology* (Gerhardt, P., ed.), pp.393-404. American Society for Microbiology, Washington D.C.
7. Marshall, B.J. and Murrell, W.G. (1970) *J. Appl. Bacteriol.* **33**, 103-129.
8. Black, S.H. and Gerhardt, P. (1962) *J. Bacteriol.* **83**, 960, 967.
9. Gerhardt, P., Beaman, T.C. and Koshikawa, T. (1982) in *Fundamental and Applied Aspects of Bacterial Spores* - personal communication.
10. Murrell, W.G. (1981) in *Sporulation and Germination* (Levinson, H.S., Sonenshein, A.L. and Tipper, D.J., ed.), pp.64-77, American Society for Microbiology, Washington D.C.
11. Marquis, R.E., Bender, G.R., Carstensen, E.L. and Child, S.Z. (1983) *J. Bacteriol.* **153**, 436-442.
12. Wilkinson, B.J., Deans, J.A. and Ellar, D.J. (1975) *Biochem. J.* **152**, 561-569.
13. Ellar, D.J. (1978) in *Relations between Structure and Function in the Prokaryotic Cell* (Stanier, R.Y., Rogers, H.J. and Ward, J.B., ed.), pp.295-325, Cambridge University Press.
14. Gould, G.W. and Dring, G.J. (1975) *Nature (London)* **258**, 402-405.
15. Warth, A.D. (1978) *Adv. Microbiol. Physiol.* **17**, 1-45.
16. Setlow, B. and Setlow, P. (1980) *Proc. Natl. Acad. Sci. U.S.A.* **77**, 474-476.
17. Bradbury, J.H., Foster, J.R., Mammer, B., Lindsay, J. and Murrell, W.G. (1981) *Biochim. Biophys. Acta* **678**, 157-164.
18. Slepecky, R.A. and Foster, J.R. (1959) *J. Bacteriol.* **78**, 119-123.
19. Eisenstadt, E. and Silver, S. (1972) in *Spores V*

(Halvorson, H.O., Hanson, R. and Campbell, L.L., eds.), pp.180-186, American Society for Microbiology, Washington D.C.
20. Seto-Young, D.L.T. and Ellar, D.J. (1981) *Microbios* **30**, 191-208.
21. Crosby, W.H., Greene, R.A. and Slepecky, R.A. (1971) in *Spore Research 1971* (Barker, A.N., Gould, G.W. and Wolf, J., eds.), pp.143-160, Academic Press, London.
22. Stewart, M., Somlyo, A.P., Somlyo, A.V., Shuman, H., Lindsay, J.A. and Murrell, W.G. (1980) *J. Bacteriol.* **143**, 481-491.
23. Nishihara, T., Ichikawa, T. and Kondo, M. (1980) *Microbiol. Immunol.* **24**, 495-506.
24. Johnstone, K., Stewart, G.S.A.B., Barratt, M.D. and Ellar, D.J. (1982) *Biochim. Biophys. Acta* **714**, 379-381.
25. Carstensen, E.L., Marquis, R.E., Child, S.Z. and Bender, G.R. (1979) *J. Bacteriol.* **140**, 917-928.
26. Aronson, A.I. and Horn, D. (1972) in *Spores V* (Halvorson, H.O., Hanson, R. and Campbell, L.L., eds.), pp.19-27, American Society for Microbiology, Washington D.C.
27. Carstensen, E.L., Marquis, R.E. and Gerhardt, P. (1971). *J. Bacteriol.* **107**, 106-113.
28. Marquis, R.E., Carstensen, E.L., Child, S.Z. and Bender, G.R. (1981) in *Sporulation and Germination* (Levinson, H.S., Sonenshein, A.L. and Tipper, D.J., eds.), pp.266-268, American Society for Microbiology, Washington D.C.
29. Bender, G.R. and Marquis, R.E. (1982) *Can. J. Microbiol.* **28**, 643-649.
30. Warth, A.D. (1980) *J. Bacteriol.* **143**, 27-34.
31. Alderton, G. and Snell, N. (1970) *Appl. Microbiol.* **17**, 745-749.

BACILLUS SUBTILIS SPORES ON SPACELAB I: RESPONSE TO SOLAR UV RADIATION IN FREE SPACE

G. HORNECK, H. BÜCKER and G. REITZ

DFVLR, FF-ME, Abt. Biophysik, Linder Höhe
D-5000 Cologne 90, Federal Republic of Germany

SUMMARY

Based on experimental data, obtained during the Apollo 16 mission and from ground-based simulation experiments, studies on the limiting factor for survival of bacterial spores in free space will be performed during the Spacelab I mission, An exposure tray on the pallet permits treatment of dry spores with space vacuum and/or selected wavelengths of solar UV. After recovery, action spectra of inactivation, mutation induction, repairability and photochemical damage in DNA and protein will be determined.

INTRODUCTION

Life on Earth has been effectively protected from most harsh and hostile environmental factors of free space. By using meteorological rockets, especially equipped with a microbe sampling device, viable microorganisms, spore-forming fungi as well as bacteria, have been isolated at altitudes up to 60-77 km (1). Therefore, an altitude of approximately 80 km has been considered as the upper boundary of the biosphere.

With the development of space technology the opportunity arose to investigate some space-related problems directly in the space environment. Examples are the following questions:

Interplanetary transfer of life. What is the chance for a small particle of the size of a microorganism to escape the gravitational forces of a planet of the size of the Earth? How realistic is the theory of Panspermia (2) of the appearance of life on Earth through infection by germs from another planet? Is a present-day microorganism capable of surviving an interplanetary journey or even a trip between different solar systems? What are the most deleterious factors in space?

Space radiation biology. What is the impact of a cosmic ray heavy ion on biological matter? What is the consequence of a combined action of cosmic radiation and low gravity on biological matter during space flight? Which radiation standards and radiation protection guidelines are required for a manned space mission?

Life on Earth. How dependent is life on Earth-bound factors? Is there a threshold for the biological response to gravity?

Bacterial spores are proper test organisms for studying several of these problems of space biology (3-4). In free space, microorganisms have to cope with an interplay of various environmental factors: ultra-high vacuum, reaching pressures down to 10^{-14} Pa; solar electromagnetic radiation, of special concern is the solar UV radiation; high energy particles of cosmic and solar origin, such as electron neutrons, protons and heavy ions; extreme temperature which may vary from very low values around 10°K, to very high values in cases of direct irradiation by the sun.

We report here on a space experiment to be flown on board Spacelab I, studying the response of *Bacillus subtilis* spores to solar UV radiation and/or to space vacuum.

METHODS

Bacterial Strains

Spores of *Bacillus subtilis* characterized by the following genetic markers are the candidate test organisms:

Strain	Genetic marker								Ref
	uvr	ssp	rec A1	pol A1	his	met	leu	trp	
168	+	+	+	+	+	+	+	+	5
Ha 101	+	+	+	+	−	−	−	+	6
Ha 101 F	+	+	+	−	−	−	−	+	6
GSY 1026	+	+	+	+	+	−	+	−	6
GSY 1025	+	+	−	+	+	−	+	−	6
TKJ 6324	+	−	+	+	−	−	+	+	7
TKJ 6312	−	−	+	+	−	−	−	+	7
TKJ 6321	−	−	+	−	−	−	+	+	7
TKJ 6232	−	−	−	+	−	−	+	+	7

For photochemical studies, in addition, cells of *E. coli* B/r thy$^-$ (5) have been used.

Facilities for Ground-based Studies

Vacuum. High vacuum chambers, equipped with an oil diffusion pump, an ion sputtering pump or a turbomolecular pump, respectively, have been used, reaching pressures down to 10^{-8} Pa.

UV irradiation. 254 nm: Hg low pressure germicidal lamp; maximum intensity: 40 J/m^2s; continuous spectrum; deuterium source, 200W (D 200 F Hanau).

Preparation

Spores were obtained on sporulation medium (16 g/l Difco nutrient broth, 2.5 x 10^{-2} M KCl, 2 x 10^{-3} M $CaCl_2$, 1 x 10^{-5} M $FeCl_2$, 2 x 10^{-5} M $McSO_4$, 1 x 10^{-4} M $MgSO_4$, 0.2% glucose) after growth at 37°C for 60 h or on Difco TAM sporulation agar. The spores were collected, washed once with water and resuspended in 0.01 M Tris HCl (pH 7.5) buffer. The spores were treated with lysozyme (100-200 μm/ml) and deoxyribonuclease I (2 μg/ml) in the presence of 1 x 10^{-4} M $MgSO_4$ for 20 min at 37°C. The suspension was heated for 10 min at 80°C and then washed six times in distilled water. The spores were further purified by sedimentation through Renografin gradient according to (8). Purified spores were stored in distilled water at 4°C. For exposure, monolayers were prepared on rough coverglass.

Analysis

Survival. Spores were counted as having survived a treatment, if they formed visible colonies on Difco NB agar after 12 h or incubation at 37°C.

Mutation induction. Mutations to azide-resistance were counted after 24 h incubation at 37°C on Difco TBAB plates containing 150 μg/ml sodium azide. Back mutations to histidine prototrophy were determined on Spizizen's minimal medium (9) supplemented with glucose (0.5%), casein hydrolysate (0.02%) and the required amino acids after incubation at 33°C for 24 h.

Photoproducts. Pyr-photoprodcuts in DNA: The DNA, labelled with ^{3}H-methyl-thymine, was extracted, hydrolysed in 98% formic acid at 175°C for 45 min and paper-chromatographed in n-butanol-acetic acid-water (80:12:30) (10). DNA protein crosslinking: After lysis with SDS, the high molecular cellular components were precipitated in 95% ethanol and resuspended in 2% SDS. Protein was precipitated by 1 M KCl and the free DNA determined in the supernatant (11).

APPROACH - FLIGHT EXPERIMENT

The joint European/US Spacelab Mission I, an Earth-orbiting flight, lasting one week, provides a flexible laboratory system installed in the Orbiter cargo for various disciplines of science, such as atmospheric physics, Earth observations, astronomy, solar physics, material sciences and technology and life sciences (12). ESA has selected our experiment "Microorganisms and Biomolecules in Space Hard Environment" 1ES 029 as one of nine European life sciences experiments. The experiment system is outlined in Table 1.

TABLE 1 *Objectives and investigators of Spacelab I experiment 1 ES 029*

Spacelab 1 - Experiment ES 029 Effects of solar UV and/or space vacuum on bacterial spores	
Phenomenon studies	Investigative Group
Action spectra of Inactivation Mutation induction Repair	DFVLR G. Horneck (Principal Investigator) G. Reitz H. Requardt
Pyr-photoproducts in DNA	University Frankfurt H.D. Mennigmann P. Weber
DNA Protein - Crosslinking	University Mainz K. Dose K.-D. Martens
Flight Hardware: Development, Integration	DFVLR G. Reitz

The experiment accommodates 400 dry samples of *Bacillus subtilis* spores, each containing approximately 10^6 organisms. They are positioned in four square-shaped, quartz-covered containers (Fig. 1). Two of them will be vented to the outside, thus allowing access of vacuum of space to the samples. Samples, which are to be exposed to UV-irradiation, are placed under a filter system, allowing exposure to UV of the following wavelengths: > 170, 200, 260 and 280 nm. A shutter will be used to determine the total exposure time. The data from post-flight evaluation of the biological flight and the ground-control samples will be compared with the findings of simulation experiments on the ground.

FIG. 1 *Experiment 1 ES 029, hardware of the exposure box assembly without filters, shutter and cover.*

RESULTS

Previous Space-flight Experiments

During the Apollo 16 space flight, in the experiment system M-191 "Microbial Response to Space Environment" (13) spores of *Bacillus subtilis* 168 were exposed to space vacuum or solar UV irradiation with a peak wavelength of 254 nm or both. It was found that a short exposure of 1.3 h to the vacuum of space around the spacecraft itself did not damage the spores detectably; that the response of wet samples to solar UV was comparable to results from ground-based studies; and that simultaneous exposure to the vacuum of space and to solar UV amplified the lethal UV effect by a factor of approximately 10 (14).

The dose-survival-curves are shown in Fig. 2. It was suggested that the reduced water content of the spores in the vacuum environment favours the production of either more or less repairable DNA photoproducts. To clarify this phenomenon of UV supersensitivity in free space, laboratory experiments in high and ultra-high vacuum have been performed (15-17).

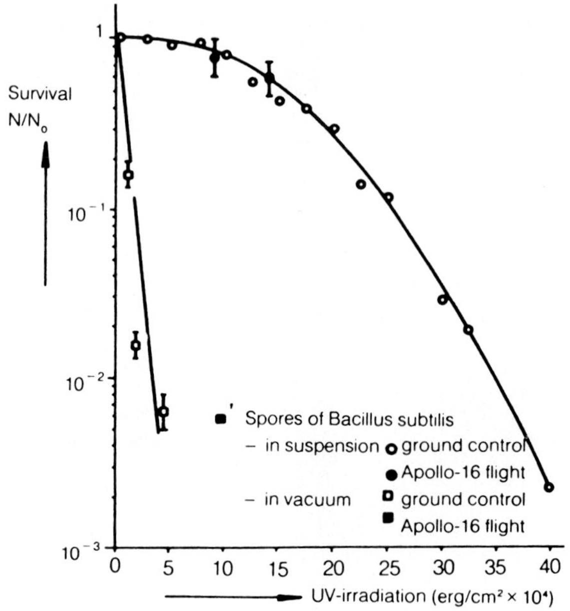

FIG. 2 *Survival of* Bacillus subtilis *spores irradiated with UV of 254 nm in suspension and* in vacuo *during space flight, and corresponding ground control.*

GROUND-BASED STUDIES

The following photoproducts have been identified in the DNA of *E. coli* cells, UV-irradiated in the laboratory: DNA-protein crosslinking, trans-syn-thymine dimer and 5-thyminyl-5,6-dihydrothymine (Fig. 3). The amounts of cis-syn-thymine dimer and cytosine-thymine dimer were reduced. In isolated DNA, UV irradiation *in vacuo* produced comparable protoproducts (18). It is very likely that at least one of these photoproducts is responsible for the increased inactivation by UV of vacuum-exposed cells.

DNA protein-cross-linking is the photochemical event that correlates with inactivation when cells are UV irradiated

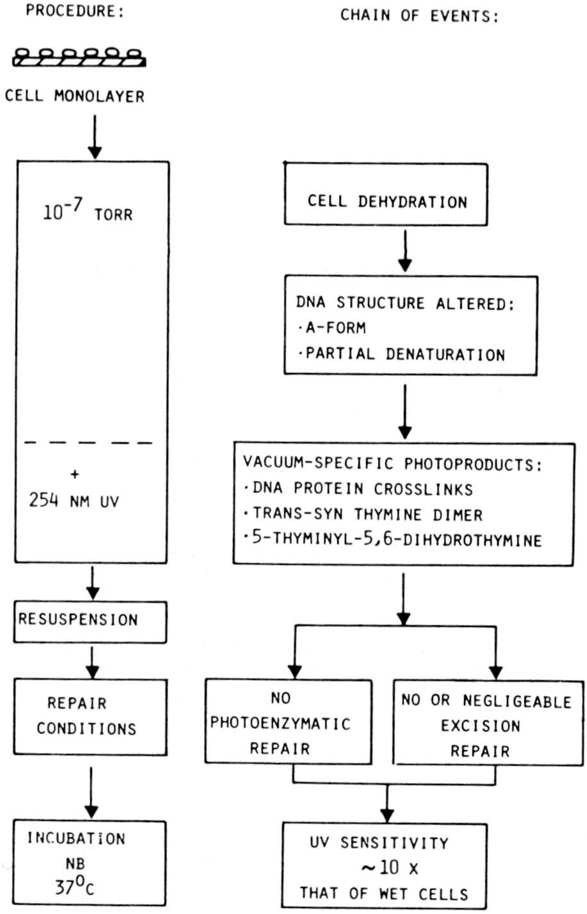

FIG. 3 *UV supersensitivity of bacterial cells* in vacuo, *observed in space and in simulation experiments on ground.*

in vacuo. The enhancement factor for the formation of this photoproduct was 12.8, the same value as for inactivation (15). Crosslinking reached saturation when approximately 80% of the DNA was rendered nonextractable, a process which requires an UV dose of 20 J/m^2.

Much higher UV doses were necessary for producing detectable amounts of thymine containing photoproducts in the DNA of *E. coli* cells irradiated *in vacuo* (19). This process was followed for doses up to 6 x 10^4 J/m^2. Whereas in wet cells, the production of cis-syn thymine dimer reaches saturation at approximately 10^4 J/m^2, in vacuum-treated cells the formation of thymine containing photoproducts continued throughout the dose range tested.

The cis-syn pyrimidine dimer of the cyclobutane-type is specifically split by the photoreactivating enzyme (20). However, the survival of cells UV irradiated *in vacuo* could not be raised by the conditions allowing photoenzymatic repair (15). It was shown that the photoenzymatic repair system itself was not affected by vacuum treatment of the cells (15). Therefore, photoproducts of the cis-syn cyclobutane pyrimidine dimer-type are certainly not responsible for the supersensitivity to UV of cells *in vacuo*.

Trans-syn thymine dimer is not a substrate for photoenzymatic repair. Since it is also formed in heat-denatured DNA (21), its production may indicate that vacuum leads, at least in part, to similar structural changes in DNA as heat treatment. This photoproduct may play a candidate role in the altered photobiological properties of cells *in vacuo*.

CONCLUSIONS AND OUTLOOK

Experimentation in space as it is offered by the NASA and ESA Space Program (22) is a new tool in biological research. This opportunity can be seized to learn more about fundamental life processes. The results obtained up to now, are sparse and sometimes contradictory (23,24). The experiment, which is reported here, will increase our understanding of the limiting factors for life. Free space has been considered as one of the most extreme and hostile environments (25).

The near-future Space Program includes: the Biorack, a Spacelab multipurpose research facility, especially designed for cell and molecular biology; its first mission will be in 1985 on the German Spacelab D1 mission (26); a passive long-duration exposure facility (LDEF), to be launched by NASA in 1984 for a nine-month orbital mission; a European retrievable carrier (EURECA), accommodating facilities for autonomous biological experimentation which is planned for 1986 for a flight lasting several months (26).

REFERENCES

1. Imshenetsky, A.A., Lysenko, S.V., Kazakov, G.A. and Ramkova, N.V. (1976) *Life Sciences and Space Research 14* Akademie-Verlag, Berlin, pp.359-362.
2. Arrhenius, S. (1908) *Worlds in the Making*. Harper and Row, New York.
3. Horneck, G., Facius, R., Enge, W., Beaujean, R. and Bartholomä, K.-P. (1974) *Life Sciences and Space Research 12*, Akademie-Verlag, Berlin, pp.75-83.
4. Spizizen, J., Isherwood, J.E. and Taylor, G.R. (1975) *Life Sciences and Space Research 13*, Akademie-Verlag, Berlin, pp.143-149.
5. Obtained from M. Brendel.
6. Obtained from J. Spizizen.
7. Obtained from N. Munakata.
8. Tamir, H. and Gilvarg, G. (1966) *J. Biol. Chem.* **241**, 1085.
9. Anagnostopoulos, C. and Spizizen (1961) *J. Bacteriol.* **81**, 741.
10. Smith, K.C. (1963) *Photochem. Photobiol.* **2**, 503.
11. Ceriotti, G. (1952) *J. Biol. Chem.* **198**, 297.
12. Craven, P.D. (ed.) (1981) *Spacelab Mission 1 Experiment Descriptions*, 2nd Edition, NASA TM-82448, ESA FSLP-EX-001.
13. Taylor, G.R., Spizizen, J., Foster, B.G., Volz, P.A., Bücker, H., Simmonds, R.C., Heimpel, A.M. and Benton, E.V. (1974) *BioScience* **24**, 505-511.
14. Bücker, H., Horneck, G., Wollenhaupt, H., Schwager, H. and Taylor, G.R. (1974) *Life Sciences and Space Research 12*, Akademie-Verlag, Berlin, pp.209-213.
15. Horneck, G. and Bücker, H. (1971) *Strahlentherapie* **141**, 732.
16. Frankenberg-Schwager, M., Bücker, H. and Wollenhaupt, H. (1974) *Raumfahrtforschung* **5**, 209-212.
17. Bücker, H., Facius, R., Reitz, G., Thomas, C., Wollenhaupt, H. (1976) *Life Sciences and Space Research* **14**, Akademie-Verlag, Berlin, pp.355-358.
18. Bücker, H., Dose, K., Horneck, G. and Thomas, C. (1979) *Life Sciences and Space Research* **17**, Pergamon Press, Oxford and New York, pp.111-115.
19. Schwager, M. (1973) Thesis, University Frankfurt/M.
20. Harm. H. (1976) in *Photochemistry and Photobiology of Nucleic Acids* (Wang, S.Y., ed), Vol.II, pp.219-263, Academic Press, London/New York.
21. Ben-Hur, E. and Ben-Ishai, R. (1968) *Biochem. Biophys. Acta* **166**, 9.
22. Bjurstedt, H. (ed.) (1979) *Biology and Medicine in Space*, European Space Agency, ESA BR-01.

23. Calvin, M. and Gazenko, O.G. (eds.) (1975) *Foundations of Space Biology and Medicine* NASA SP 374.
24. Klein, H.P. (1981) *Acta Astronautica* **8**, 927-938.
25. Horneck, G. (1981) *Adv. Space Res.* **1**, issue 14, pp.39-48.
26. Further information from: Dr H. Oser, ESA, 8-10 rue Mario Nikis, 75738 Paris, Cedex 15, France.

EFFECT OF ULTRASONIC WAVES ON THE HEAT RESISTANCE OF *BACILLUS STEAROTHERMOPHILUS* SPORES

B. SANZ, P. PALACIOS, P. LÓPEZ and J.A. ORDÓÑEZ

Departamento de Higiene y Microbiologia de los Alimentos, Facultad de Veterinaria, Universidad Complutense, Madrid-3, Spain

SUMMARY

The heat resistance of *Bacillus stearothermophilus* spores in distilled water markedly decreases after an ultrasonic treatment (20 KHz, 120 W, 12°C) which is, by itself, unable to kill any spores. The heat sensitizing effect of this ultrasonic treatment is greater as the ultrasonication time increases. The ultrasonic treatment does not seem to affect the z-values. The heat sensitizing effect of the ultrasonic waves is not affected by prolonged storage times of the spores maintained at 1°C.

INTRODUCTION

Many efforts have been made to increase the efficiency of sterilization processes. A combination of heat with other physical or chemical agents has proven to increase the killing effects of some treatments (1,2,3). Ultrasonic waves alone have only a very low (4,5) if any (6,7) sporicidal power but the later reports showed that they may potentiate the antimicrobial activity of some chemical agents, such as hydrogen peroxide. Other workers have observed that some strains of *B. cereus* whose spores have been ultrasonically treated maintain their heat resistance (8,9) while in other strains of *B. cereus*, *B. licheniformis* and *B. subtilis*, the heat resistance of spores decreases significantly (9,10). Recently, in our laboratory (11), we verified that ultrasonic treatments applied prior to heating reduced the D-values of two strains of

B. subtilis spores about 2-fold. This reduction was even higher when the spores were simultaneously subjected to ultrasonic and heat treatment.

Because the *B. stearothermophilus* spores are highly resistant to heat, many of the sterilization treatments applied to foods take into consideration their heat-resistance parameters. Any method able to increase the killing effect of heat could be very useful in the food, and other industries. The present work reports on the usefulness of ultrasonic waves in this context.

METHODS

Organisms and Culture Conditions

Spores of *B. stearothermophilus* NCA-1518 and *B. stearothermophilus*-L were used as the test organisms. The later strain was isolated by Dr. J. Burgos (Leon University, Spain) from spoiled sterilized milk.

The spores were obtained after growth of the bacteria at 55°C during 5 days in a manganese-containing nutrient agar. The crude spore suspensions were washed twice by centrifigation (5,000 g) in sterile distilled water and submitted to a papain (0.4 mg/ml) treatment (48 h at 10°C) in 0.05 M phosphate buffer (pH 7) containing sodium thioglycolate (0.05 mg/ml) and 1:1 toluene/chloroform mixture (0.5%) as recommended by Murrel and Warth (12). The spores were washed three times in sterile distilled water at 5°C and stored at 1°C. Before use the spores were activated by heating at 100°C for 15 min.

Ultrasonic Treatment

Aliquots of 40 ml of the purified spore suspensions were subjected to an ultrasonic treatment of 20 KHz, 120 W using a Heat Systems Ultrasonic Inc wave generator. The temperature was kept constant by holding the cells throughout the ultrasonic process in a jacketed glass vessel connected to an ice-water bath. A thermocouple linked to a digital thermometer allowed continuous monitoring of the actual temperature of the suspension. When the ultrasonic waves were applied under these conditions the medium temperature quickly increased from 0.5°C to 12°C after about 1 min. This temperature was then maintained and was considered as that of the ultrasonic treatment.

To determine the killing effect of the ultrasonic treatment, aliquots of the spore suspensions were periodically removed and the number of survivors estimated by colony-counts on starch milk agar incubated at 55°C for two days.

Heat Resistance Determination

The technique used was the capillary tube method of Stern and Proctor as described by Stumbo (13). Heating was carried out in a thermostatically-controlled glycerol bath and the capillary tubes were washed after cooling by immersing them in sterile water.

Initial and survivor spores were estimated by plating, in duplicate, suitable dilutions of the heated suspensions on starch milk agar at 55°C.

For each temperature a survival curve was obtained by plotting the logarithm of the number of survivors at each time of sampling. The D-value (decimal reduction time) was calculated as the time required for the survival curve to traverse one logarithmic cycle. The z-value (temperature necessary to bring about a 10-fold change in the D-value) was calculated from linear regression equations of the logarithm of D obtained at the corresponding temperatures.

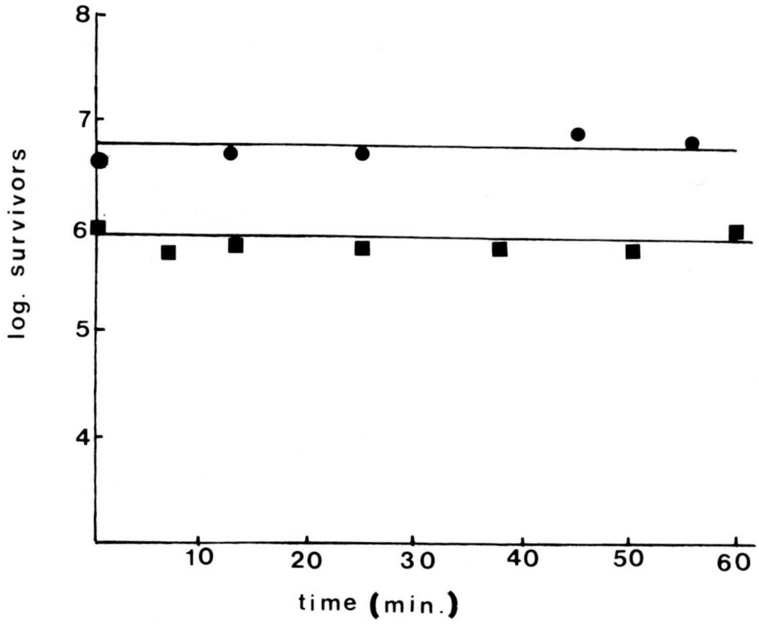

FIG. 1 *Effect of ultrasonic treatment (20 KHz, 120 W, 12°C) on* B. stearothermophilus *NCA-1518 (■) and* B. stearothermophilus-*L (●) spores.*

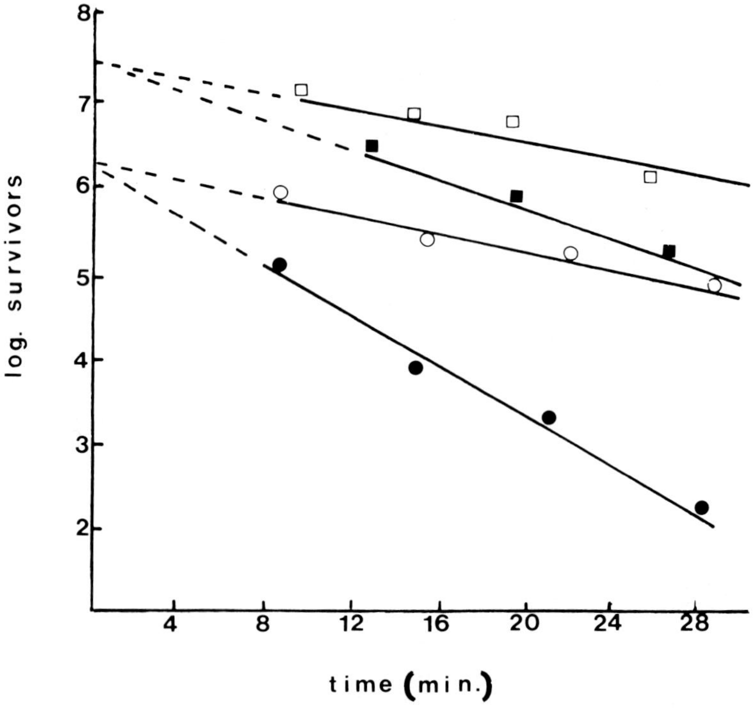

FIG. 2 *Heat survival curves of* B. stearothermophilus *NCA-1518 (circles) and* B. stearothermophilus-*L (squares) at 115°C without (empty symbols) and after (full symbols) ultrasonic treatment (20 KHz, 120 W, 12°C, 60 min).*

RESULTS AND DISCUSSION

Effect of the Ultrasonic Treatment on th- Survival of *B. stearothermophilus* Spores

Aliquots of 40 ml of spore suspensions of *B. stearothermophilus* NCA-1518 and *B. stearothermophilus*-L were subjected to the above mentioned ultrasonic treatment for periods ranging from 5 to 60 min. Figure 1 shows the effect on spore viability. The number of survivors was not affected after 60 min of ultrasonic wave application. These results are in agreement with others (6,7) previously published showing the great resistance of bacterial spores against ultrasonic treatment. The *B. stearothermophilus* spores seem to be more resistant than spores of mesophilic species of the genus *Bacillus*, such as

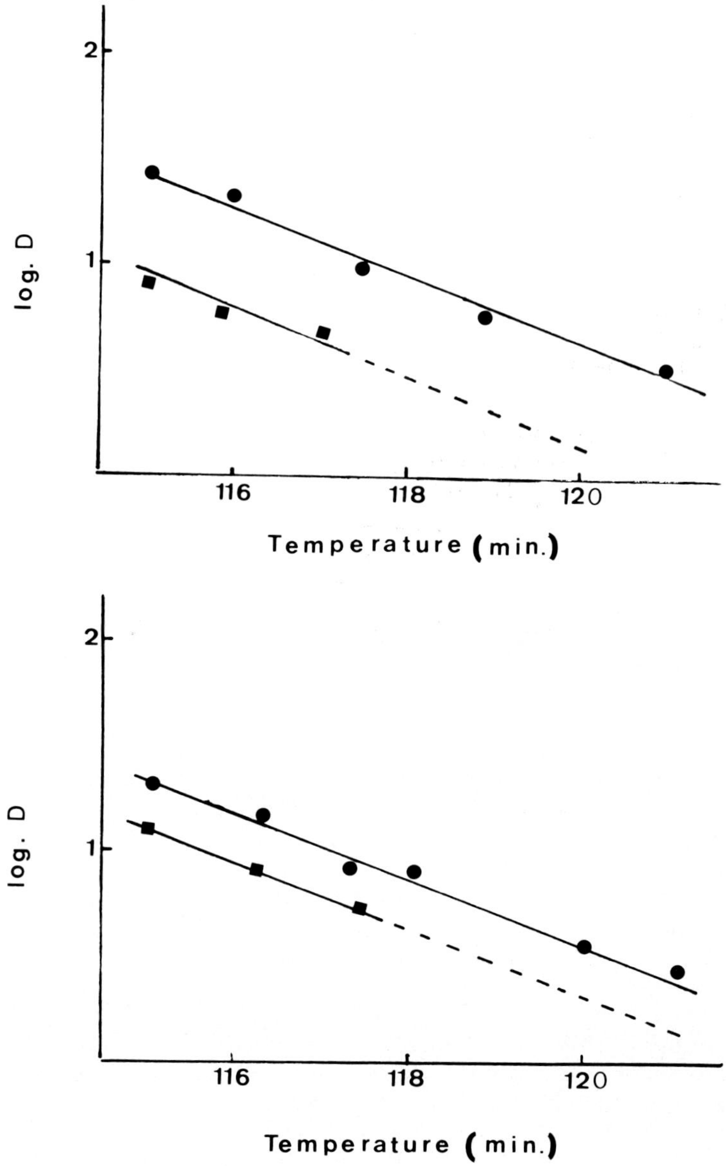

FIG. 3 *Thermal destruction curves of* B. stearothermophilus *NCA-1518 (upper) without (● Z = 6.49°C) and after (■ Z = 6.19°C) ultrasonic treatment (60 min) and* B. stearothermophilus-*L (lower) without (● Z = 6.34°C) and after (■ Z = 6.55°C) ultrasonic treatment (60 min).*

B. cereus. A disruption rate coefficient of 0.045 min^{-1} was calculated for *B. cereus* (5), whilst for *B. stearothermophilus* the coefficient is less than 2 x 10^{-5} min^{-1} as estimated from the results reported here.

Effect of the Ultrasonic Treatment on Heat Resistance

Aliquots of 40 ml of spore suspensions of *B. stearothermophilus* NCA-1518 and *B. stearothermophilus*-L were subjected to the above mentioned ultrasonic treatment for 60 min at 12°C. Immediately after this treatment heat resistance of the spores at different temperatures was determined. The heat resistance of spores without previous ultrasonic treatment was assessed as control. The results are shown in Table 1. Figure 2 shows

TABLE 1 *D-values (min) of* B. stearothermophilus *spores at several temperatures without and after ultrasonic treatment (20 KHz, 120W, 12°C, 60 min)*

	B. stearothermophilus NCA-1518			*B. stearothermophilus*-L		
Temp. (°C)	D	D_s	D/D_s	D	D_s	D/D_s
115.0	25.0	8.6	2.91	21.7	12.8	1.69
115.8	19.5(+)	6.7	2.91	-	-	-
116.2	-	-	-	14.0(+)	8.1	1.73
117.0	12.8(+)	4.3	2.98	-	-	-
117.3	-	-	-	9.4(+)	5.5	1.71

(+) Calculated from the corresponding linear regression equation of the thermal destruction curve (see Fig. 3).
D Decimal reduction time of heat treatment alone.
D_s Decimal reduction time of heat treatment after ultrasonic treatment.

the effect of ultrasonic waves on the heat resistance of *B. stearothermophilus* species at 115°C. It may be seen that the slope of the survival curves increases when the spores were ultrasonically treated before heating. The D-value at 115°C showed a decrease of 2.91-fold for *B. stearothermophilus* NCA-1518 and of 1.69 for *B. stearothemophilus*-L (Table 1). The D-values were reduced and similarly affected in both strains at all temperatures tested as may be observed in the table. The effect of the ultrasonic treatment on the z-value was also studied. In both strains, this parameter was constant irrespective of the treatments at which the spores were submitted (Fig. 3).

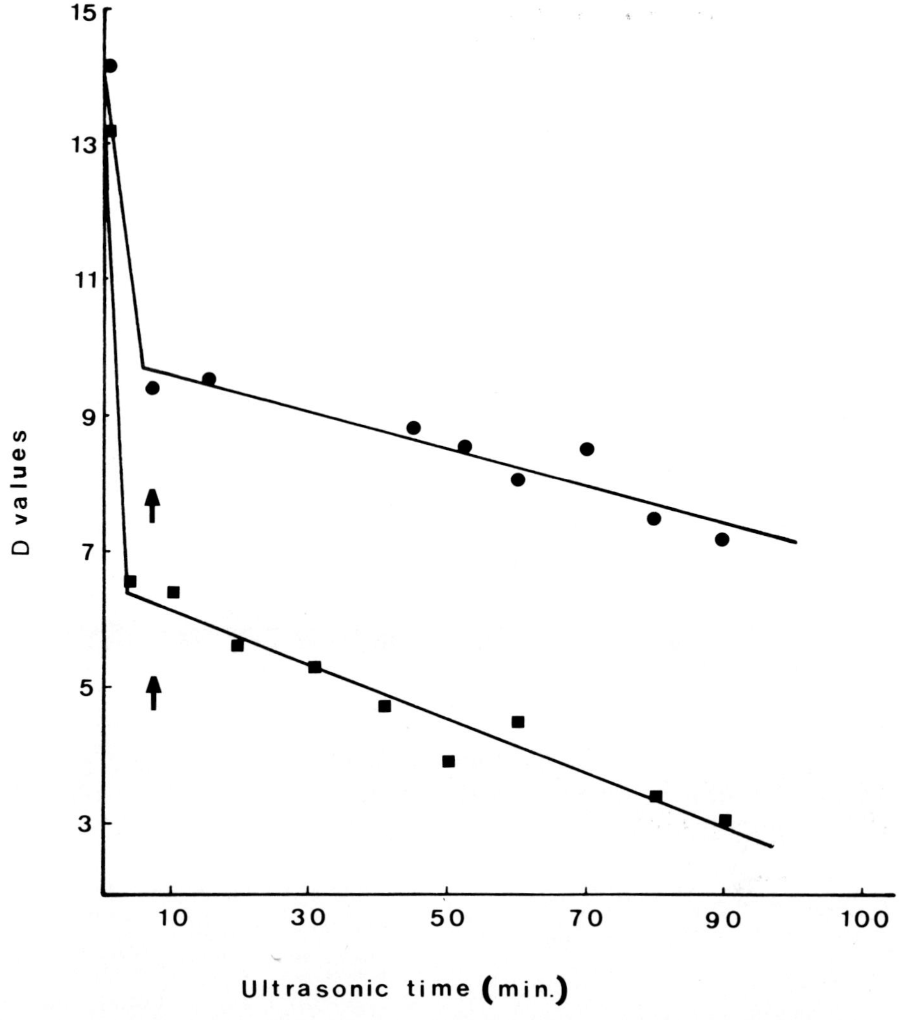

FIG. 4 *Effect of ultrasonication time on heat resistance of* B. stearothermophilus *NCA-1518 (■) at 116.9°C and* B. stearothermophilus-*L (•) at 116.3. The arrows indicate the beginning of the ultrasonication prior to heating.*

Further experiments were carried out over a wide range of ultrasonication times (from 3-90 min). The results shown in Fig. 4 demonstrate that the heat sensitizing effect of ultrasonic waves was greater as the ultrasonication times increased. The $D_{116.9°C}$ decreased from about 2- to 4-fold for *B. stearothermophilus* NCA-1518 and the $D_{116.3°C}$ from about 1.5- to

2-fold for *B. stearothermophilus*-L.

Finally, as shown in Fig. 5, the original heat resistance of ultrasonically-treated spores of *B. stearothermophilus*-L was not restored after prolonged storage times at 1 C.

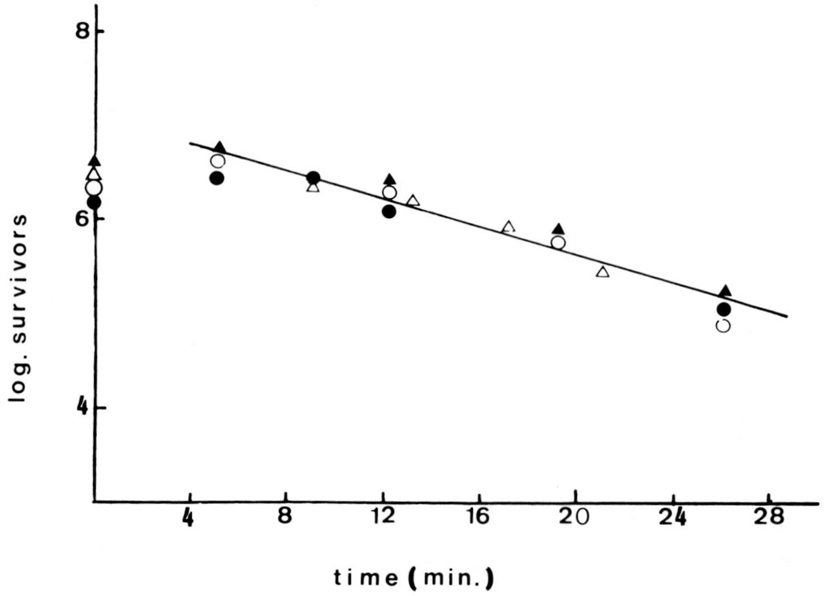

FIG. 5 *Effect of storage time (● = 2 h; o = 7 h; ▲ = 20 h and Δ = 72 h) at 1°C of ultrasonically treated (60 min)* B. stearothermophilus-*L spores on their heat resistance at 115°C.*

The mechanisms underlying the high resistance of the spores to heat remains unsolved but it seems that various factors contribute to this phenomenon, the dehydration state of the spore core being a predominating factor. The cortex plays an important role in the core-dehydration process and several theories have been proposed to explain it (14). Dipicolinic acid (DPA) and Ca^{++} have also been related (14,15) with heat resistance and, in fact, it has been demonstrated that *Bacillus* spore mutants unable to synthesize DPA (16) produce heat sensitive spores. However, this picture is obscured by the fact of the isolation of DPA-less heat resistant mutants (17) and by studies involving the effect of osmotic stabilizers on the heat resistance of heat-sensitive DPA-less spores (18). In this context, it is difficult to explain the origin of the heat sensitizing effect of ultrasonic waves. It could be possible that ultrasonic treatments induce a permeability change in spore coats or a disturbance in the dehydration mechanism so that the core undergoes partial rehydration. On the other hand, it has been reported that ultrasonication of *B. cereus* spores produces an exponential release of DPA depending on the ultrasonication time (5). This could also be related to the heat-sensitizing effect of ultrasonic waves, and could explain the phenomenon illustrated in Fig. 4. Further studies on the sensitization process are in progress.

REFERENCES

1. Russell, A.D., Jenkins, J. and Harrison, I.H. (1967) *Adv. Appl. Microbiol.* **9**, 1-38.
2. Xezones, H. and Hutchings, I.J. (1965) *Food Technol.* **19**, 113-115.
3. Fisher, D.A. and Pflug, I.J. (1977) *Appl. Environ. Microbiol.* **33**, 1170-1176.
4. Davies, R. (1959) *Biochim. Biophys. Acta* **33**, 481-493.
5. Berger, J.A. and Marr, A.G. (1960) *J. Gen. Microbiol.* **22**, 147-157.
6. Ahmed, F.I.K. and Russell, C. (1975) *J. Appl. Bacteriol.* **39**, 31-40.
7. Boucher, R.M.G. (1979) *Can. J. Pharmac. Sci.* **14**, 1-12.
8. Heiligman, F., Desrosier, N.W. and Broumand, H. (1956) *Food Res.* **21**, 63-69.
9. Ordoñez, J.A. and Burgos, J. (1976) *Appl. Environ. Microbiol.* **32**, 183-184.
10. Burgos, J., Ordoñez, J.A. and Sala, F. (1972) *Appl. Microbiol.* **24**, 497-498.
11. Garcia, M.L., Ordonez, J.A., Dela Hoz, L. and Lopez, P. (1981) *VIII Cong. Nac. Microbiol.* p.324, Madrid, Spain.
12. Murrell, W.G. and Warth, A.D. (1965) in *Spores III* (Campbell, L.L. and Halvorson, H.O., eds.), pp.1-24, American Society for Microbiology, Michigan.
13. Stumbo, C.R. (1973) *Thermobacteriology in Food Processing* p.105, Academic Press, New York.
14. Gerhardt, P. and Murrell, W.G. (1977) in *Spores VII* (Chambliss, G. and Vary, J.C., eds.), pp.18-20, American Society for Microbiology, Washington D.C.
15. Keynan, A. (1977) in *Spores VII* (Chambliss, G. and Vary, J.C., eds.), pp.43-53, American Society for Microbiology, Washington D.C.
16. Fukuda, A. and Gilvarg, C. (1968) *J. Biol. Chem.* **243**, 3871-3876.
17. Hanson, R.S., Curry, M.V., Garner, J.V. and Halvorson, H.O. (1972) *Can. J. Microbiol.* **18**, 1139-1143.
18. Bhothiapaksa, K. and Busta, F.F. (1978) *Appl. Environ. Microbiol.* **35**, 800-808.

HEAT RESISTANCE OF PA 3679 (NCIB 8053) AND OTHER ISOLATES OF *CLOSTRIDIUM SPOROGENES*

S.J. ALCOCK and K.L. BROWN

The Campden Food Preservation Research Association
Chipping Campden, Gloucestershire GL55 6LD, England

SUMMARY

The heat resistance of over 200 spore crops of *Clostridium sporogenes* PA 3679 (NCIB 8053) grown in trypticase broths ranged from $D_{121°C}$ values of 0.1-1.5 min, with a median value of 0.3 min. Modifications to the sporulation medium rarely affected spore heat resistance which was similar for spores produced in trypticase and in beef heart casein broths. $D_{121°C}$ values for 33 other isolates of *Cl. sporogenes* grown in trypticase broth ranged from ca 0.1 to 1.2 min, with a median value of 0.1 min.

INTRODUCTION

Spore suspensions with specific heat resistance characteristics are often required for heat resistance experiments and inoculated pack trials. Thermal resistance characteristics may vary between spore crops and certain factors are known to influence sporulation and heat resistance. For example, calcium is concerned in spore heat resistance (1) and the presence of manganese ions in the sporulation medium improves the heat resistance of *Bacillus megaterium* (2). Cadmium has subtle effects on bacterial physiology e.g. inhibition of transformation by DNA and of synthesis of RNA and protein (2) and may interfere with spore development.

Other substances which inhibit sporulation may be counteracted by the addition of adsorbents such as starch or activated carbon (4). Stirring is reported to increase spore yield of *Cl. parabotulinum* in 5% polypeptone (5), while sporulation of

Cl. botulinum is influenced by an agar phase (6).

However, it is difficult to compare data obtained in different laboratoties since strains of each bacterial species and growth and recovery conditions are not constant. Thus a systematic study of factors affecting sporulation and heat resistance of *Cl. sporogenes* was carried out in order to establish the range of variability within the species. The history of isolates was ascertained wherever possible since some "strains" are in fact subcultures from original collections. Since highly heat resistant spores of PA 3679 have been reported to be produced in beef heart casein broth at the National Food Processors Association (NFPA; K. Ito, personal communication) this was included in the experiments.

METHODS

Isolates. The origin of all of the isolates used is given in Table 3. All isolates were subcultured into liver broth (Oxoid CM77), incubated at 37° until gas production occurred, and stored at 4°C until required.

Sporulation. Spore crops were generally produced in trypticase broth (g/1: Trypticase peptone (BBL) 50.00, Mn $SO_4.4H_2O$ (Analar) 0.082; Bacteriological peptone (Oxoid L37) 5.00, K_2HPO_4 (GPR) 1.250, $(NH_4)_2SO_4$ (Analar) 20.00); pH adjusted to 7.5 with NaOH, autoclaved at 115°C for 30 min. Beef heart casein (BHC) broth was prepared according to the NFPA method (7.8). The spore crops were produced either in Buchner flasks (usually 0.51) equipped with an air filter, or modular fermenters (LH Engineering series III).

The pH of the sporulation medium was controlled in appropriate experiments by automated addition of sterile HCl or NaOH. Trypticase broth was variously supplemented before inoculation with $Cd(NO_3)_2$, $MnSO_4$ or soluble starch (Table 1). Culture fluid (20 or 50% by volume) from earlier crops D_{121} = 0.2 min) was added to certain cultures 48 h after inoculation. The agar phase in other flasks was formulated as trypticase broth with (g/l) yeast extract (Oxoid L21) 1.0 and agar (Difco) 20.0, and used at a ratio of ca 9 cm^2 agar surface to 100 ml broth.

An aliquot (1 ml) of a subculture in 10 ml sterile distilled water was pasteurised for 10 min at 80°C, 5 ml was then transferred into 20 ml liver broth and after incubating for 1 day at 37°C an aliquot comprising 5% by volume (of the next culture) was transferred into trypticase broth and incubated for 6 h. A single bottle of inoculum was used for all appropriate flasks within a batch: 2% inoculum by volume was added to sporulation broth equilibrated at 37°C and the head space was flushed with nitrogen, followed by incubation at 37°C.

usually for 7 days but a few cultures were incubated for up to 21 days. The NFPA method (8) was followed for the evaluation of BHC broth: four daily transfers in BHC or trypticase broth resulted in a final inoculum of 10% and incubation was at 25°C for 17 days.

Heat resistance determination. Spores were harvested from the growth medium by centrifuging at 8000 rpm for 10 min at 7°C, washed, resuspended in sterile distilled water and stored at 4°C until required. D-values were estimated using 0.1 ml

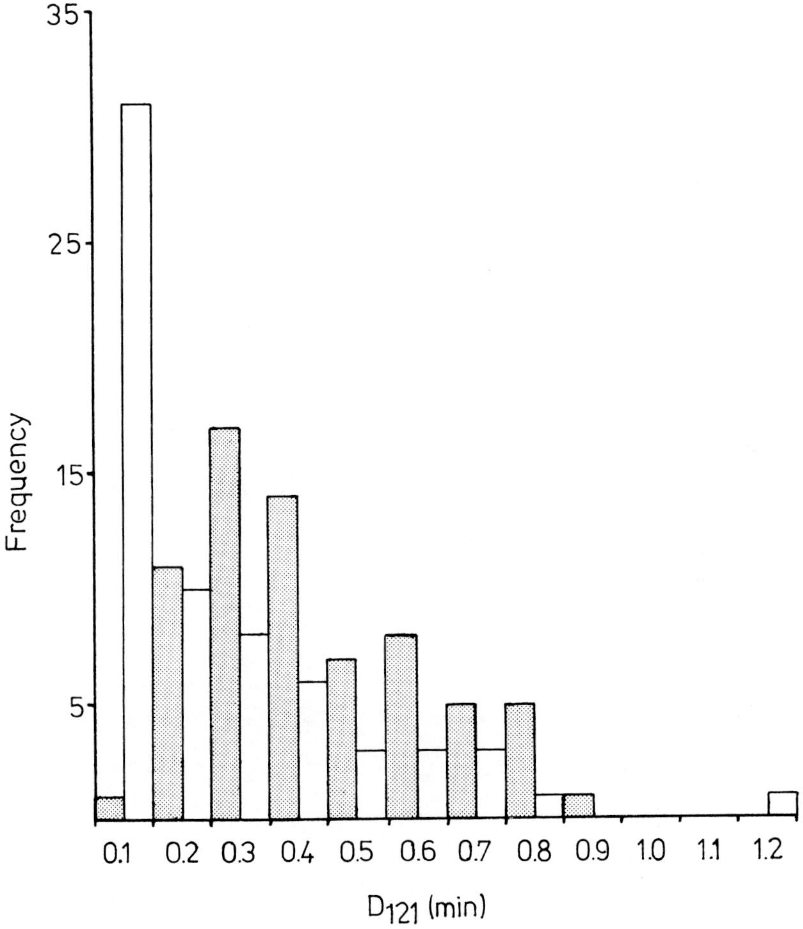

FIG. 1 *Heat resistance of* Cl. sporogenes. *Shaded bars represent spore crops of NCIB 8053, unshaded bars are duplicate crops of 33 other isolates.*

aliquots sealed into glass spheres blown from melting point tubes. The spheres were heated in glycerol at 121.1°C (± 0.1°C) and cooled in water. Each sphere was crushed into sterile distilled water and survivors were recovered on Eugonagar (BBL) + 0.01% soluble starch pH 7.4 (ES) with 7 days incubation at 30°C.

RESULTS

Variability in heat resistance of replicate spore crops. Spore crops produced under similar conditions in trypticase broth were highly variable in their heat resistance characteristics (see Fig. 1). For example, in one batch of eight flasks, D-values ranged from 0.3 to 0.8 min. The distribution of D-values obtained from many crops grown under standard conditions ranged from 0.2 to 0.9 min, and spore crops with a D of 0.3 min occurred most commonly.

Effect of incubation period on heat resistance. Spores grown in trypticase broth were most resistant after ca 7 days incubation at 37°C (Fig. 2), and prolonged incubation resulted in fewer, less heat resistant spores.

Effect of sporulation medium and cultural conditions on heat resistance. Addition of cadmium, starch or high levels of

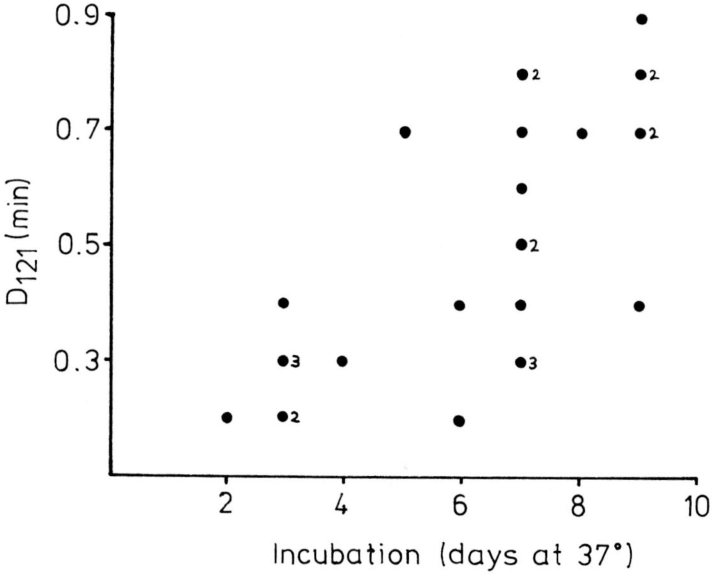

FIG. 2 *Effect of incubation time on heat resistance of NCIB 8053.*

TABLE 1 *Effect of addition of Mn^{2+}, Cd^{2+} and starch to trypticase sporulation broth on heat resistance of PA 3679*

Component	Concentration	D_{121}	(min)
Mn^{+}	0	0.2,	0.5
	3.7×10^{-4}M	0.2,	0.6
	7.4×10^{-4}M	0.1,	0.5
	3.7×10^{-3}M	0.1,	0.2
Cd^{+}	0	0.8	
	4.6×10^{-6}M	0.6	
	4.6×10^{-5}M	0.3	
	4.6×10^{-4}M	0.5	
	4.6×10^{-3}M	0.1	
	4.6×10^{-2}M	0.1	
Starch	0	0.8,	0.5
	0.1%	0.3,	0.2

TABLE 2 *Effect of certain constituents of trypticase broth on heat resistance of PA 3679 spore crops*

Trypticase	Bacteriological peptone	$(NH_4)_2SO_4$	K_2HPO_4	D_{121} (min)
+	+	+	+	0.3,0.4,0.4,0.2,0.6
−	+	+	+	<0.1
−	+*	+	+	0.2
+	−	+	+	0.3
+	+	−	+	0.3
+	−	−	+	0.2
−	+	−	+	<0.1
+	+	+	−	0.7,0.4

*Trypticase replaced by equal weight of bacteriological peptone.

TABLE 3 (cont'd)

Isolate	Source	Comments	D_{121} (min)	
Other isolates:				
FRA 1089*	BFMIRA	Labelled CN 644	0.1,	0.1
NCTC 276	NCTC	R60 X V steam and formaldehyde sterilizer test strain	0.1,	0.1
PA 174*	MRI	From McClung who obtained it from NFPA San Francicso	0.7,	0.8
NCIB 532	NCIB		0.1,	0.1
MRI 12 PA-R	MRI	From DMRI	0.1,	0.1
MRI 13 PA-93R	MRI	From DMRI	0.4,	0.3
MRI 14 PA-CP	MRI	From CP	0.6,	0.2
MRI 15 PA-Unox	MRI	From CP	0.1,	0.1
PA $1077\text{-}S^{+}$	MRI	From McClung who obtained it from JM. Sporulation variant	0.2, 0.3	0.3
MRI 61	MRI ?		0.3,	0.4
CG5	MRI ?		0.1,	0.1
NCIB 8243	DSRTC		0.2,	0.5

*Strains which are thought to derive from Cameron's PA.
+Sub from single spore.
NCTC = National Collection of Type Cultures, UK: FRI = Food Research Institute, UK; BFMIRA = FRA = British Food Manufacturing Industries Research Association; NCIB = National Collection of Industrial Bacteria, UK; DSRTC = Dalgety Spillers Research and Technical Centre, UK; NFPA = National Food Processors Association, USA; MB = Metal Box, UK; MRI = Meat Research Institute, UK; LARS = Long Ashton Research Station, UK; DMRI = Danish Meat Research Institute; CP = Canada Packers; JM = J. Morrell and Co., UK.
++ Mean of 4 crops.

manganese to the trypticase medium appeared to reduce spore heat resistance (Table 1). Other factors had no obvious effect on resistance; these included pH maintained at 6 to 7.5, incubation at 27, 30, 33, 35 or 37°C, addition of culture fluid from an earlier crop, stirring (ca 300 rpm), and inclusion of an agar phase.

The effect of individual constituents of trypticase broth on growth and heat resistance is shown in Table 2. Elimination of trypticase reduced growth and heat resistance, and replacement of trypticase by an equal weight of bacteriological peptone favoured vegetative growth but sporulation and heat resistance were poor. In the absence of $(NH_4)_2SO_4$ sporulation was reduced, but thermal resistance did not appear to be affected. Omission of K_2HPO_4 had no obvious effect on growth or heat resistance, and elimination of $MnSO_4$ seemed to reduce sporulation but heat resistance was not altered.

Cell numbers in BHC broth were higher than in comparable trypticase cultures, but sporulation was reduced. A high proportion (ca 90%) of spores were mature in each crop and this seems to be a result of prolonged incubation at 25°C. However, there was no evidence for any beneficial effect of BHC or the sporulation technique on thermal resistance. D-values for NCIB 8053 grown in BHC broth were 0.2, 0.3 and 0.2 min and corresponding values for spores produced in trypticase broth were 0.2, 0.2 and 0.1 min. Spores of the NFPA isolate of PA 3679 (D = 0.1, 0.2 min) and the MRI isolates of strain 1075 (D = 0.2 min) produced in BHC broth did not show exceptional heat resistance.

Variation in heat resistance between strains. A comparison was made of the growth and heat resistance characteristics of 33 isolates of *Cl. sporogenes*. The isolates were grown in batches which included a control (NCIB 8053).

Isolates or strains did not appear to differ in growth characteristics except that McClung strain no. 1075 yielded fewer cells than other strains. Colony morphology in ES differed between isolates but failed to correlate with individual strains. Results of heat resistance determinations (Table 3) tended to be similar for duplicate crops of individual isolates, and the range of D-values (ca 0.1 - 1.2 min) was similar to that of NCIB 8053 (Fig. 1). However, crops of low resistance (ca 0.1 min) were produced more frequently from other isolates than from NCIB 8053.

DISCUSSION

Trypticase, a peptone derived from casein by pancreatic digestion, provides various nutrients including potassium, sulphur and phosphorus which may be required for sporogenesis, but supplementation with additional quantities improved growth of PA 3679. The addition of manganese or ammonium sulphates favoured sporulation, but trypticase peptone was the only constituent of the sporulation broth showing an obviously beneficial effect on spore heat resistance. The variability of heat resistance characteristics of spore crops grown under similar conditions may be attributable to heterogeneity of inoculum and alteration of the medium during growth. D-values for crops derived from single spore isolates were similar for duplicates, however three crops of isolate PA 174 grown subsequently were less resistant (D = 0.2 min).

The apparently reduced resistance of spore crops grown in a medium incorporating starch may result from adsorption of a substance(s) important for development of heat resistance since starch is known to remove the effect of sporulation inhibitors (4). Many proteins require metal co-factors for their activity but high concentrations are often inhibitory (9), and it may be that cadmium or high levels of manganese inhibit enzymes involved in development of heat resistance.

Modifications to trypticase broth and variation in cultural conditions often had little or no effect on heat resistance; D-values for 200 such crops ranged from 0.1 to 1.5 min with a median of 0.3 min. The range is wider than for control crops, probably due to the greater number of observations. There was no apparent correlation between heat resistance and pH, E_h, cell concentration or sporulation.

Spore crops are often harvested at maximum sporulation, but spore crops produced in trypticase were most resistant after incubation for ca 7 days at 37°C. Prolonged incubation for several weeks is sometimes recommended but yield and heat resistance of spores in the present experiments decreased with increasing incubation beyond ca 7 days, probably due to germination of spores and less favourable growth conditions.

The NPFA technique is reported to produce heat resistant spores, but using the same or different strains of PA 3679 we obtained spores with resistance similar to crops produced in trypticase medium. The difference in results may be due to the heart tissue used to prepare the medium, and/or to insufficient replication since some NFPA crops produced by this method are of low heat resistance (K. Ito, personal communication). However, a more likely explanation appears to be differences in the method of recovery of the heated spores. The pork-pea infusion agar used by the NFPA for recovery would give better resuscitation of heat-damaged spores than ES agar (10), which was selected here because of its more reproducible nature. The choice of recovery medium, together with the use of the most probable number technique (with its inherently large error) would be expected to give higher estimates of numbers of survivors, particularly after longer heating times. Thus the higher heat resistance reported by the NFPA is more likely to be a function of better recovery rather than specifically associated with the production of spores in BHC broth.

The D-values for spore crops of various isolates of *Cl. sporogenes* ranged from ca 0.1 to 1.2 min, which is similar to replicate crops of NCIB 8053. These spore crops were generally less resistant (D = 0.1 min) than NCIB 8053 which most frequently has a D-value of 0.3 min ($t = 4.682$, $p < 0.001$). Further analysis showed that heat resistance characteristics of isolates of PA 3679 differed from NCIB 8053 ($t = 3.164$, $p < 0.01$) whereas the San Francisco sub-strain did not differ significantly from NCIB 8053. Strain NCIB 10696 differed from NCIB 8053 ($t = 14.241$, $p < 0.001$), and also from the San Francisco sub-strain and from McClung strain 1075 at the 5% level. The results also support the contention that the San Francisco strain (which has similar heat resistance to NCIB 8053) is more heat resistant than other strains studied although individual spore crops may have relatively low resistance. However, further replication would be required to assign individual isolates to particular strains on the basis of heat resistance.

ACKNOWLEDGEMENTS

This work was supported by the Ministry of Agriculture, Fisheries and Foods. We thank Mrs P. Banks and Miss L. Sellors for technical assistance and Dr T. Roberts, MRI, for initial discussions.

REFERENCES

1. Rajan, K.S. and Grecz, N. (1977) in *Spore Research 1976* (Barker, A.N., Wolf, J., Ellar, D.J., Dring, G.J. and Gould, G.W. eds.), Vol. 2, pp.527-543, Academic Press, London, U.K.
2. Aoki, H. and Slepecky, R. (1974) in *Spore Research 1973* (Barker, A.N., Gould, G.W. and Wolf, J., eds.), pp.93-102, Academic Press, London, U.K.
3. Babich, H. and Stotzky, G. (1978) *Adv. Appl. Microbiol.* **23**, 55-110.
4. Murrell, W.G. (1961) *Symp. Soc. Gen. Microbiol.* **11**, 100-150.
5. Perkins, W.E. (1965) *J. Appl. Bact.* **28**, 1-16.
6. Bruch, M.K., Bohrer, C.W. and Denny, C.B. (1968) *J. Fd Sci.* **33**, 108-109.
7. Anon. (1968) *National Canners Association Laboratory Manual*

for Food Canners and Processors, Vol. 1, p.11, AVI Publishing Company, Inc., Westport, Conn.
8. Anon. (1966) *Preparation of Spore Suspensions (Putrefactive Anaerobes)*. National Food Processors Association, Washington, D.C.
9. Williamson, F.B. (1977) in *Spore Research 1976* (Barker, A.N., Wolf, J., Ellar, D.J., Dring, G.J. and Gould, G.W., eds.), Vol. 1, pp.55-67, Academic Press, London, U.K.
10. Wheaton, E., Pratt, G.B. and Jackson, J.M. (1958) *Fd. Res.* **24**, 134-145.

HEAT RESISTANT THERMOPHILIC ANAEROBE ISOLATED FROM COMPOSTED FOREST BARK

K.L. BROWN

The Campden Food Preservation Research Association, Chipping Campden, Gloucestershire, GL55 6LD, UK

SUMMARY

An extremely heat resistant strain of *Clostridium thermosaccharolyticum* was isolated during investigation of a spoilage outbreak in canned mushrooms. Spores of this organism withstood processes over F_0 30, which are above those normally employed in the canning industry. Resistant spores were found in composted forest bark used as a component of mushroom compost. Samples of composted forest bark were heated for different time intervals at temperatures between 115 and 126°C and surviving spores enumerated using a Most Probable Number (MPN) technique. The D-value at 121.1°C was 68 min and the z-value was 11°C.

INTRODUCTION

Spoilage of canned foods by thermophilic anaerobes can occur when heat resistant spores of strains of *Clostridium thermosaccharolyticum* (1) survive the heat process and the cans are either inadequately cooled so that the pallet loads remain at an elevated temperature (44-55°C) for some time or samples are incubated. The organism grows rapidly producing large volumes of gas, mainly carbon dioxide and hydrogen, which swell and eventually burst the can. The product has a lowered pH and frequently a strong butyric or "cheesy" odour (2,3).

In a recent survey (4), of the types of spoilage investigated by the Continental Can Co., 14% involved thermophilic spoilage. This included, as well as thermophilic anaerobic spoilage, flat sour spoilage caused by growth of *Bacillus stearothermophilus* and sulphur stinker spoilage caused by *Desulfotomaculum nigrificans* (formerly *Cl. nigrificans*).

Thermophilic anaerobe and thermophilic flat sour spoilage account for most of the thermophilic spoilage today and sulphur stinker spoilage is now relatively uncommon. Segner (4) reports one outbreak in 18 years, while Campden FPRA has investigated only one in the last 10 years. However, in the 1920s an entire season's production of sweet corn could be lost (5) due to sulphur stinkage spoilage.

A routine investigation by the author in 1978 of blown and burst cans of mushrooms showed spoilage was due to a thermophilic anaerobe. The high heat resistance of the spores was of particular interest. In experimental trials, spores were found to survive processes of over F_0 30 (a process equivalent to 30 min at 121°C). The spores appeared to be carried forward on small particles of compost adhering to the mushrooms even though they had been washed several times. Similar findings have been reported elsewhere (6).

Mushroom compost usually contains high numbers of thermophiles (7,8) and some very resistant spores of a thermophilic anaerobe were found in the composted forest bark used in the mushroom beds as a casing material. This is a layer, usually composed of soil or peat and lime, which is spread over the compost in which the mushroom mycelium has grown to induce fruiting body formation. Recently composted forest bark has been advocated as an alternative casing material (9,10).

Fresh, pulverised conifer bark contains a number of volatile substances which apparently inhibit the growth of both plants and mushrooms (11). By composting the bark for several weeks the volatile oil content can be reduced. During this time the temperature of the bark rises to 50°C or more and the pH rises from 4.5 to 5.5 and it is likely that the heat resistant strain of *Cl. thermosaccharolyticum* studied here grows and sporulates during this composting procedure.

Spores of a thermophilic anaerobe showing similar heat resistance to those isolated from the composted forest bark used for mushroom casing have also been isolated from a bag of composted forest bark obtained from a garden centre.

In view of the obvious commercial significance of such heat resistant spores, as well as their academic interest a series of heat resistance experiments were carried out on the naturally occurring spore population in composted forest bark.

METHODS

Heat Resistance

Samples of composted forest bark obtained from the mushroom farm were weighed into 3 sets of 5 x 150 ml round bottles in quantities of 10, 1 and 0.1 g per bottle for each set respectively. Distilled water (25 ml) was added to each bottle. The pH of the bark suspension ranged from 6.7 to 6.9. Each batch of 15 bottles was autoclaved for different time intervals at temperatures of 115.6, 118.3, 121.1, 123.9 and 126.7°C in either a Fraser pilot scale retort or a small Baird and Tatlock laboratory autoclave. Both heating vessels were direct steam heated and could be brought to temperature within 5 min. No attempt was made to correct for come-up time of the bottles in view of the long heating times involved.

After the heat treatment, approximately 50 ml of molten Eugonagar (BBL) + 0.1% soluble starch (BDH Analar) pH 7.4 cooled to 50°C was added to each bottle. When this had solidified, an overlayer of double strength agar (Difco) was poured to a depth of approximately 1 cm.

Bottles were incubated at 55°C until no new positives were observed (35-39 days). Gas production and/or appearance of irregular-shaped colonies or general growth in the medium was recorded as a positive result. Gas production was variable with a tendency for the cultures heated for the longer times to lose the ability to produce gas.

Growth Characteristics

Cultures of both the bark isolate and the type strain of *Cl. thermosaccharolyticum* (NCIB 9385) were set up in 1 oz (29.35 g) bottles containing freshly steamed liver broth (Oxoid CM77) which was then overlayered with double strength agar (Difco). The cultures were incubated at 25, 30, 37, 44, 55, 60 and 70°C for up to 2 weeks.

In addition, a limited range of carbohydrate fermentation tests were carried out in trypticase agar base (BBL) to which was added the appropriate filter sterilized carbohydrate to a final concentration of 10 g/l.

RESULTS

Heat Resistance

The MPN of survivors and 95% confidence limits were calculated as described by de Man (12). The decimal reduction (D) value at each temperature was calculated by plotting log MPN heating time and fitting a line by linear regression to the straight line portions of the survivor curves. At 121.1°C the D-value was 68 min as shown in Fig. 1. At 115.6°C the D-value was calculated to be 204 min after experiments with heating times up to 5 h.

A z-value of 11°C was obtained on plotting log D-value against temperature (Fig. 2).

Growth Characteristics

Both the bark isolate and NCIB 9385 grew and produced gas in liver broth between 30 and 60°C, although growth was rather slow and poor at 30 and 37°C. No growth of either strain was recorded at 25°C. In addition the bark isolate was able to grow at 70°C. These results are more in accord with the data in the manual of Prevot and Fredette (13) than that presented in Bergey's Manual (14).

Both strains were able to ferment arabinose, lactose, maltose and mannose.

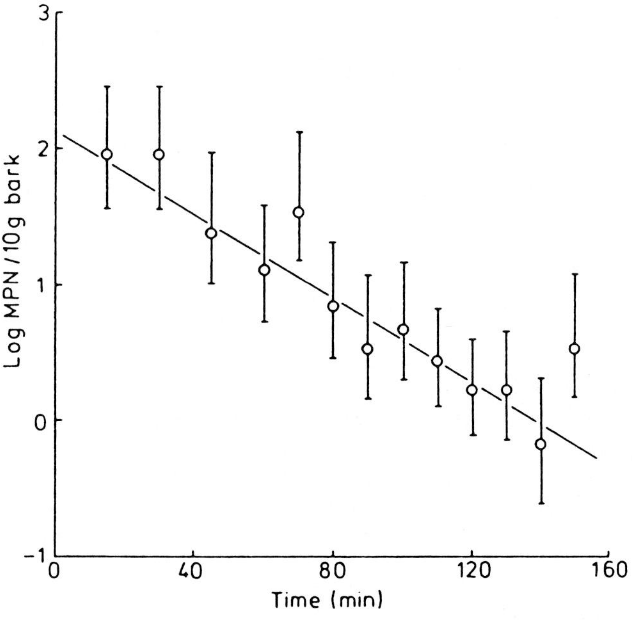

FIG. 1 *Heat resistance of* Cl. thermosaccharolyticum *spores from forest bark heated in bark/distilled water at 121.1°C. (Vertical lines indicate 95% confidence limits).*

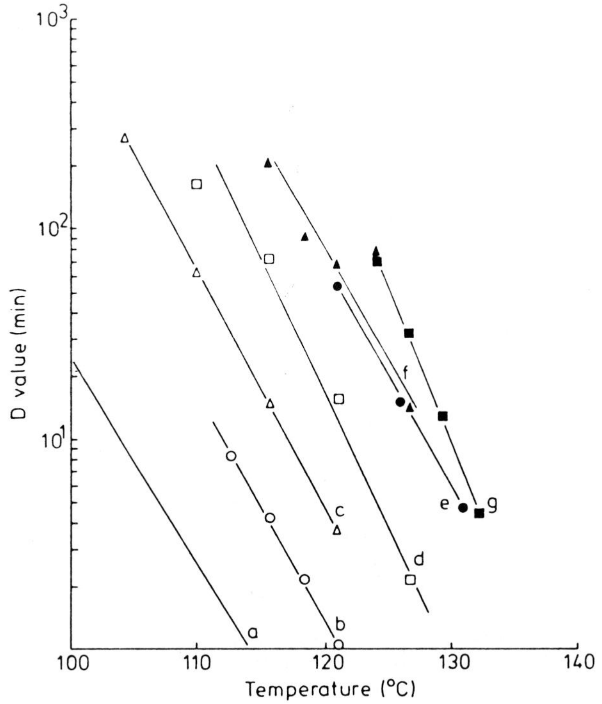

FIG. 2 *Heat resistance of* Cl. thermosaccharolyticum *spores from forest bark compared with the resistance of other resistant sporeformers. a, classical* Cl. botulinum *resistance (18); b,* Cl. sporogenes *(24, o); c,* Cl. thermosaccharolyticum *(19, Δ); d,* B. stearothermophilus *NCIB 8919 (22, □); e,* D. nigrificans *(8, ●); f, bark isolate (▲); g,* Cl. thermosaccharolyticum *(17, ■).*

DISCUSSION

The spores of the thermophilic sporeformers are usually much more resistant to wet heat than those of the mesophilic sporeformers (15,16). Probably the highest wet heat resistance reported is that for spores of a strain of *Cl. thermosaccharolyticum* isolated from molasses by Xezones, Segmiller and Hutchings (17) which had a D-value of 72.5 min when heated in distilled water at 123.9°C. With a z-value of 6.9°C this equates to a D-value of 195 min at 121.1°C (Fig. 2). The spores of the strain of *Cl. thermosaccharolyticum* isolated from composted forest bark were less resistant with a D-value of 68 min at 121.1°C and a z-value of 11°C (Fig. 2).

Resistances of this order are considered exceptional for *Cl. thermosaccharolyticum* (3,18), a more usual order of resistance is that reported by Gillespy (19) with a D value of 3.7 min at 121° and a z value of 8.9° (Fig. 2). Addition of 2.5% soil to the sporulation medium has been found to increase the thermal resistance of *Cl. thermosaccharolyticum.*

Hardly any data are published on the resistance of *D. nigrificans*. Speck (20) reports a D-value at 121°C of 2 min and a z-value of 9°C while Brown and Thorpe (21) recorded a D-value of 2.6 min and a z-value of 6.8°C. However spores of this organism produced in a 40% infusion of spent mushroom compost have been reported (8) to have D-values as high as 54.4 min at 121°C and a z-value of 9.5°C (Fig. 2).

A noticeable feature of these spores of exceptionally high heat resistance was the length of the incubation time required for recovery of survivors of heat treatments, 2-3 weeks for *D. nigrificans* (8) and up to 6 weeks for *Cl. thermosaccharolyticum* (17).

Another resistant thermophilic sporeformer is *B. stearothermophilus*, the resistant strain NCIB 8919 having D-values at 121°C around 16 min (22) and a z-value of 7.7°C (Fig. 2) while strain TH24 has been reported to have a D-value of 121°C at 11.9 min and z-value of 7.3°C (23).

In comparison with the organisms mentioned above, the heat resistance of the spores of other canned food spoilage organisms are somewhat lower. Pflug and Esselen (24) reported a D121°C value for *Cl. sporogenes* of 1.06 and a z-value of 9.3°C (Fig. 2) while the generally accepted resistance for the most resistant strains of *Cl. botulimum* (18), is D121°C, 0.2 min and z-value 10°C (Fig. 2). Other sporeformers with resistances in this range include *B-coagulans* D121°C, 3 min (25), *B. subtilis* D121°C, 1.5 min (26), and *B. cereus* D121°C, 2.3 min (27).

Spores of the thermophilic anaerobes are clearly much more resistant than those of the mesophilic sporeformers and may cause commercial spoilage problems in temperate climates if inadequate cooling of the product occurs after processing. However, such heat resistant spores may also have a useful role if, for example, they could be used to estimate processes in the same way as spores of *B. stearothermophilus* have been used, (28,29) or alternatively, they may provide some useful information on the mechanisms of spore heat resistance.

ACKNOWLEDGEMENTS

J.R. Aaron, Forestry Commission, Alice Holt Lodge, Wrecclesham, Farnham, Surrey, U.K. for providing information on the composting of forest bark, and the Ministry of Agriculture, Fisheries and Food for their financial support.

REFERENCES

1. McClung, L.S. (1935) *J. Bacteriol.* **29**, 189-202.
2. Ashton, D.H. (1981) *J. Fd. Prot.* **44**, (2), 146-148.
3. Hersom, A.C. and Hulland, E.D. (1980) *Canned Foods, Thermal Processing and Microbiology*. 7th Edn. Churchill Livingstone, Edinburgh.
4. Segner, W.P. (1979) *Fd. Technol.* **33**, 55-59, 80.
5. Werkman, C.H. and Weaver, H.J. (1927) *Iowa State J. Science* **2**, 57-67.
6. Riviere, J., Denys, C. and Cassegrain, Y. (1969) *Ann. Technol. Agric.* **18** (2), 75-91.
7. Fordyce, C. (1970) *Appl. Microbiol.* **20** (2), 196-199.
8. Donnelly, L.S. and Busta, F.F. (1980) *Appl. Env. Microbiol.* **40** (4), 721-725.
9. Aaron, J.R. (1973) *J. Inst. Wood Sci.* No. 33, **6** (3), 22-27.
10. Stewart, N. (1979) *Horticulture Industry* Feb. 1979, 26, 30.
11. Aaron, J.R. (1976) *Conifer bark, its properties and uses.* Forestry Commission Record 110. HMSO.
12. de Man, J.C. (1975) *Eur. J. Appl. Microbiol.* **1**, 67-78.
13. Prevot, A. and Fredette, C. (1966) *Manual for the classification and determination of the anaerobic bacteria.* Lea and Febiger, Philadelphia.
14. *Bergey's Manual of Determinative Bacteriology.* 8th Edn. (1977) Williams and Wilkins, Baltimore.
15. Denny, C.B. (1981) *J. Fd. Prot.* **44** (2), 144-145.
16. Warth, A.D. (1978) *J. Bacteriol.* **134** (3), 699-705.
17. Xezones, H., Segmiller, J.L. and Hutchings, I.J. (1975) *Fd. Technol.* **19**, 1001-1003.
18. Stumbo, C.R. (1973) *Thermobacteriology in Food Processing.* 2nd Edn. Academic Press, London.
19. Gillespy, T.G. (1947) *Ann. Rept.* pp.40-54. Fruit and Veg. Presn. Resn. Stn., Chipping Campden.
20. Speck, R.V. (1981) *J. Fd. Prot.* **44** (2), 149-153.
21. Brown, K.L. and Thorpe, R.H. (1978) *Tech. Memo. 24*, Campden Fd. Pres. Res. Ass., Chipping Campden.
22. Brown, K.L. (1974) in *Aspetic Packaging of Vegetable Products, Progress Report for Agricultural Market Development Executive Committee*, Campden Fd. Presn. Res. Ass., Chipping Campden.
23. Davies, F.L., Underwood, H.M., Perkin, A.G. and Burton, H. (1977) *J. Fd. Technol.* **12**, 115-129.
24. Pflug, I.J. and Esselen, W.B. (1954) *Fd. Res.* **19** (1), 92-97.
25. Murrell, W.G. (1964) *Aust. J. Pharm.* S40-S46.
26. Put, H.M.C. and Aalbersberg, W.I.J. (1967) *J. Appl. Bact.* **30**, 411-419.

PROTEIN DEGRADATION DURING BACTERIAL SPORE GERMINATION

P. SETLOW

Department of Biochemistry,
University of Connecticut Health Center,
Farmington, Connecticut 06032, USA.

SUMMARY

Up to 20% of the protein of dormant spores of various species is degraded to amino acids in the first minutes of spore germination. These amino acids are essential for rapid protein synthesis early in germination and outgrowth. The proteins degraded are a group of low molecular weight (small) acid soluble species (SASP), which are found in the spore core associated with spore DNA and are synthesized midway through sporulation; synthesis of the SASP is under transcriptional control. The proteolysis during germination is initiated by a serine-type endoprotease which acts only on the SASP, since it cleaves only within a specific pentapeptide sequence found in the SASP. The activity of this protease is regulated in part by its synthesis during sporulation as a zymogen, which is only activated later in development.

INTRODUCTION

There are a number of unsolved problems concerning the system of sporulation and germination in *Bacillus* species. These problems include the identification and description of mechanisms for: the regulation of synthesis of spore and/or sporulation specific gene products; causing spore heat and/or radiation resistance; bringing about and maintaining spore dormancy; and triggering spore germination. One approach to these individual problems is not to attack each one separately, but to examine in detail one particular aspect of sporulation and germination to which all of the above problems may apply in the hope that the specifics of the one system will provide generalizations about sporulation and germination as a whole. The system we have chosen for study involves the various components which interact to give the rapid degradation of dormant spore protein during spore germination, and these studies have indeed provided insight into a number of the problems noted above. The majority of these studies have used *B. megaterium* as the experimental organism and, in the discussion below, data cited and conclusions drawn are from work with this species. However, the more limited studies with *B. cereus*, *B. sphaericus*, and *B. subtilis* as well as *Clostridium bifermentans* have indicated that the proteolysis of dormant spore protein during germination is relatively similar in these other organisms with one major exception as noted below.

RESULTS

Proteolysis During Germination

Kinetics, products and function. The first 20 to 30 minutes of spore germination are accompanied by the degradation of up to 20% of the dormant spore's protein, as well as rapid turnover of the proteins first synthesized during germination (1). While the latter process requires metabolic energy, degradation of dormant spore protein during germination is unaffected by: inhibition of macromolecular synthesis or energy metabolism; high levels of exogenous amino acids; and use of non-metabolizable germinants versus metabolites (1). These data indicate that not only the substrate(s) but also the enzyme(s) involved in the rapid proteolysis of dormant spore protein are present in the spore.

In *B. megaterium* the products of the degradation of dormant spore protein are free amino acids and at least 16 of the common free amino acids are generated by this process (2). Of the four not found two (tryptophan and cysteine/cystine) may be produced but moderate levels would not have been detected in the analytical procedure used; the other two, asparagine and glutamine, may be produced but then rapidly deamidated (2,3). While the degradation of dormant spore protein generates a large supply of amino acids for the germinating and outgrowing spore, the free amino acids are rapidly reutilized, about 50% directly for protein synthesis with the remainder either for other biosynthetic pathways (i.e. nucleotide metabolism) or catabolized (2). The latter process is particularly important for the young germinated spore which contains many enzymes of amino acid interconversion and catabolism (2). Thus the high energy compounds (i.e. NADH, acetyl CoA and ATP) formed from catabolism during germination of proteolytically derived amino acids greatly exceeds those derived from 3-phosphoglyceric acid, another major energy reserve stored in the dormant spore (4,5,6,7).

The reutilization of the proteolytically derived amino acids for protein synthesis is also very important, since this reutilization is essential for rapid protein synthesis during germination (1,2). Surprisingly, the dormant and young germinated bacterial spore is a functional multiple amino acid auxotroph (even if it is genetically prototrophic), since it lacks enzymes for synthesis of at least seven amino acids (2). These enzymes are synthesized at defined times during germination and outgrowth using amino acids generated by proteolysis, and only then does *de novo* amino acid biosynthesis begin allowing the developing spore to escape from its dependence on degradation of dormant spore protein as a source of amino acids. Consequently, a second function of this proteolytic system is to generate amino acids rapidly in spore germination to allow rapid protein synthesis.

Dormant Spore Proteins Degraded During Germination

Identification, purification and characterization. Given that a large fraction of dormant spore protein is degraded during spore germination, it is of obvious interest to know the identity of the protein(s) degraded. While some spore membrane protein may be degraded during germination, this can account for only a small percentage of the total degraded (8). Similarly, of seven spore enzymes examined, none showed significant

TABLE 1 *Comparative properties of Class A and B SASP*[a]

	Class A	Class B
Molecular weight	5,900 - 9,000	10,300 - 12,000
Percentage of met, tyr, leu, ile and pro	13.5 - 19.8	1 - 5
Isoelectric point	> 6 but always less than that of Class B protein from same organism	> 7.5 - > 9.5 and always greater than that of Class A protein from the same organism

[a]Data taken from references 11, 12, 13 and 14.

degradation early in germination (1). Strikingly, work with spores of both *Bacillus* and *Clostridium* species has established that the great majority ($\geq$ 85%) of the dormant spore proteins degraded during germination are a group of low molecular weight (6-12,000) (small) acid soluble proteins (SASP) (9,10,11,12,13). Studies with SASP from spores of four *Bacillus* species (*B. cereus*, *B. megaterium*, *B. sphaericus* and *B. subtilis*) and one *Clostridium* species (*C. bifermentans*) have shown that the SASP consist of 2-3 major proteins ($\sim$ 80% of the total) and 5 or more minor ones; almost all the SASP have isoelectric points > 7 (11,12,13,14). Most of the major proteins have been purified and characterized from spores of the four *Bacillus* species studied and all species contain two types of major proteins - one or two representatives of class A and one of class B (11, 12,13,14). While the general physical and chemical properties of the proteins within the same class are conserved across species, the two classes differ in amino acid composition, molecular weight and isoelectric point (13) (Table 1). A number of the minor SASP have also been purified from *B. megaterium* spores; these proteins are different from the Class A and B proteins, and are not derived from the latter (15).

Localization of SASP in vitro. Initial work with *B. megaterium* spores demonstrated that both major and minor SASP were located in the spore core (9,15). More recent work has refined this location through analysis of spores in which various stable cross-links have been induced *in situ* using high doses of ultraviolet light. This work has shown that there is no detectable crosslinking of dipicolinic acid to the major SASP (16). However, SASP are crosslinked to DNA *in vivo* with a quantum yield similar to that for SASP-DNA complexes prepared *in vitro*; very little SASP-DNA cross-linking was detected *in vivo* (17). This data, plus the binding of purified *B. megaterium* SASP to DNA *in vitro* (14), has strongly suggested that the majority of the SASP are bound to spore DNA *in vivo*. Unfortunately, some of these experiments have not been repeated with spores of other species, in particular *B. subtilis* in which the isoelectric point of the two major SASP is < 7.

Amino acid sequences of SASP. The complete primary sequence of the three major *B. megaterium* proteins has been established (18,19,20). The two class A proteins (termed proteins A (61 residues) and C (71 residues)) have 80% sequence identity, and are also very closely related immunologically (17,18,19). The *B. megaterium* class B protein (termed protein B) is much less related, showing only two small regions of homology with the A and C proteins (19). The B protein (96 residues) also has a large internal repeating region of 34 residues which only shows

5 differences in the two repeats (19).

One striking finding concerning the primary sequence of the major SASP concerns the conservation of primary sequence across species. As noted above, general physical and chemical properties of the class A or B proteins from spores of different species are conserved. However, the class A or B proteins from different species are very different based on immunological criteria (13). Similarly, there are a number of differences (6 out of 15 residues) between the amino terminal sequences of the *B. cereus* and *B. megaterium* class A proteins and even less homology between the *B. cereus* and *B. megaterium* B class proteins (13). Strikingly, the amino termini of the *B. megaterium* and the *B. subtilis* B class proteins show essentially no homology. However, very high degrees of homology have been found between the amino termini of the *B. subtilis* B class protein and internal (and in some cases almost carboxy-terminal) regions of the *B. megaterium* B class protein (13). These findings have led to the suggestion that there is only limited constraint (but see below) on the primary sequence of major SASP, and that consequently they may evolve rapidly by using mechanisms such as intragenic recombination (13).

Regulation of SASP synthesis. With the knowledge that the SASP (both major and minor) disappear during germination, an obvious question is - are these species also synthesized late in sporulation i.e. are they spore specific gene products? The answer to both questions is yes. The major SASP are present in vegetative or young stationary phase cells at a level <0.01% that in the dormant spore, with the upper limit for the minor SASP at the 1% level (15,21). The SASP appear only in the developing forespore during sporulation, about 2 to 3 h before dipicolinic acid synthesis (9,11,15,22). While the majority of the different SASP appear at approximately the same time, a few forms may show slightly different kinetics (9,11,15). The major SASP accumulate in parallel with the acquisition of resistance to ultraviolet light by the developing spore (9,11), and mutants of *B. sphaericus* blocked in spore cortex formation accumulate SASP and become UV resistant (12). This, plus the data on DNA binding *in vivo* and *in vitro* has led to the suggestion that the SASP may be involved in the resistance of bacterial spores to ultraviolet light (14).

The major SASP which have been studied are synthesized *in vivo* coincident with the appearance of the mature proteins (11,21). At the time of maximum rates of SASP synthesis, these proteins may comprise as much as 7% of the protein being made in the sporulating cell (21). There is no evidence that SASP translated *in vitro* are made as large precursors (11,23). Indeed, the size of the mRNAs for the A and C proteins is such that they must be mono-cistronic (24), although they could have been processed from a polycistronic mRNA. Since the SASP appear at a defined time in sporulation and only in the forespore, these proteins would appear to represent genes which are turned on during sporulation in a defined manner both temporally and spatially. How is regulation of expression of these genes mediated? Studies in several laboratories have now provided strong evidence that synthesis of these proteins is regulated at the transcriptional level (21,23). Synthesis of the proteins *in vivo* is shut off almost immediately by inhibition of RNA synthesis, and levels of mRNA for the major SASP (as determined by *in vitro* translation) parallel rates of synthesis of the major SASP *in vivo*.

The Protease Initiating the Degradation During Germination

Identification of the enzyme. As in most intracellular proteolytic reactions the rate limiting step in degradation of dormant spore protein during spore germination is the first proteolytic cleavage (1). This strongly suggests that the enzyme which initiates the proteolysis during germination is an endoprotease. While sporulating cells of various *Bacillus* species contain high levels of a number of non-specific proteases, these are not found in the spore (25,26). Indeed the endoprotease specific activity on typical protease substrates (haemoglobin, casein, etc.) is only slightly higher in spores than in vegetative cells (26). However, spores do contain a protease active on the major SASP which is not found in other stages of growth (26,27). This enzyme has been purified to homogeneity, and is specific for cleavage of the major SASP; it has no activity on other amide, ester, peptide or protein substrates (27,28). This enzyme, called spore protease, is the enzyme responsible for initiating degradation of the major SASP *in vivo* during germination. This has been proven by the isolation of four independent mutants (at least one of which is in the structural gene for the enzyme) with decreased spore protease enzyme activity *in vitro* (29). In all of these mutants the rate of degradation of the major SASP during germination is decreased (although it still takes place), even though the rates of initiation of germination are normal (29). The process of germination and outgrowth (as well as growth and sporulation) in these mutants is also normal, suggesting that degradation of the major SASP is not rate limiting for outgrowth. Whether the degradation of the SASP in the protease mutants is due to a leaky mutation or a second unrelated protease is not clear.

Characterization and sequence specificity. The spore protease as purified from germinated spores of *B. megaterium* is a tetramer of identical 40,000 mol wt subunits (28). Only the tetramer is active, and the tetrameric conformation is stabilized by divalent cations. The enzyme is a serine type endoprotease which cleaves the major SASP (as well as many of the minor ones (15)) from various *Bacillus* species at one or two specific sites generating large (20-35 residue) oligopeptides (13,27,30). These oligopeptides can then be degraded to free amino acids *in vitro* (and presumably *in vivo*) by spore peptidases (25,26). The latter enzyme(s) has little activity on the major SASP without prior protease cleavage (27).

The remarkable specificity of the spore protease is due to its requirement for a specific sequence for recognition and cleavage. The recognition sequence is at least a pentapeptide, and is probably larger (20,30,31). The pentapeptide sequence required is Glu ↓ X Y Z Glu, where X is Ile or Phe, Y is Gly or Ala and Z is Ser, Thr or Gln, with cleavage at the first glutamyl bond (arrow). A synthetic heptapeptide with the sequence NH_2-Thr-Glu↓Phe-Ala-Ser-Glu-Phe is a substrate for the spore protease and is cleaved at the first Glu-Phe bond (arrow) with a V_{max} similar to that for the protein substrates (31). However, the Km for the heptapeptide is $>10^3$ higher than that for the protein substrates (31).

The required pentapeptide sequence is found not only in the *B. megaterium* proteins, but also in the major SASP of *B. cereus* and *B. subtilis* and is the site of spore protease cleavage in these proteins (13). Thus this sequence, which has a defined function, is conserved in the major SASP from different *Bacillus* species. Presumably the strict sequence requirement of the spore protease allows initiation of rapid and massive intracellular proteolysis during germination without the possibility that proteins needed later in development will be degraded, as might be the case with a non-specific protease.

Enzyme level and localization during development. Use of a radioimmunoassay against the spore protease has shown that like its SASP substrates, the protease is also a spore specific gene product. Spore protease antigen accumulates $\sim$3 hr before dipicolinic acid synthesis and only within the developing forespore (28). Again like its substrate, the spore protease appears localized within the spore core, although its distribution within the core is not known. During spore germination the spore protease antigen disappears in parallel with loss of its enzymatic activity with a half life of about 40 min (28). The latter process requires metabolic energy, and is presumably the energy dependent proteolysis of the spore protease (28). The loss of the spore protease during spore germination makes functional sense, since the protease's only potential substrates are lost during germination with a t 1/2 of 5 to 8 min.

Regulation of the spore protease. Several of the findings concerning the spore protease and its substrates strongly indicate that the enzyme must be regulated in some fashion. Thus the spore protease antigen appears during sporulation at about the same time and in the same compartment as the SASP (22,28). However, there is no interaction between enzyme and substrate in the developing forespore or dormant spore, as the SASP are neither degraded nor turned over (12). On the other hand, in the first minutes of spore germination the protease and its substrates interact rapidly resulting in degradation of the SASP within 30 min. Consequently, the spore protease must be regulated in *some* fashion such that it does not act in the developing or dormant spore yet acts rapidly in the germinated spore. One well established method for regulating proteolytic enzymes is by zymogen formation, and this appears to be a major mechanism for regulating the spore protease (32,33).

The spore protease is synthesized (presumably in the forespore) during sporulation ∿3 h before dipicolinic acid synthesis and as a polypeptide of mol wt 46,000 (P_{46}) (32). P_{46} is *catalytically inactive* (33). Approximately 3 h later P_{46} is processed probably by proteolysis to a form with a 41,000 mol wt polypeptide (P_{41}). P_{41} can also form tetramers, and has full catalytic activity (33). Strikingly, this processing step takes place in parallel with accumulation of dipicolinic acid by the developing spore, an event which probably causes a drastic change in the spore core environment (32). Perhaps this then precludes any protease action at this time. P_{41} is the predominant form of the protease in the dormant spore, with small amounts of unprocessed P_{46}.

During spore germination P_{41} is rapidly (t 1/2 3 to 4 min) processed (again probably by proteolysis) for a form with a 40,000 mol wt polypeptide (P_{40}); P_{40} is the form in which the protease has been purified, and P_{40} has the same specific activity as P_{41} (32,33). The conversion of P_{41} to P_{40} is energy dependent; however, the P_{40} once formed then disappears (and no smaller forms are found) with a t 1/2 of 40 min in an energy dependent process. Any P_{46} present in the dormant spore is not processed further during germination, again suggesting that special intracellular conditions are required for P_{46} to P_{41} conversion.

This identification of an inactive zymogen form of the spore protease explains in part the regulation of this enzyme, and implicates conversion of P_{46} to P_{41} as a key regulatory step. The significance of the P_{41} to P_{40} conversion is less clear, as this seems to have no effect on the catalytic activity of the enzyme; perhaps it simply targets the enzyme for eventual complete proteolysis.

DISCUSSION

The system of protein degradation during bacterial spore germination is but one example of the rapid degradation of a protein store during the return to active metabolism of a dormant or quiescent stage of growth. Other systems, such as fungal spore or plant seed germination, exhibit many of the general properties of the bacterial system such as a unique group of proteins degraded and a unique protease. Thus the bacterial spore system is not unique, but is only the first (in terms of evolution) in a long line of other organisms which have adopted this type of strategem. Presumably, it is advantageous for a dormant or quiescent system to store amino acids to be used for protein synthesis when rapid development begins, rather than to synthesize them at that time. Again, presumably storage of amino acids as protein is more sparing of storage space and safer in terms of potential loss of stored material.

While the proteolysis during spore germination may not be a unique system in nature, detailed studies of this system have made (and one hopes will continue to make) significant contributions to our understanding of some fundamental problems concerning bacterial sporulation and germination, as posed in the introduction of this article.

Mechanisms for regulation of gene expression during sporulation. The SASP as well as the spore protease are spore specific gene products which are almost certainly expressed only in the forespore. In this latter regard they differ from many other sporulation genes which are expressed in the mother cell. Clearly, isolation and analysis of the SASP and spore protease genes could provide insight into mechanisms for regulation not only of these genes during sporulation but possibly many others. While a number of sporulation genes have now been cloned and analysed, in most cases the gene product and/or function are not known; this is not the case with the SASP and spore protease.

Mechanisms for causing spore heat and radiation-resistance. While it is clear that the SASP are not involved in spore heat resistance, there is significant circumstantial evidence that they are involved in spore resistance to ultraviolet light.

It is clear, however, that the SASP are not the sole cause of ultraviolet light resistance, for the spore protease mutants all lose ultraviolet light resistance at the same rate as wild type spores despite decreased rates of loss of SASP (28). Thus if the SASP are involved in spore ultraviolet light resistance they must act synergistically with some other factor, possibly partial dehydration. While this suggestion has not yet been proven, isolation of mutants lacking one or more SASP (possibly by using appropriate cloned genes) could answer this question.

Mechanisms for bringing about and maintaining spore dormancy. The synthesis of the spore protease as a zymogen is a classic example of a mechanism for maintaining enzymatic dormancy. This finding certainly explains the lack of interaction of spore protease and substrate prior to P_{46} to P_{41} conversion. While we do not yet know the details of the regulation of P_{41} itself, the findings that the P_{46} to P_{41} conversion is needed to activate the protease, and that the latter process requires unusual intracellular conditions (31) strongly suggests that understanding of the P_{46} to P_{41} conversion may explain, at least in part, the enzymatic dormancy of the spore.

Mechanisms for triggering of spore germination. While degradation of dormant spore protein is an early event in spore germination, it is by no means the earliest. Indeed, slowing proteolysis by mutation has no effect on the triggering of germination (28). Thus the proteolytic system is not rate limiting for initiation of germination. However, it is clear that the triggering of spore germination activates the spore protease and primes it for conversion from P_{41} to P_{40}, yet does not allow the P_{46} to P_{41} conversion to take place. Again contrasting the P_{41} to P_{40} conversion with the P_{46} to P_{41} conversion *in vitro* may allow us to draw conclusions about the environment inside the spore at the time of these different processes.

From the results presented in this review it is clear that we now know many of the details of the proteolytic system involved in proteolysis during spore germination. However, it should be equally clear that many other details are still obscure. It is to be hoped that the elucidation of the latter will provide us not only with important information about proteolysis during germination, but also with new insight into sporulation and germination themselves.

ACKNOWLEDGEMENTS

The work from the author's laboratory has been supported by

grants from the National Institute of Health and the Army Research Office.

It is a pleasure to acknowledge the participation of many colleagues in the work from the author's laboratory including Susan Dignam, Mitchell Dunn, Craig Gerard, Susan Goldrick, Renecca Hackett, Charles Loshon, Deborah Miller, Juris Ozols, Cynthia Postemsky, Grace Primus, Barbara Setlow, Ravendra Singh, Bonnie Swerdlow, William Waites, Scott Wood and Katherine Yuan.

REFERENCES

1. Setlow, P. (1975) *J. Biol. Chem.* **250**, 631-637.
2. Setlow, P. and Primus, G. (1975) *J. Biol. Chem.* **250**, 623-630.
3. Wade, H.E., Robinson, H.K. and Philipps, B.W. (1971) *J. Gen. Microbiol.* **69**, 299-312.
4. Setlow, P. and Kornberg, A. (1970) *J. Biol. Chem.* **245**, 3637-3644.
5. Setlow, B. and Setlow, P. (1977) *J. Bacteriol.* **129**, 857-865.
6. Setlow, B. and Setlow, P. (1977) *J. Bacteriol.* **132**, 444-452.
7. Setlow, B., Shay, L.K., Vary, J.C. and Setlow, P. (1977) *J. Bacteriol.* **132**, 744-746.
8. Seto-Young, D.L.T. and Ellar, D.J. (1979) *Microbios* **26**, 7-15.
9. Setlow, P. (1975) *J. Biol. Chem.* **250**, 8159-8167.
10. Setlow, P. and Waites, W.M. (1976) *J. Bacteriol.* **127**, 1015-1017.
11. Johnson, W.C. and Tipper, D.J. (1981) *J. Bacteriol.* **146**, 972-982.
12. Setlow, B., Hackett, R.H. and Setlow, P. (1982) *J. Bacteriol.* **149**, 494-498.
13. Yuan, K., Johnson, W.C., Tipper, D.J. and Setlow, P. (1981) *J. Bacteriol.* **146**, 965-971.
14. Setlow, P. (1975) *J. Biol. Chem.* **250**, 8168-8173.
15. Setlow, P. (1978) *J. Bacteriol.* **136**, 331-340.
16. Setlow, P., unpublished results.
17. Setlow, B. and Setlow, P. (1979) *J. Bacteriol.* **139**, 486-494.
18. Setlow, P. and Ozols, J. (1979) *J. Biol. Chem.* **254**, 11938-11942.
19. Setlow, P. and Ozols, J. (1980) *J. Biol. Chem.* **255**, 8413-8416.
20. Setlow, P. and Ozols, J. (1980) *J. Biol. Chem.* **255**, 10445-10459.
21. Dignam, S.S. and Setlow, P. (1980) *J. Biol. Chem.* **255**, 8417-8432.

22. Singh, R.P., Setlow, B. and Setlow, P. (1977) *J. Bacteriol.* **130**, 1130-1138.
23. Leventhal, J.M., Johnson, W.C., Tipper, D.J. and Chambliss, G.H. (1981) in *Sporulation and Germination* (Levinson, H.S., Sonenshein, A.L. and Tipper, D.J., eds.), pp.209-214, Amer. Soc. for Microbiol. Washington, D.C.
24. Setlow, B. and Setlow, P., unpublished results.
25. Maurizi, M.R. and R.I. and Switzer, R.L. (1980) *Curr. Top. in Cellular Regulation* **16**, 163-224.
26. Setlow, P. (1975) *J. Bacteriol.* **122**, 642-649.
27. Setlow, P. (1976) *J. Biol. Chem.* **251**,7853-7862.
28. Loshon, C.A. and Setlow, P. (1982) *J. Bacteriol.* **150**, 303-311.
29. Postemsky, C.J., Dignam, S.S. and Setlow, P. (1978) *J. Bacteriol.* **135**, 841-850.
30. Setlow, P., Gerard, C. and Ozols, J. (1980) *J. Biol. Chem.* **255**, 3624-2638.
31. Dignam, S.S. and Setlow, P. (1980) *J. Biol. Chem.* **255**, 8408-8412.
32. Loshon, C.A., Swerdlow, B.M. and Setlow, P. (1982) *J. Biol. Chem.* **257**, 10838-10845.
33. Hackett, R.H. and Setlow, P. (1983) *J. Bacteriol.* **153**, 375-378.

ACTIVATION OF *BACILLUS CEREUS* SPORES WITH CALCIUM

H.A. DOUTHIT and R.A. PRESTON

*The University of Michigan,
Ann Arbor, Michigan, USA.*

SUMMARY

Treatment of spores with calcium solutions was shown to activate all strains of *Bacillus cereus* T tested, some more than others. The activation was greatest with calcium. Strontium and manganese were moderately effective. Magnesium, zinc, barium, cobalt, cadmium, and copper were ineffective. The kinetics of activation were slow at 4°C, occurred with a pH optimum of about 8.5, and saturated at a calcium concentration of about 5 mM. Calcium was effective as an activator whether chelated by dipicolinic acid or present as a chloride. It is suggested that calcium may be central to the control of dormancy in bacterial spores via mechanisms that control its distribution at critical sites.

INTRODUCTION

Bacterial spores generally require an activation treatment, for example by sublethal heat, to optimize their response to physiological germinating agents. The molecular details of the activation mechanisms are still obsecure, but are assumed to involve configurational changes in spore macromolecules, particularly those in the outer layers (see Keynan and Evenchik (1) for review). In reviewing the literature on activation, we were intrigued by a report indicating that calcium dipicolinate (CaDPA), a well-known germinating agent at higher concentrations, seemed capable, at low concentrations of activating *Bacillus subtilis* spores for germination by L-alanine (2). Paradoxically, calcium and DPA are also major components of bacterial spores, and it seemed reasonable to speculate that insights into mechanisms of activation in general might be obtained through a study of the details of activation by this particular chemical agent. Subsequently, during the course of an investigation of the capacity of CaDPA to activate spores of *B. cereus* for germination by L-alanine and inosine, we discovered that calcium by itself was an effective activating agent. We report here on an initial characterization of these phenomena.

METHODS

Bacterial Strain

The strain of *B. cereus* T used for most of these experiments was obtained from H.O. Halvorson, via Neil McCormick, and is referred to as NEIL. The strain of NAK is a spontaneous, non-capsulated variant of NEIL. Strains HD1, HD4, HD7, and HD13 are our designations of *B. cereus* T kindly supplied by Elliot Juni, Larry Sacks, James Vary and Richard Hanson respectively.

Spore Production

Spores were produced in the G-medium of Church *et al.* (3) in the following manner. A suspension of dormant spores (A660 = 1.0 Gilford 240 spectrophotometer with a 1 cm cuvette) was heated for 60 min at 65°C in distilled water to activate and pasteurize. These were inoculated into tryptic soy broth (Difco), and growth in a shaking water bath was followed spectrophotometrically to A660 = 0.5. A ten per cent inoculum from this culture was made into G-medium, and grown similarly to A660 = 0.5. Serial growth in G-medium from a 10% inoculum was repeated one more time, and finally, a 500 ml volume of G-medium in a 1 l Erlenmeyer flask was inoculated with 10 ml of culture having A660 = 0.5. This last culture was grown on a rotaty shaker at 200 rev/min, 30°C for 20 to 24 h, by which time release of mature spores from their sporangia was essentially complete. Harvesting and washing were accomplished by centrifugation (9,200 x g, 5 min, 0°C) and sequential suspension of the spore pellets in cold distilled water (3 washes), 1.0 M KCl (3 washes) and distilled water (4 washes). Washed spores were stored in aqueous suspension on ice at A660 = 30 for no more than 2 weeks.

Pretreatment

Typically 0.5 ml of spores were centrifuged at 4°C in an Eppendorf table-top centrifuge (model 5412) for 2 to 5 min.

The supernatant was decanted, and spores suspended in the test solution with the aid of a Pasteur pipette. Incubation, except where noted, was 60 min on ice. Spores were centrifuged again, washed once with either cold distilled water or cold buffer, resuspended and stored on ice until used within a few hours.

Germination

Germination was assayed spectrophotometrically at 450 nm. Spores were added to 3.0 ml reaction mixtures containing 20 mM Tris-(hydroxymethyl)-aminomethane (Tris), 10 mM NaCl, 1.0 mM inosine, and 1.0 mM L-alanine, mixed well, and absorbance recorded semi-continuously until germination was essentially complete.

Calculations

Estimated best straight lines were drawn along the initial, relatively unchanging part of germination curves, and along

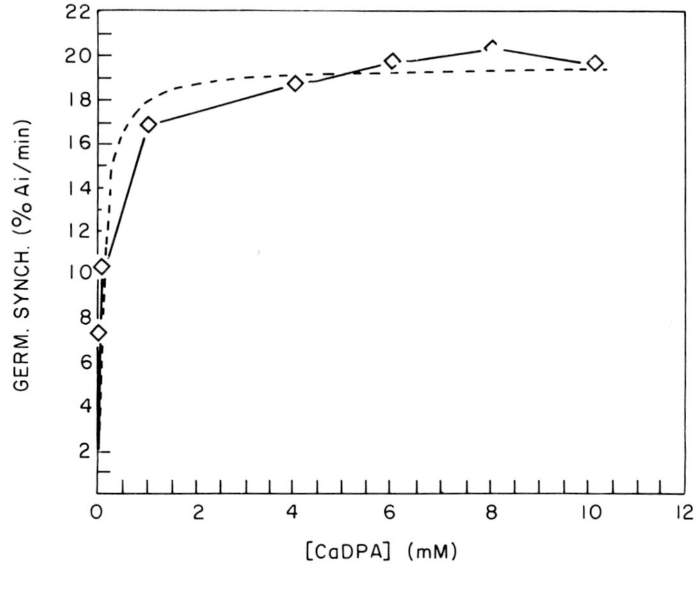

FIG. 1 *Activation by CaDPA. Spores were incubated for 10 min at 30°C in the indicated concentrations of CaDPA in 20 mM Tris, harvested, washed once with distilled water, and germinated under standard conditions. The dotted line is a Michaelis-Menten approximation of the data points, assuming no endogenous CaDPA.*

of CaDPA below 17 mM, a stimulated response to InAl was observed (Fig. 1), the effect saturating at about 4 mM CaDPA. The dashed line in Fig. 1 illustrates a Michaelis-Menten approximation to the data points obtained by a non-linear regression program (apparent V_{max} = 19.5% Ai/min, apparent Km = 0.93 mM CaDPA). Again, there was no evidence of germination in this experiment prior to the addition of spores to the InAl germination solution.

It was possible that either of the two components of CaDPA had been responsible for the activation, rather than the chelated pair, and for this reason, control experiments were conducted. The surprising result was that calcium chloride was as effective as CaDPA in causing activation of NEIL spores for InAl germination. An example of such an experiment is shown in Fig. 2, which details the effect of calcium, without DPA, on germination synchrony. Saturation kinetics was obtained (apparent V_{max} = 15.9% Ai/min, apparent Km = 0.26 M calcium), and again, the effect saturated at quite low calcium concentrations.

The activation reactions shown above were conducted in

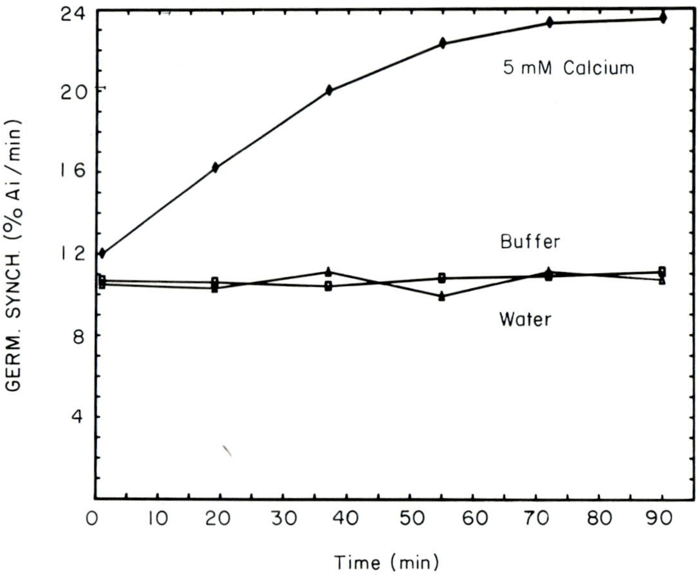

FIG. 4 *Activation kinetics. Spores were mixed with water, buffer (20 mM Tris, pH 8.5) or buffer plus 5 mM calcium chloride, incubated at 4°C for the time indicated, harvested, washed once with 20 mM Tris, and germinated under standard conditions.*

Tris buffer at the pH (8.5) which is near optimum for InAl germination. Since it was possible that activation by calcium could have quite different pH characteristics than those for germination, the effect of pH on calcium activation was assessed as shown in Fig. 3. Calcium was ineffective as an activator below pH 6 and above pH 10, having an optimum at about pH 8.5.

An unexpected result of the experiments on pH dependence was the indication that the buffers themselves had a capacity to activate *B. cereus* T spores in a pH-sensitive manner, though not nearly as effectively as calcium (Fig. 3, upper panel). Although these "buffer activations" are not yet well characterized, their mediation by an endogenous calcium source remains an interesting possibility. Whatever their mechanism, the activation with Tris, at least, could be avoided almost entirely by activating with calcium at low temperature (see below), thereby simplifying comparative analysis of activations by various ions.

Detailed kinetic studies of the calcium activation mechanism are in progress, but the results for 5 mM calcium at 4°C

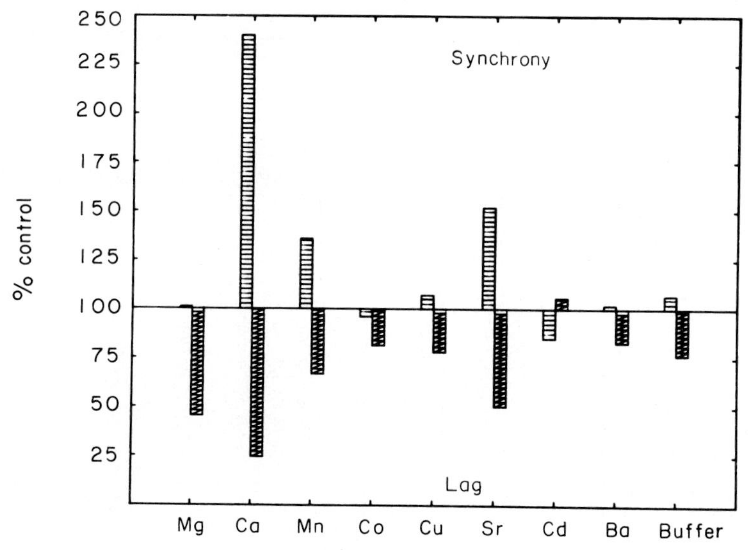

FIG. 5 *Effect of other cations. Spores were mixed with buffer (20 mM Tris, pH 8.5) containing one of the indicated cation chlorides (5 mM), incubated at 4°C for 90 min, harvested, washed once with cold buffer, and germinated under standard conditions. Data are calculated as percent of values obtained from untreated spores (synchrony = 5.5% Ai/min, Lag = 6.2 min).*

show the activation is quite slow at low temperature, requiring about 80 min to attain a maximum level of activation (Fig. 1). As mentioned above, no activation by the pH 8.5 Tris buffer, without calcium, occurred at 4°C.

As a test of the specificity of the activation for calcium ions, we compared chlorides of the group IIa elements magnesium, calcium, strontium, and barium, and the transition metals manganese, iron, cobalt, copper, zinc, and cadmium for their effects on germination lag and synchrony. Figure 5 shows results of these tests, except for iron and zinc, the hydroxides of which were too insoluble to allow spectrophotometric assays of germination. The results showed that specificity for calcium is not complete, with Sr and Mn ions both causing some synchrony increase and lag decrease over that caused by the buffer treatment alone. The results of magnesium

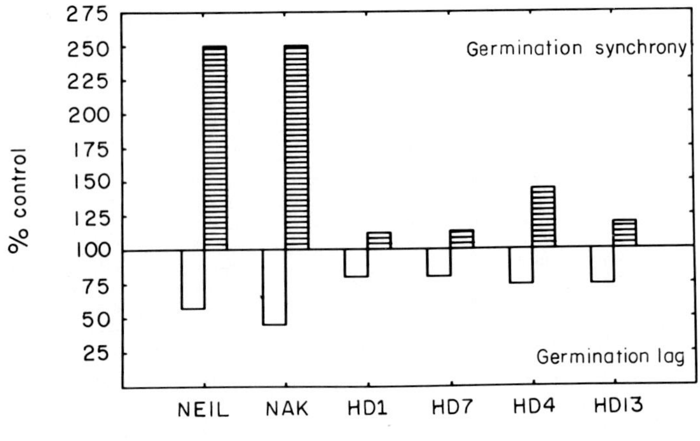

FIG. 6 *Response of several strains. Spores of several strains of* Bacillus cereus *T were activated by incubating in 20 mM Tris, pH 8.5 containing 5 mM calcium chloride, in the cold for 60 min. They were harvested, washed once with distilled water, and germinated under standard conditions. Data are calculated as a percent of the values obtained from untreated spores of the appropriate strain.*

treatment were anomalous, in that it caused a rather large reduction in lag without affecting synchrony at all. This provides an example of a situation in which the "germination rate" apparently is increased uniformly for all spores in the sample, while germination synchrony remains unaffected.

We tested several strains of *B. cereus* T to begin to assess the generality of the phenomenon of calcium activation. All of the strains tested were grown in the same G-medium (divided prior to inoculations) at the same time, harvested and washed at the same time, and stored for the same length of time. The results (Fig. 6) show that, although the lags of all strains were reduced, and the germination synchrony of all strains increased somewhat, only strain NEIL and its variant NAK were activated substantially by the treatment.

DISCUSSION

The present experiments serve to establish the broad outlines of the phenomenon of activation by calcium of *B. cereus* T for InAl germination. The pH dependence of the activation, with a peak at pH 8.5 (Fig. 3) is consistent either with an enzymic process or with non-enzymic, metal coordination chemistry, or both. The appreciable rate of the activation reaction at 4°C (Fig. 4) might argue against enzymic processes, but without more extensive data, for example, on temperature relationships, such conclusions are premature. The lack of activation at moderately low and high pH values might more reasonably be attributed to competitions by protons for relevant calcium binding ligands, or by hydroxyl ions for calcium, either of which could exclude calcium from essential binding sites.

The slow kinetics of the activation shown in Fig. 4 might indicate a permeability barrier(s) external to the essential binding site, as, for example, the exosporium, spore coat or membranes. Simple mechanisms involving binding to an easily accessible ligand seem unlikely, in view of the slow kinetics, but more complex mechanisms, dependent on slow reactions following rapid binding, are certainly possible.

Various lines of evidence suggest that the phospholipids of spore membranes might be binding sites for calcium relevant to the activation reaction described. Calcium is known to be specifically required for maintaining the integrity of spore protoplast membranes, unlike the requirement by vegetative cell membranes for magnesium as a divalent counterion (6). Recent results with *B. megaterium* spores implicate a protein in the inner spore membrane as the essential receptor site for the germinant L-proline, and fluorescence depolarization studies have revealed changes in membranes associated with the binding of L-proline at this receptor site (7). "Non-physiological" germination induced by long-chain aliphatic amines has been correlated with the ability of these amines to displace calcium from isolated membrane systems and intact spores (8). It is tempting to speculate that exogenously supplied calcium may promote structural rearrangements of spore membranes that could unmask or otherwise activate receptors for physiological germinants.

The stability constant series for chelation of divalent cations by DPA (Cu, Ni, Zn, Co, Cd, Ca, Mn, Sr, Ba, and Mg in order of decreasing stability constant) appears often in discussions of ionic phenomena in spores (see (9) for general review). In the present study, the first five of these ions (except for Zn for which data were not presented) were shown to be ineffective as activators. Of the second five in the series, Ca, Mn, and Sr were shown to be effective as activators in the order Ca>Sr>Mn. Neither of the least strongly chelated of this series was effective as an activator. In nitrate-induced germination of heat activated *B. megaterium* spores (10) the same series was found to be effective as counterion for nitrate (Ca, Sr, Mn), although, in that study, barium was found to be about as effective as strontium. Furthermore CaDPA, and SrDPA, and to a lesser extent MgDPA have been shown to be germinative in appropriate concentrations (11). However, while there is a general similarity in all these systems with regard to their cationic specificities and the chelation properties of DPA, considering the complexity of the activation/germination system of bacterial spores, we consider it unwise at this time to assume that DPA is involved directly in all of them.

Clearly there are genetic differences in the strains of *B. cereus* T we tested with regard to activation by calcium. Our laboratory strain NEIL and its derivative NAK were the only ones of the several tested to show substantial activation by calcium. All six of these strains are closely related, in a taxonomic sense, but their exact pedigrees and methods of maintenance in stock cultures are unknown. At this point it remains unclear whether isolates of this taxon from nature would be activated by the calcium treatment described here. In any case, this genetic potential exhibited by NEIL and NAK are seen as useful attributes in studies designed to help understand the activation phenomenon. Studies in progress will determine whether this genotype allows greater permeation by exogenous calcium, which in turn should allow assessment of the importance of low free calcium concentrations in the maintenance of dormancy.

ACKNOWLEDGEMENTS

This investigation was supported in part by grant No. 403961 from the LSA Development Fund of the University of Michigan, USA.

REFERENCES

1. Keynan, A. and Evenchik, Z. (1968) in *The Bacterial Spore* (Gould, G.W. and Hurst, A., eds.), pp.359-396, Academic Press, London and New York.
2. Freeze, E. and Cashel, M. (1965) in *Spores III* (Halvorson, H.O. and Campbell, L.L., eds.), pp.144-151, American Society for Microbiology, Ann Arbor, Michigan.
3. Church, B.D., Halvorson, H. and Halvorson, H.O. (1954) *J. Bacteriol.* **69**, 393-399.
4. Vary, J.C. and Halvorson, H.O. (1965) *J. Bacteriol.* **89**, 1340-1347.
5. Lee, W.H. and Ordal, Z.J. (1963) *J. Bacteriol.* **85**, 297-217.
6. Fitz-James, P.C. (1971) *J. Bacteriol.* **105**, 1119-1136.
7. Racine, F.M., Skomurski, J.F. and Vary, J.C. (1981) in *Sporulation and Germination* (Levinson, H.S., Sonenshein, A.L. and Tipper, D.J., eds.), pp.224-227, American Society for Microbiology, Washington D.C.
8. Ellar, D.J., Eaton, M.W. and Posgate, J. (1974) *Biochem. Soc. Trans.* **2**, 947-948.
9. Murrell, W.G. (1968) in *The Bacterial Spore* (Gould, G.W. and Hurst, A., eds.), pp.215-273, Academic Press, London and New York.
10. Levinson, H.S. and Feeherry, F.E. (1975) in *Spores VI* (Gerhardt, P., Costilow, R.N. and Sadoff, H.L., eds.), pp.495-505, American Society for Microbiology, Washington D.C.
11. Jaye, M. and Ordal, Z.J. (1965) *J. Bacteriol.* **89**, 1617-1618.

THE KINETICS OF BACTERIAL SPORE GERMINATION

G.M. LEFEBVRE and A.F. ANTIPPA

Université du Québec a Trois-Rivières, Trois-Rivières, Québec, Canada G9A 5H7.

SUMMARY

The entire absorbance versus time curve of a sample of germinating bacterial spores can be reproduced by a model which is qualitatively consistent with the currently accepted description of this process. Its quantitative predictions are also in very good agreement with the rather precise data now available. The model makes a clear distinction between the intrinsic aspects of the germination process on the one hand, and the observable changes consequent to germination on the other. The intrinsic processes consist of two successive random transitions between three spore states. Each observable change has a characteristic time lag and a characteristic rate of evolution controlled by a function $\alpha(t)$.

INTRODUCTION

To describe the kinetics of a process in physically meaningful terms requires that its mathematical expression incorporate its qualitative as well as quantitative features. The model of bacterial spore germination which we propose not only allows the time-dependence of the observed index of this process to be reproduced, but also gives expression to generalizations based on past experiments in terms of its parameters.

Past studies of the germination process have culminated in its description as a time-ordered sequence of four distinct phases [1,2]:

Activation - Triggering - Initiation - Outgrowth

Activation results in an increase in both the rate and extent of germination, as well as a reduction in the observed lag time. Triggering is a direct consequence of contact of the spores with the germinant [3,4] and is considered to constitute an irreversible commitment to completion of the germination process [5,6]. During initiation one observes the characteristic manifestations of germination, such as release of Ca^{2+} and DPA, loss of heat resistance, uptake of stains, loss of refractility and loss of absorbance in suspensions. These events have been shown to occur in a time-ordered sequence [7]. Moreover, some of these indices are known to occur over a finite interval of time in the single spore [8,9]. Finally, outgrowth is a period of synthesis of new materials, the outcome of which is a vegetative cell.

The model which we develop [10] is shown in Fig. 1. The rate of entry of spores into the phase "initiation" of germina-

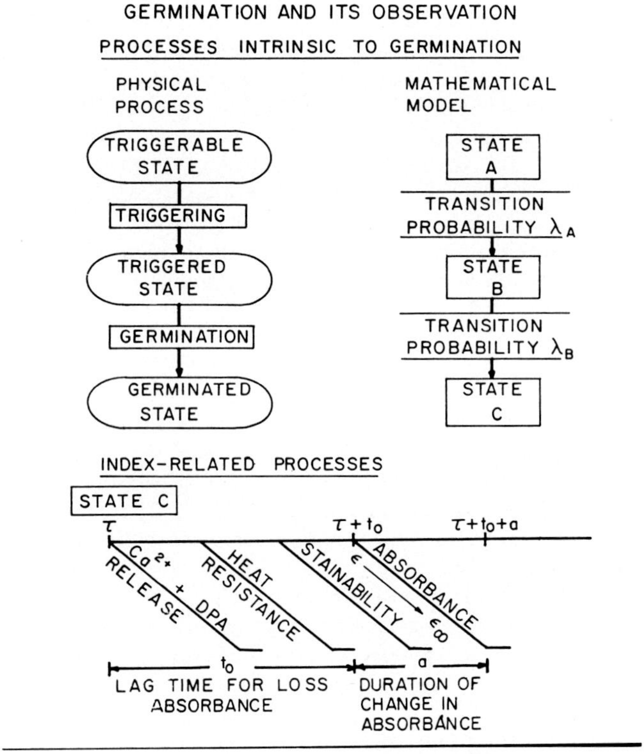

FIG. 1 *Schematic diagram of the germination process and its observation according to the model proposed.*

tion is determined by transitions between three spore states, denoted A, B, and C, as well as by the concentration of triggerable spores, A_0. The transitions between spore states are postulated to occur randomly in time, somewhat analogous to nuclear decay. In the presence of a germinant, the triggerable spores (those in state A) eventually undergo triggering to state B (the triggered state). These triggered spores are committed to eventually undergo germination to state C. A spore in state C is irreversibly engaged in the sequence of degradative events called initiation.

The transitions between spore states are intrinsic to the germination process. The same values of λ_A and λ_B should be obtained whatever the index chosen for observation of this process. Asynchrony in the sample of germinating spores is specified completely in terms of the transition probabilities λ_A and λ_B.

Each index-related process, i.e. the progress of an individual spore through a chosen observable event, can be described by an appropriate function $\alpha(t)$. Though a complete parametric description of an $\alpha(t)$ may be quite complex, it can be considered to have a lag time, t_0, and a finite duration, a. The lag time of a given observable event is the interval from the moment a spore enters state C to the beginning of the observed change. Thus, t_0 fixes the position of the observable event in the initiation sequence. The duration of the change a, is the time required by the individual spore to alter its relevant coefficient from its initial value, ε_0, to its final value, ε_∞.

METHODS

The experimental data used in testing our model were obtained by observing the absorbance of a sample of germinating spores as a function of time. Preparation of the *B. megaterium* (ATCC 14581) spores was as previously described [11].

A spore sample consisted of a monolayer film of these cells dried from a water suspension over a circular area 1.5 cm in diameter on a transparent metalized glass slide. The sample could be heat-activated by plunging the slide into a bath of pure water maintained at the desired temperature, transferring to a second bath at ice temperature to terminate the interval of activation. The slide was then air dried and placed in a suitable temperature-controlled mounting in the sample beam of a Coleman 124 spectrophotometer, the lower end of the slide resting in a shallow volume of germinant solution. A second clear slide was placed facing the first, with the two in contact at the solution end. At t = 0 the clear slide was pushed against the first, trapping a thin film of solution against the monolayer of spores. Absorbance was then recorded at 560 nm until no further change was observed.

The output of the spectrophotometer was interfaced to a microcomputer (Ohio Scientific model C8P-DF) equipped with a 12-bit analog to digital converter. With programming in BASIC, the sampling rate was 34.5 sec^{-1}. Each recorded data point was the average of 500 observations, with a total of 200 data points accumulated per experimental curve. The data were then transferred electronically to a CDC 3600 computer for analysis.

THEORY

Ley y(t) represent the evolution in time of a characteristic index of germination, and $\varepsilon(t)$ the contribution of a single spore to this index. The coefficient $\varepsilon(t)$ maintains its initial value ε_0 until the spore enters state C, whereupon its time-dependence is described by

$$\varepsilon(t) = \varepsilon_0 + (\varepsilon_\infty - \varepsilon_0) \int_0^t \alpha(\tau)\, d\tau \qquad (1)$$

As well as determining the rate of evolution of $\varepsilon(t)$, $\alpha(t)$ also fixes its position in the sequence of degradative events. The corresponding expression

$$y(t) = y_0 + (\varepsilon_\infty - \varepsilon_0) \int_0^t C(t - \tau)\alpha(\tau)\, d\tau \qquad (2)$$

where

$$C(t) = A + \frac{A_0}{\lambda_B - \lambda_A} \left[\lambda_A e^{-\lambda_B t} - \lambda_B e^{-\lambda_A t} \right] \qquad (3)$$

is the concentration of spores having entered state C up to the time t_0 [10], and $y_0 = \varepsilon_0 N_0$, N_0 being the total spore concentration. In the context of our model, the specific rate of germination is

$$\frac{1}{A_0} \frac{cD(t)}{dt} = \frac{\lambda_A \lambda_B}{\lambda_B - \lambda_A} \left[e^{-\lambda_A t} - a^{-\lambda_B t} \right] \qquad (4)$$

If N is the total spore concentrations of the sample, and A_0 the concentration of spores which will germinate in the experiment, then the extent of germination is A_0/N_0. Thus,

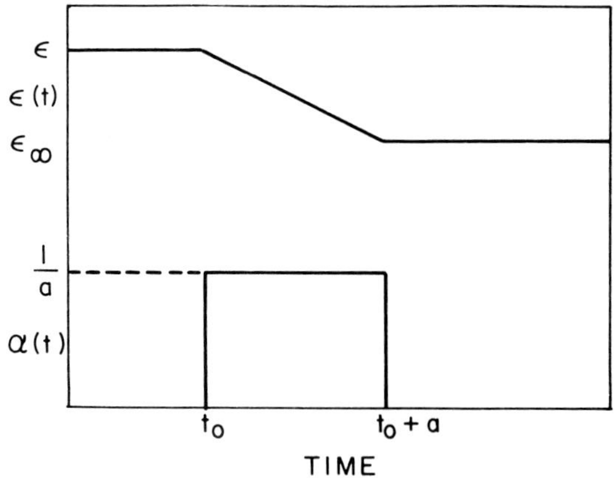

FIG. 2 *A plot of the rectangular pulse approximation for $\alpha(t)$, together with the corresponding function $\varepsilon(t)$.*

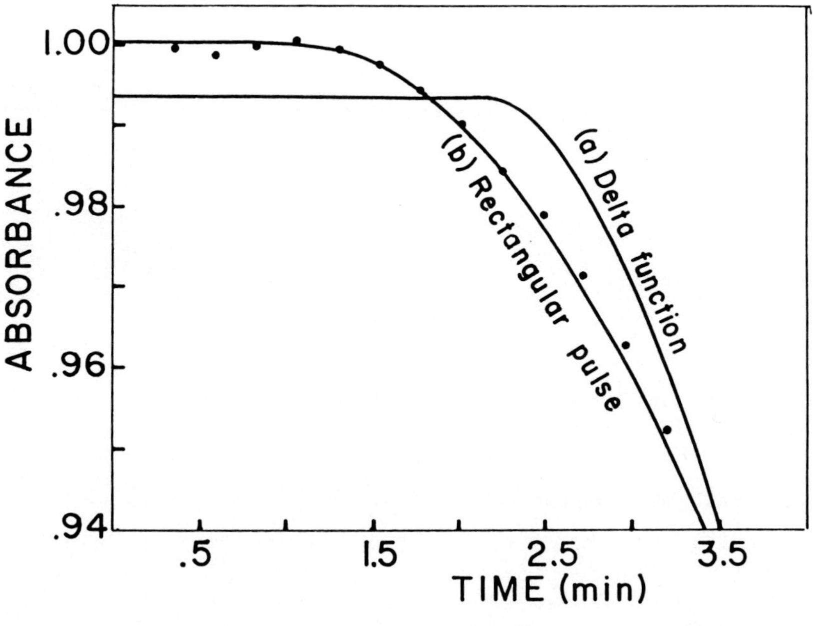

FIG. 3 *An enlarged view of the "shoulder" region of an absorbance vs. time curve. The spores were heat-activated at 70°C for 3 min and placed in contact with β-D-glucose (10 mM) in phosphate buffer (50 mM), pH 7.0, at 28°C. The data points are shown individually while the curve predicted from analysis using equations (5)(b) and that when $\alpha(t) = \delta(t-t_0)$ (a), are shown as solid traces.*

extent and rate of germination are shown to be separable quantities.

The function $\alpha(t)$ can be determined experimentally by observation of germination in the single spore [8]. In the absence of a complete parametric description, $\alpha(t)$ can be approximated to varying degrees by functions which recognize that $\Delta\varepsilon$ has a finite duration, and that it occupies a definite position in the sequence of degradative events. The simplest function meeting these requirements is the rectangular pulse, shown in Fig. 2. Substitution of this approximation of $\alpha(t)$ into equation (2) yields

$$y_a(t) = y_0, \qquad t < t_0 \tag{5a}$$

$$y_a(t) = y_0 + A_0(\varepsilon_\infty - \varepsilon_0)\left[\frac{(t-t_0)}{a} + \frac{\lambda_A}{\lambda_B-\lambda_A}\left(\frac{1-e^{-\lambda_B(t-t_0)}}{a\lambda_B}\right) - \frac{\lambda_B}{\lambda_B-\lambda_A}\left(\frac{1-e^{-\lambda_A(t-t_0)}}{a\lambda_A}\right)\right], \qquad t_0 \le t \le t_0 + a \tag{5b}$$

$$y_a(t) = y_\infty + A\,(\varepsilon_\infty - \varepsilon_0)\left[\frac{\lambda_A}{\lambda_B-\lambda_A}\left(\frac{e^{a\lambda_B}-1}{a\lambda_B}\right)e^{-\lambda_B(t-t_0)} - \frac{\lambda_B}{\lambda_B-\lambda_A}\left(\frac{e^{a\lambda_A}-1}{a\lambda_A}\right)e^{-\lambda_A(t-t_0)}\right], \qquad t > t_0 + a \tag{5c}$$

where t is the spore's lag time, and y_∞ is the final value of the chosen index.

RESULTS AND DISCUSSION

In Fig. 3 are shown, as individual points, part of the data accumulated from the germination of spores heat activated at 70°C for 3 min. These data were analyzed for best fit of equations (5) and again in the limit as a approaches zero, i.e. for $\alpha(t) = \delta(t-t_0)$. The resultant predicted curves are shown as solid traces. The fit with the rectangular pulse is clearly superior in this initial portion, while both curves are essentially coincident for all later times. The ability of equations (5) to reproduce experimental results is illustrated in Fig. 4.

Perhaps the most elusion aspect of our model is the inter-

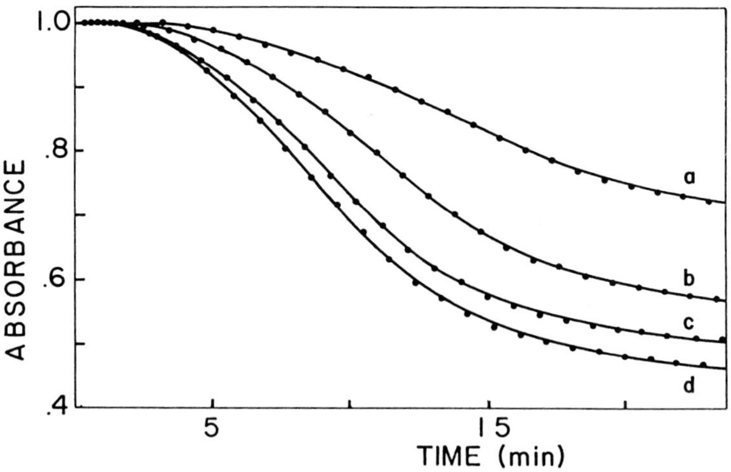

FIG. 4 *Absorbance versus time curves predicted using equations (5) corresponding to a rectangular pulse approximation for α(t). The experimental conditions were as in Fig. 3, except that heat-activation was at 60°C for 0(a), 1(b), 3(c) and 5 min (d). The data are shown as individual points.*

mediate state B, or triggered state. Experimental support for this state is found in a study of germination of temperature-sensitive mutants [12], which germinated normally at 30°C, but not at 46°C or greater. Exposure to the germinant at the permissive temperature followed by shift-up to 46°C allowed a fraction of the spores capable of germinating to do so. The dependence of the extent of germination on the duration of exposure to the germinant at 30°C can be explained on the basis of a three-state model, but not on a two-state model. When coupled to the results of our theoretical considerations [20], the three-state model is the more realistic.

We have identified the transition from the triggered state into state C as germination. In principle, germination can be observed by utilizing any one of the many manifestations which together comprise the phase of initiation. However, since these indices occur in a time-ordered sequence, the extent of germination in a sample of spores at a given moment would appear to be greater on an early event than on a later one [7]. It is reasonable to consider that the germination event cannot depend in this way on the index chosen for its observation. For this reason we believe that the spore is germinated on reaching the earliest event in the sequence, that is, entry into state C itself.

Considering the qualitative agreement with past experiments on the one hand, and the degree of good fit of absorbance versus time curves on the other, we feel that our model provides a valid description of the germination process, and this in terms of parameters that are physically meaningful. In formulating the model we were guided by a principle of maximum uncertainty, i.e. the model is the simplest compatible with experimental evidence. Further experimentation is needed to tie down details, such as the shape of the function $\alpha(t)$ for the several indices of germination, the dependence of λ_A and λ_B on environmental variables, etc. Also, further theoretical development is necessary in order to formulate other verifiable consequences of the model.

REFERENCES

1. Keynan, A. (1978) in *Spores VII* (Chambliss, G. and Vary, J.C., eds.), pp.43-53, American Society for Microbiology, Washington D.C.
2. Vary, J.C. (1979) *Spore Newslett.* **6**, 191.
3. Harrell, W.K. and Halvorson, H. (1955) *J. Bacteriol.* **69**, 275-279.
4. Halmann, M. and Keynan, A. (1962) *J. Bacteriol.* **84**, 1187-1193.
5. Scott, I.R., Stewart, G.S.A.B., Koncewicz, M.A., Ellar, D.J. and Crafts-Lighty, A. (1978) in *Spores VII* (Chambliss, G. and Vary, J.C., eds.), pp.95-103, American Society for Microbiology, Washington D.C.
6. Rossignol, D.P. and Vary, J.C. (1979) *J. Bacteriol.* **138**, 431-441.
7. Levinson, H.S. and Hyatt, M.T. (1966) *J. Bacteriol.* **91**, 1811-1818.
8. Hashimoto, T., Frieben, T. and Conti, S.F. (1969) *J. Bacteriol.* **100**, 1385-1392.
9. Vary, J.C. and Halvorson, H.O. (1965) *J. Bacteriol.* **89**, 1340-1347.
10. Lefebvre, G.M. and Antippa, A.F. (1982) *J. Theor. Biol.* **95**, 489-515.
11. Lefebvre, G.M. (1978) *J. Theor. Biol.* **75**, 307-326.
12. Vary, J.C. (1975) *J. Bacteriol.* **121**, 197-203.

COAT STRUCTURE AND MORPHOGENESIS OF BACTERIAL SPORES IN RELATION TO THE INITIATION OF SPORE GERMINATION

H.-Y. CHEUNG* and M.R.W. BROWN

Department of Pharmacy, Aston University, Birmingham, UK.
**Laboratory of Molecular Biology, National Institute of Neurological and Communicative Disorders and Stroke, NIH, Bethesda, Maryland 20205, USA.*

SUMMARY

The surface layers of nutrient depleted spores as revealed by microelectrophoretic mobility studies and electron microscopy were structurally different. The mean surface charges of nitrogen-, sulphur- and carbon-depleted spores were respectively -2.87 and -4.13 μ/s V/cm. Replica electron micrographs showed that sulphur-depleted spores containing low manganese (67.3 ng/10^7 spores) had a smooth, thin coat and were all oval in shape. Spores with high manganese (158 nm/10^7 spores) were angular in shape and had a rough and incomplete layer. Spores which germinated rapidly had a complete but rough and angular coat and plentiful pits on the coat surface while slow germinating spores did not. Thus, the lack of pits on the spore surface and the incomplete coat layers may be the cause for defective germination in sulphur-depleted spores.

INTRODUCTION

Sporulation of *Bacillus* is induced when cells are grown in a medium limited for metabolizable carbon, nitrogen, phosphate or sulphate (1). The depletion of any one of these nutrients will lead to the formation of spores with different germination properties. Previous studies on *B. megaterium* (2) and

B. stearothermophilus (3) have shown that spores produced after glucose depletion of the medium germinated faster and to a greater extent than those produced after sulphate depletion.

In combination with specific nutrient depletion conditions, the concentration of manganese is also known to have a strong effect on germination. Spontaneous and fast germination takes place in spores from medium supplemented with high manganese concentrations (3,4,6). Although these observations have been reported, no current study is devoted to an explanation of the reasons behind the differences. This paper reports the result of our study on the structural details of nutrient-depleted spores with special attention to coat layers. Correlations between germination dormancy and the morphological structure of spore coats are shown. Close parallels are demonstrated for the existence of granular coats and spores which germinate spontaneously and extensively. Implications of these results for the mechanism of triggering of spore germination are discussed.

MATERIALS AND METHODS

Organisms and Spore Preparation

Bacillus stearothermophilus NCTC 10,003 was used throughout this study. The cultivation of carbon- and sulphur-depleted spores in a chemically defined medium (CDM) has been described previously (3). When preparing nitrogen-depleted spores, the NH_4Cl concentration was 1.5 mM and manganese concentration was 62.5 μM. The concentrations of other nutrients were not changed. Cells were incubated for 60 h at 60°C with aeration at 800 ml/min, and spores were harvested by centrifugation. The pellet was washed in cold sterile deionized distilled water, resuspended in phosphate buffered saline (5 mM, pH 7.2) and stored at 4°C. In each batch, 1 to 4% of spores appeared non-refractile under the phase-contrast microscope after more than 2 months storage.

Spore Germination

To incude germination spores were resuspended in pre-warmed glucose/glutamate CDM (3). Germination was assessed by monitoring the decrease in turbidity at 470 nm (A_{470}) of appropriately diluted spore suspension. The optical density of spore suspensions germinating at 60°C was followed for 70 min by using an automatic recording spectrophotometer equipped with a temperature controller (Gilford Instrument Lab. Inc.). Confirmation of germination was checked by observing changes in refractility using a phase-contrast microscope.

Measurement of Electrophoretic Mobility (EPM)

EPM of nutrient depleted spores was measured in a cylindrical cell micro-electrophoresis apparatus, Rank Mark II model (Rank Bros. Bottisham, Cambridge, UK). Measurements were carried out at 25°C in constant ionic strength sodium chloride solution (1 mM, pH 6.8) containing 0.013 M KCl. The true electrophoretic velocity was measured only at the "stationary" level of the cell, where the occurrence of electro-osmosis near tube walls cancels with the compensating return flow of electrolytes at the centre of tube. EPM of spores at the stationary layer was consequently determined using a cytopherometer (Carl Zeiss, Oberkochen/Wuertt). The movement of individual spores was timed over a fixed distance on the eyepiece scale which was magnified on a television screen. Spore velocity was determined over one graticule square (=70 μM). The current was adjusted to give transit times of 5 to 15 seconds. The EPM of nutrient depleted spores was calculated from the following equation:

$$\mu = V \cdot E^{-1}$$

where μ was electrophoretic mobility, V was electrophoretic velocity (calculated from an average of the reciprocals of more than 100 graticule migration readings of different spores) and E was field strength which is the voltage gradient between two electrodes.

Electron Microscopy

Carbon replicating. The one-phase replica method was used in which drops of spore suspension were applied to a slide and dispersed evenly on the glass surface with the aid of 2% albumin as dispersing agent. The spore smear was allowed to dry in the air and then coated with a layer of evaporated carbon. The carbon film was cut into 3 mm squares and the spores were dissolved by immersing the slide in 10 N NaOH for 4 days. After washing, the carbon replica films left a carbon shell exactly conforming to the original topography of the spore surface. The replica films were then mounted on a grid and dried in a vacuum desiccator. The replicas were shadowed with gold-palladium at an angle of ∿30° and used for electron microscopy.

Thin sectioning. Spore samples were fixed in 1% osmium tetroxide buffer and immediately centrifuged. The prefixed spores were resuspended in 1 ml of fresh standard fixative (at pH 6.1 in Veronal buffer containing 0.01 M calcium chloride) of Kellenberger *et al.* (6) overnight at room temperature. After about 16 h of fixation the suspension was diluted with 8 ml of the buffer solution above and centrifuged. The resulting pellet was resuspended in 2 ml of 2% warm agar. After cooling, the agar was cut into half millimeter blocks. These were soaked in 0.5% uranyl acetate (aqueous) for 2 h at room temperature, dehydrated through a series of alcohol strengths up to absolute alcohol and then transferred through 25 and 75% mixtures of TAAB embedding resin (TAAB Laboratories, Reading, UK) and acetone. The blocks were retained overnight in the pure TAAB embedding resin, containing both activator and initiator, at room temperature and were put into size 0 gelatin capsules in fresh resin the following day. Polymerization and curing were obtained in an oven at 60°C for 3 to 5 days. Sections were cut onto distilled water with a glass knife held in a LKB automatic "Ultratome" and in the Sewall Porter-Blum microtome. They were picked up on grids filmed with Formvar. Preparation contrast was augmented by staining the cut sections with lead hydroxide using the method of Dalton and Zeigel (7). Electron microscopy was performed using a JEM 100B Electron Microscope.

RESULTS

Germination and Viable Counts of Nutrient Depleted Spores

The germination response of washed spores which had experienced different nutrient depletion is shown in Fig. 1. There were significant differences between germination rates of sulphur-depleted and of carbon- or nitrogen-depleted spores. Sulphur-depleted spores were characteristically superdormant compared to other nutrient-depleted spores. The dormancy of sulphur-depleted spores was not affected significantly by the amount of manganese ions added to the sporulation medium (data not shown. However, the germination rate of carbon-depleted spores was greatly enhanced by higher concentrations of manganese in the medium when cells were induced to sporulate (Fig. 1). Viability of both carbon- and sulphur-depleted spores was checked. The data in Table 1 show that both kinds of spores were equally capable of giving rise to colonies after 24 h incubation following plating on glucose tryptone agar.

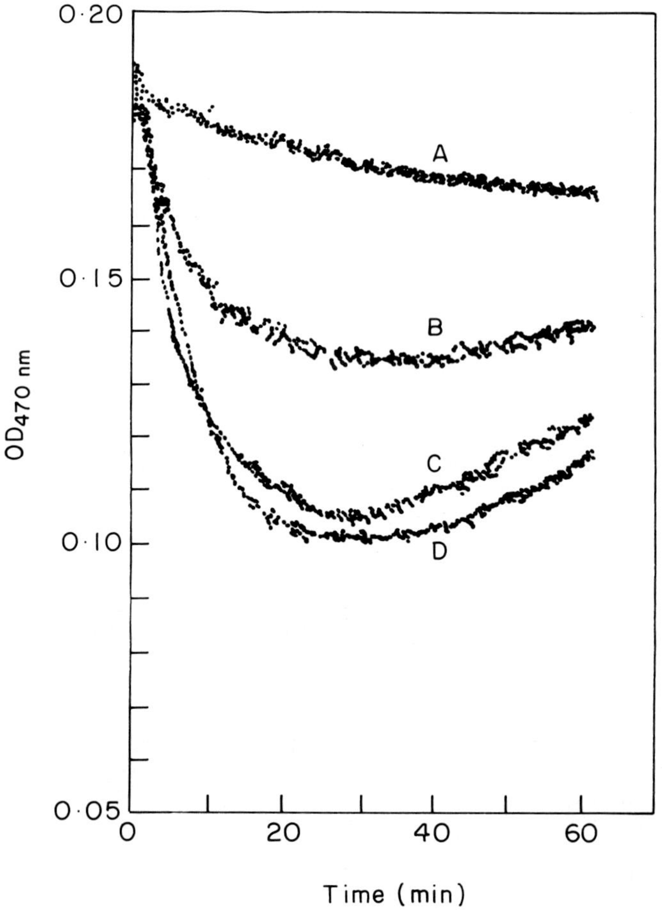

FIG. 1 *Germination of nutrient depleted spores in chemically defined medium. (A) sulphur-depleted spores produced from sulphate limited culture. (B) carbon-depleted spores produced from glucose limited culture with low manganese content (67 ng/10^7 spores). (C) carbon-depleted spores produced from glucose limited culture with high manganese content (186 ng/10^7 spores). (D) nitrogen-depleted spores produced from ammonium ion-limited culture.*

EPM of Nutrient Depleted Spores

The above results suggest that the structure of spores may differ when they have arisen from cells which have experienced different nutrient depletions. The electrophoretic mobility, being a function of average net surface charge density,

TABLE 1 *Colony formation of carbon-depleted and sulphur-depleted spores of* B. stearothermophilus *NCTC 10,003*

$[Mn^{2+}]$ (μM) in sporulation media	% spores which formed colonies* carbon-depleted	sulphur-depleted
5.0	--	73.6
10.0	72.0	76.4
62.5	90.5	89.6
100.0	88.6	66.3

*Spore suspensions were diluted in phosphate buffer (pH 7) and plated on glucose-tryptone agar. Colonies were counted after 24 h incubation at 60°C.

TABLE 2 *Electrohporetic mobilities of nutrient depleted spores at pH 6.9*

Nutrient depletion	Intra-sporal Mn^{2+} content[a]	Mean of mobilities[b]	Standard deviation
Nitrogen	--	-2.87	0.27
Sulphur	67	-3.31	0.33
	158	-3.47	0.43
Carbon	67.3	-3.69	0.71
	186	-4.13	0.44

a. ng/10^7 spores
b. mobilities unit: μ/s/V/cm

reflects the effect of surface components on the surface charge of spores (8). Therefore, an EPM study might give initial information about the change in spore surface structure under different sporulation conditions. Table 2 shows that of three kinds of spores studied, nitrogen-depleted spores had the lowest mobility rate and carbon-depleted spores had the highest mobility rate with sulphur-depleted spores in between. These results indicated that carbon-depleted spores had the most negative net charge on the spore surface and nitrogen depleted spores the least. The mobilities of both sulphur- and carbon-depleted spores containing 67 ng of manganese per 10^7 spores

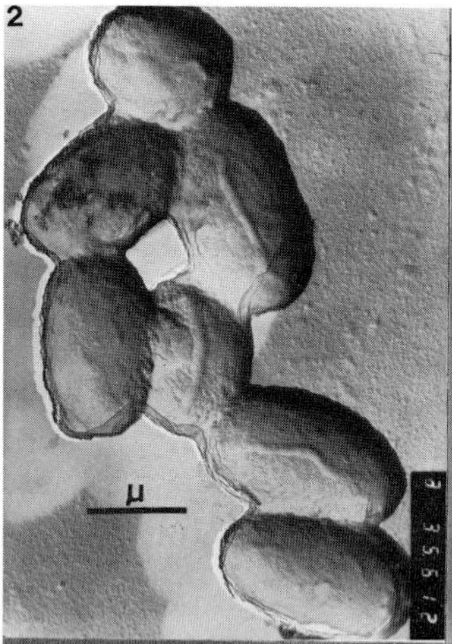

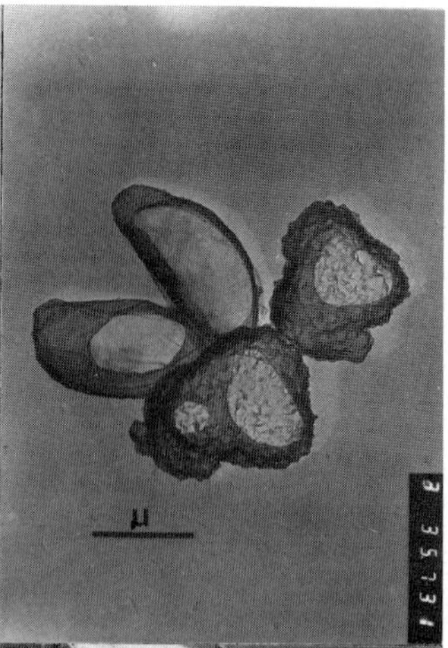

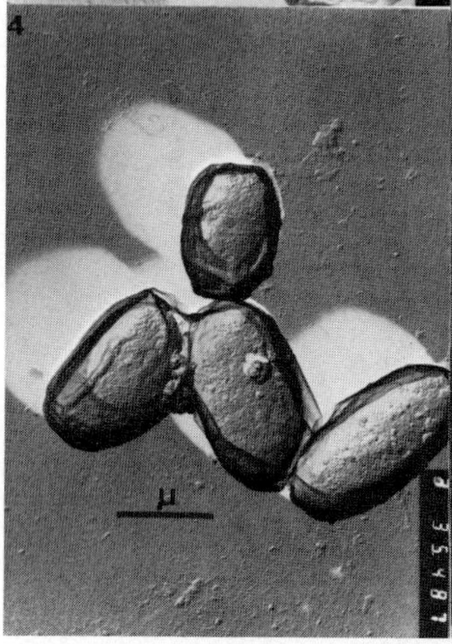

FIG. 2 *Carbon replica micrograph of sulphur-depleted spores containing 67 ng of Mn^{2+} per 10^7 spores, showing smooth spore surface.*

FIG. 3 *Carbon replica micrograph of sulphur-depleted spores containing 158 ng of Mn^{2+} per 10^7 spores. Note angular and rough surface type and smooth surface type of spores.*

FIG. 4 *Carbon replica micrograph of carbon-depleted spores containing 67.3 ng of Mn^{2+} per 10^7 spores, showing fewer and insignificant granular particles on spore surface.*

FIG. 5 *Carbon replica micrograph of carbon-depleted spores containing 186 ng of Mn^{2+} per 10^7 spores, showing the rough and granular particles on spore surface.*

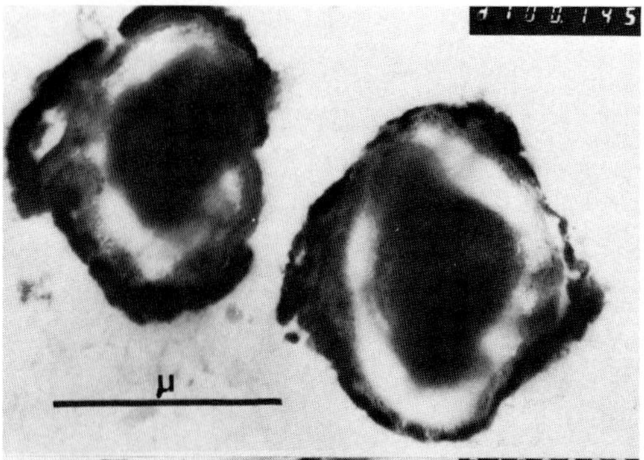

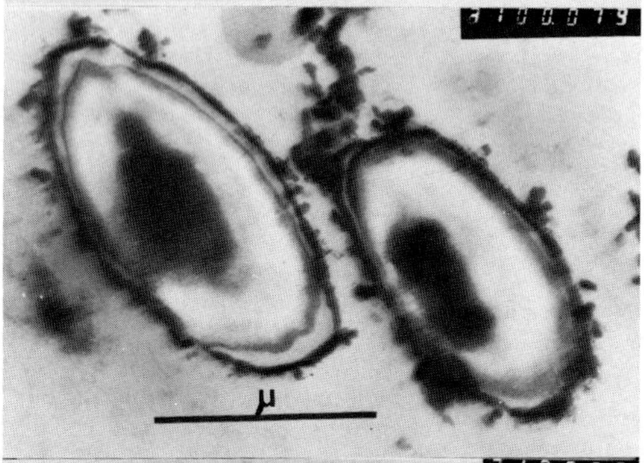

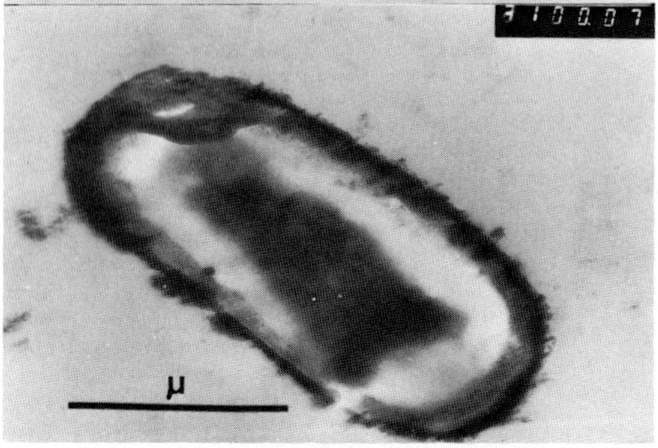

FIG. 6 *Section of sulphur-depleted spores of* B. stearothermophilus *NCTC 10,003 produced from CDM with 10 µM Mn^{2+} added. These spores contain 67 ng of Mn^{2+} per 10^7 spores.*

FIG. 7 *Section of sulphur-depleted spores of* B. stearothermophilus *NCTC 10,003 produced from CDM with 0.1 mM Mn^{2+} added. These spores contain 158 ng of Mn^{2+} per 10^7 spores.*

FIG. 8 *Section of carbon-depleted spores of* B. stearothermophilus *NCTC 10,003 showing thick and complete coat layers. These spores contain 186 ng of Mn^{2+} per 10^7 spores.*

were all lower than those containing higher manganese levels. The observed increase in mobilities with increasing intrasporal manganese content therefore suggests that the net negative charge on the spore surface was related to the sporal content of divalent metal ions. The relationship however, was not revealed by this EPM study.

Electron Microscopy

In order to reveal the structural differences between nutrient depleted spores, two kinds of electron microscopy were used. The carbon replica pictures of sulphur-depleted spores containing low manganese (67.3 ng/10^7 spores) and high manganese (158 ng/10^7 spores) are illustrated in Fig. 2 and 3 respectively. The electron micrograph of the former show that these spores had smooth surfaces and were fine in structure. They were almost all cylindrical in shape (Fig. 2) in contrast to those with high manganese which were heterogeneous. There were two distinct morphological types in the spore crop (Fig. 3). Nearly 90% of the spores were smooth but smaller in size than sulphur-depleted spores containing low manganese. These spores were neither cylindrical nor oval, but kidney shaped. The other 10% had ribbing and were very rough on the surface. They were nearly angular in shape. The differences between carbon-depleted spores containing low (67.3 ng/10^7 spores (Fig. 4) and high (186 ng/10^7 spores (Fig. 5)) manganese content were not as big as in sulphur-depleted spores. Carbon-depleted spores containing high manganese content had fewer coat pits than spores containing low manganese. Nitrogen-depleted spores also had a lot of such pits on the spore coat, similar to carbon-depleted spores containing high manganese. The appearance of both carbon- and nitrogen-depleted spores was unique and they were oval in shape.

The ultrathin sections of each type of nutrient depleted spore is shown in Figs. 6,7 and 8). All kinds of spores, except sulphur depleted ones (Figs. 6 and 7) had complete and integral coat layers. The section of carbon depleted spore shown in Fig. 8 is typical. The volume of interior compartments of each spore type however was different, depending on manganese concentrations during sporulation. Although sulphur-depleted spores containing low manganese (67.3 ng/10^7 spores (Fig. 6)) had an integral coat, the coat of these spores was less than two thirds the thickness of those of carbon- and nitrogen depleted spores. Sulphur-depleted spores prepared from medium-containing high manganese were apparently unable to form a perfect coat.

DISCUSSION

Cheung *et al.*(9) have noted that the growth of sulphur-limited cultures was more sensitive than other kinds of nutrient limited culture to high manganese concentrations. When nutrient depletions are varied, the spores produced have different germination characteristics. Since sulphur in the form of the sulphydryl group is known to be an important component for the formation of the insoluble disulfide coat fraction (10), its depletion in the sporulation medium would affect the formation of coat. The poor germination of sulphur-depleted spores reported by Cheung *et al.*(3) is confirmed here. As illustrated in Fig. 7, their difficulty in germination correlates with incomplete formation of coat layers. There is growing evidence that the coat layers of a spore are relevant to the initiation of spore germination (11,12). Spores of *B. cereus* mutants selected for slow response to germinants and sensitivity to lysozyme, although heat-resistant and containing the same quantity of dipicolinic acid as the wild-type, have been found to be deficient in coat (13). Our results confirm other reports on the inability, or difficulty of germination of chemically stripped spores (14,15), as well as in coatless mutants (16,17,18). However, coatless spores can not directly indicate where spores are triggered to germinate. As we have shown above, spores which have spore coats can be reluctant to germinate. These spores include sulphur- and carbon-depleted spores produced in CDM containing low manganese concentrations (<62.5 μM). By comparing their morphology, we found that the only differences among nutrient depleted spores is the absence of coat pits on poor germinating spores. Moberly *et al.* (19) have reported that heat-activated *B. anthracis* spores showed a mottled appearance on the coat, and they germinated rapidly. Their description may be the same as we report here. In both cases, the germination rate and extent of germination are related to the presence and the number of spore coat particles. Therefore, we conclude that the initiation of spore germination may depend on coat pits and perhaps the conformational changes of these pit proteins, after being triggered, result in releasing signals from outside inwards to initiate germination.

ACKNOWLEDGEMENTS

We thank Dr. E. Freese for his permission to repeat part of this work by using the equipment in his laboratory.

REFERENCES

1. Lee, Y.H. and Brown, M.R.W. (1975) *J. Pharm. Pharmacol.* **27**, suppl. 22p.
2. Hodges, N.A. and Brown, M.R.W. (1975) in *Spores VI* (Gerhardt, P., Costilow, R.N. and Sadoff, H.L., eds.), pp. 550-554. American Society for Microbiology, Washington, D.C.
3. Cheung, H.Y., Vitkovic, L. and Brown, M.R.W. (1982b) *J. Gen. Microbiol.* **128**, 2403-2409.
4. Pelcher, E.A., Fleming, H.P. and Ordal, Z.J. (1963) *Can. J. Microbiol.* **9**, 251.
5. Gruft, H., Buchman, J. and Slepecky, R.A. (1965) *Bact. Proc.* **37**.
6. Kellenberger, E., Ryter, A. and Sechaud, J. (1958) *J. Biophys. Biochem. Cytol.* **4**, 671-678.
7. Dalton, J. and Zeigel, R.R. (1958) *J. Biophys. Biochem. Cytol.* **7**, 409-410.
8. Richmond, D.V. and Fisher, D.J. (1973) in *Advances in Microbial Physiology* **9**, 1-29. (Rose, A.H. and Tempest, D.W., eds.), Academic Press, New York.
9. Cheung, H.Y., Vitkovic, L. and Brown, M.R.W. (1982a) *J. Gen. Microbiol.* **128**, 2395-2402.
10. Aronson, A.I. and Fitz-James, P. (1976) *Bacteriol. Rev.* **40**, 360-402.
11. Watabe, K., Ichilawa, T. and Kondo, M. (1974) *Japan J. Microbiol.* **18**, 173-180.
12. Crafts-Lighty, A. and Ellar, D.J. (1981) *J. Appl. Bacteriol.* **48**, 135-145.
13. Stelma, Jr. G.N., Aronson, A.I. and Fitz-James, P. (1980) *J. Gen. Microbiol.* **116**, 173-185.
14. Duncan, C.L., Labbe, R.G. and Reich, R.R. (1972) *J. Bacteriol.* **109**, 550-559.
15. Shibata, H., Murakami, H. and Tani, I. (1980) *Microbiol. Immunol.* **24**, 291-298.
16. Stelma, Jr. G.N., Aronson, A.I. and Fitz-James, P. (1978) *J. Bacteriol.* **134**, 1157-1170.
17. Sacks, L.E. (1981) *Biochim. Biophys. Acta* **674**, 118-127.
18. Moir, A. (1981) *J. Bacteriol.* **146**, 1106-1116.
19. Moberly, B.J., Shafa, F. and Gerhardt, P. (1966) *J. Bacteriol.* **92**, 220-228.

GRAMICIDIN S AND SPORE GERMINATION AND OUTGROWTH

S. NANDI*, M. FRANGOU-LAZARIDIS, I. LAZARIDIS*** and B. SEDDON**

Department of Developmental Biology, University of Aberdeen, Aberdeen AB9 2UE, UK
**Present address: Department of Molecular Biology, Madurai Kamaraj University, Madurai, India*
***Present address: Department of Biological Chemistry, University of Ioannina, Ioannina, Greece.*
****Present address: Department of General Biology, University of Ioannina, Ioannina, Greece.*

SUMMARY

Transmission electron microscopy of germinating wild-type and gramicidin S-negative (GS^-) mutant spores of *Bacillus brevis* revealed that when gramicidin S (GS) inhibited outgrowth it blocked the much earlier event of chromatic aggregation towards the centre of the spore protoplast. When outgrowth was not inhibited by GS, spores did not exhibit this block. These findings supported *in vivo* studies on RNA synthesis and inhibition by GS. Although toluene-permeabilised cells failed to establish such inhibition, *in vitro* studies using native DNA and purified RNA polymerase showed GS inhibition of transcription by GS-DNA complex formation.

INTRODUCTION

The possible role and mode of action of the cyclic decapeptide antibiotic gramicidin S (GS) in the developmental cycle of the producer organism *Bacillus brevis* is under investigation in several laboratories (1-4). Using the GS producing strain Nagano and selected GS^- mutants (E-1,B1-7,B1-9) we have shown that GS is not connected with sporulation or spore properties

(DPA content, heat resistance, spore density) but that outgrowth characteristics are regulated by GS (5). Contrary to our findings on spore properties, Marahiel and co-workers, repeating our studies on *B. brevis* strain ATCC 9999 and its GS^- mutants, reported that lack of GS in the mutants gives rise to abnormal spores both with respect to heat resistance and DPA content (6). At the same time, however, these workers did confirm our findings on outgrowth and its inhibition by GS. More recently these workers have also claimed that extensive studies showed no complex formation between GS and the native DNA and that GS does not inhibit *in vitro* transcription using purified RNA polymerase (4).

Considering the earlier contradictions between our findings and those of Marahiel and co-workers the present investigation was carried out in an attempt to determine whether or not, in the Nagano strain, outgrowth control by GS could be related to an effect at the level of transcription.

METHODS

Organisms and Culture Conditions

B. brevis Nagano wild-type and GS^- mutants (E-1,B1-7,B1-9) were used (7,8). Spores were produced on the surface of agar plates using a nutrient growth and sporulation (NGS) medium. Mature spores were harvested and washed in glass-distilled water at 4°C until at least 95% were phase-bright, then stored at -20°C. For germination, outgrowth and growth studies a nutrient broth (NB) medium was used. Details have been given previously (8,9).

Electron Microscopy

Aliquots of organisms at various stages of development were fixed by the method of Kellenberger *et al.* (10). Thin sections (100 nm thick) were cut on an LKB ultramicrotome, stained with uranyl acetate and lead citrate and viewed with a Philips model 201 electron microscope. Photographs were taken at instrumental magnifications from 15,000-70,000.

In vivo Incorporation of Radioactive Precursors into RNA

The method used for initial studies on ^{14}C-uracil uptake into acid precipitable material during germination and outgrowth and for cells in the growth phase was as described previously (9). In order to detect the very early stages of RNA synthesis during germination where incorporation was low, higher levels of radioactivity were needed together with continuous labelling of spores. 0.2 - 1.0 μCi/ml [2-^{14}C] uracil (50-60 mCi/mmol) was used, depending on the spore concentration (10^7-10^8 spores/ml for wild type; 10^7-10^9 spores/ml for GS^- mutants). The precursor was added at zero time of incubation and duplicate 0.5 ml samples were withdrawn at 15 min intervals and incorporation into trichloroacetic acid precipitable material was monitored as before (9). When the effect of GS was tested 0.5 ml duplicate samples were removed every 30 s after GS addition. In certain experiments 0.2 μCi/ml [5,6-^{3}H] uridine (28 Ci/mmol) replaced ^{14}C-uracil.

Incorporation of Radioactive Precursor into RNA using Toluene-Permeabilised Cells

To make cells permeable to the substrates of RNA polymerase the method described by Fisher *et al.* (11) was used as modified by Rothstein *et al.* (12). It was also found necessary to incubate both the reaction mixture and the toluene-treated cells separately at 37°C for 5 min prior to mixing. The assay was performed at 37°C and contained: 0.04 Tris-HCl pH 7.9; 0.3 mM GTP, UTP CTP; 1.5 mM ATP; 0.2 mM dithiothreitol; 0.1 M KCl; 0.01 M $MgCl_2$; 0.1 mM EDTA; 2 μCi [5,6-^{3}H] UTP (40-60 Ci/mmol) and, after mixing, toluenised cells (4-8 x 10^9 cells); total volume 1 ml. Duplicate 100 μl samples were removed every 2.5 min and added to 2 ml 5% trichloroacetic acid containing 0.01 M $Na_4P_2O_7$. Samples were then processed for counting as before (9).

Isolation of *B. brevis* DNA

Native DNA was isolated by the method of Marmur (13). For the preparation of ^{3}H-DNA cells were grown in the presence of 4 x 10^{-3} μCi/ml [6-^{3}H] thymidine (28 Ci/mmol) for several generations before DNA extraction. The specific activity of the extracted ^{3}H-DNA was found to be 10^4 cpm/mg DNA.

Purification and *in vitro* Assay of RNA Polymerase

RNA polymerase was purified to near homogeneity from both wild-type and a GS^- mutant of *B. brevis* Nagano. We attempted to follow the method of Paulus and co-workers who have successfully purified RNA polymerase from *B. brevis* ATCC 8185 (14-16). All procedures were carried out at 4°C. Several modifications, however, were found necessary. Cells were suspended in four volumes of buffer A (14) except that Tris-HCl at a pH of 8 was used and phenylmethane sulphonyl fluoride (PMSF) at a concentration of 1 mM was added. Cells were broken in an MSE 150 W ultrasonic disintegrator with 10 x 5 s sonication pulses and a cooling interval of 30 s. The extract was centrifuged at

120,000 g for 30 min. The supernatant was taken through stepwise precipitation using ammonium sulphate and the fraction (designated P60) precipitating between 45% and 60% saturation was taken up in buffer A and chromatographed on DEAE-cellulose (14). Active fractions were precipitated with 70% ammonium sulphate and the pellet, collected by centrifugation, dissolved in buffer B (0.05 M Tris-HCl pH 7.9); 0.1 mM EDTA; 0.1 M dithiothreitol and 15% glycerol) was applied to an agarose colume (Bio-gel A-5m, 200-400 mesh) and eluted with an upward flow rate of 10 ml/h with buffer B. Active fractions were pooled, concentrated by dialysis against buffer A containing 70% ammonium sulphate and the precipitate collected by centrifugation. This procedure was repeated and the final pellet dissolved in buffer A plus 50% glycerol and stored at -20°C at a concentration of 4-8 mg protein/ml. The enzyme was estimated to be about 70% pure as estimated by polyacrylamide gel electrophoresis in the presence of sodium dodecyl sulphate and compared favourably with the enzyme isolated from *B. brevis* ATCC 8185 (16).

RNA polymerase was assayed at 37°C for 15 min in the presence of 30 mM Tris-HCl pH 7.9; 1 mM (each) ATP, GTP and CTP; 10 mM $MgCl_2$, 0.1 mM 5,6-^3H UTP (10 μCi/mmol); 2 mM $MnCl_2$; 10 mM 2-mercaptoethanol; 0.4 mM potassium phosphate pH 7.5; 0.5 mg/ml bovine serum albumin and appropriate amounts of enzyme and DNA in a final volume of 0.2 ml. When *E. coli* RNA polymerase was assayed 0.15 M KCl was added (16). The reaction was terminated by the addition of 2 ml ice-cold trichloroacetic acid (5%) containing $Na_4P_2O_7$. Precipitates were collected on glass-fibre filter papers (Whatman GF/C) previously soaked in trichloroacetic acid (2%) containing 0.01 M $Na_4P_2O_7$ and washed three times with the same mixture and three times with ethanol. The radioactivity was counted in a liquid scintillation counter. Antibiotics and GS to be tested were added to parallel assays keeping the solvent volumes less than 1%.

Complex Formation between *B. brevis* DNA, Peptide Antibiotics and RNA Polymerase

These studies were performed by the filter binding technique of Riggs *et al.* using nitrocellulose filters (17).

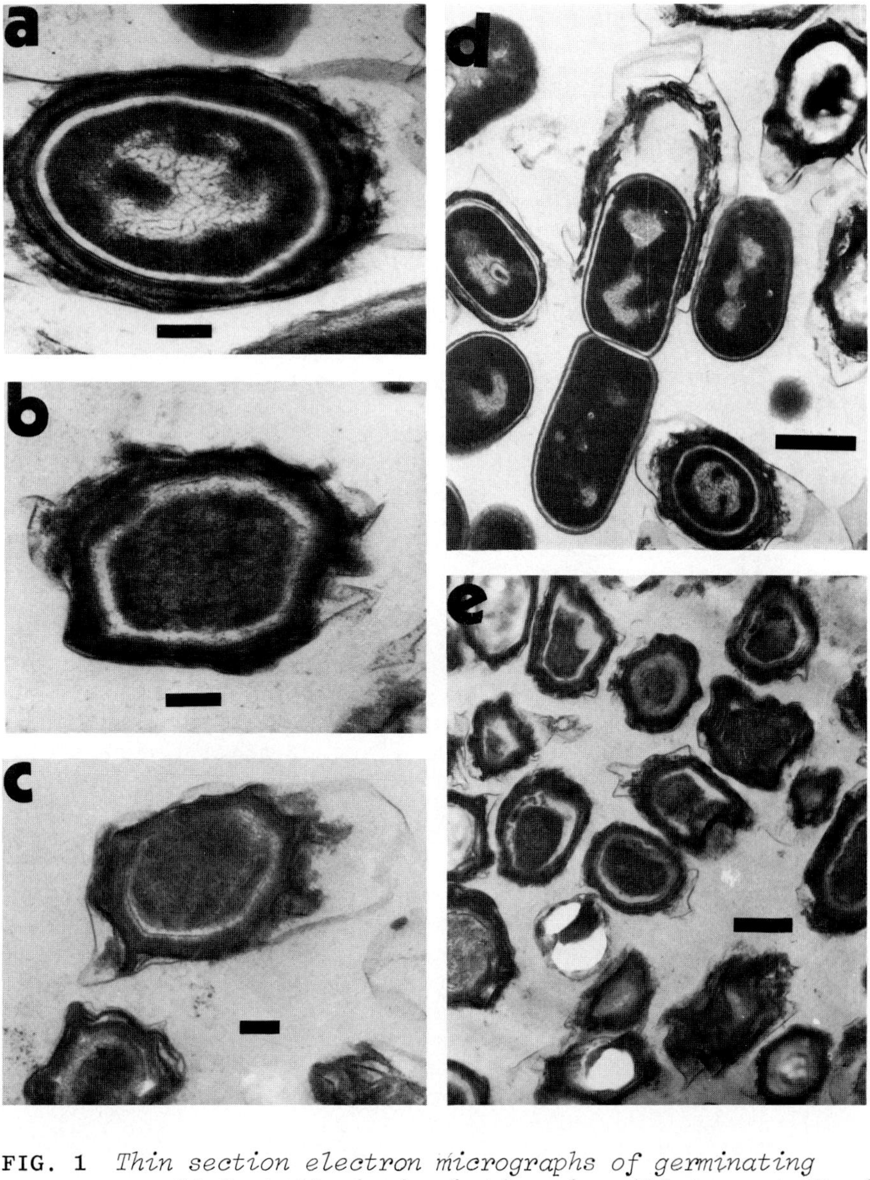

FIG. 1 *Thin section electron micrographs of germinating spores. a, B1-9 at 60 min incubation; b, wild-type at 60 min incubation; c, B1-9 at 60 min incubation in GS (10 μg/ml) supplemented medium; d, B1-9 at 110 min incubation; e, wild-type at 120 min incubation (E-1 showed similar characteristics to B1-9). Bar marker represents 0.25 μm (a,b,c) and 0.75 μm (d,e).*

MATERIALS

Actinomycin D, rifampicin, gramicidin S and PMSF were obtained from Sigma. *E. coli* RNA polymerase was purchased from Boehringer Corporation and all radioactive chemicals were obtained from the Radiochemical Centre, Amersham.

RESULTS AND DISCUSSION

The process of germination, as observed by darkening of spores under phase microscopy, is similar in wild-type and GS^- mutants (1,5,6,8). However, observations on thin sections of germinating spores using electron microscopy clearly indicated differences in the germination characteristics of wild-type and GS^- mutant spores (Fig. 1). Mutant spores (which outgrow normally without a population effect; 5,8) showed the normal sequence of germination events as described by Rousseau *et al.* (18). At 60 min of incubation stage III of germination (as described by electron microscopic studies; 18) was reached in all germinating spores with chromatin aggregation towards the centre of the spore core and complete breakdown of the cortex material (Fig. 1a). Sections taken at various time intervals of incubation showed that this aggregation of chromatin could be detected in some spores as early as 10 min into the germination period and the proportion of spores showing chromatic aggregation steadily increased in the population in parallel to darkening of spores under the phase microscope. At 80-120 min of incubation all of the spores germinated showed chromatic aggregation (Fig. 1d) and outgrowth of spores had commenced (Fig. 1d; centre). Not all spores, however, germinated as can be seen by the presence of the dormant spore (Fig. 1d; top right).

Wild-type spores, on the other hand, when incubated at population densities greater than 10^8 spores/ml failed to reach stage III of germination. There was no observed appearance of chromatin aggregation within the spore core at 60 min of germination (Fig. 1b) and on close examination some residual cortex material was observed in some spores although this was not always clearly defined. Even at 80-120 min of incubation (the normal timing of outgrowth in GS^- spores) no signs of chromatin aggregation could be observed and there was no sign of outgrowth (Fig. 1e). The spores appear to be blocked at or before stage II of germination as defined by Rousseau *et al.* (18). GS^- mutant spores to which GS had been added back at inhibitory concentrations behaved phenotypically like wild-type spores in that they appeared to be blocked at the stage of germination prior to chromatin aggregation (Fig. 1c). In contrast to this, wild-type spores incubated at population densities permissive for outgrowth (less than 10^8 spores/ml) showed characteristics

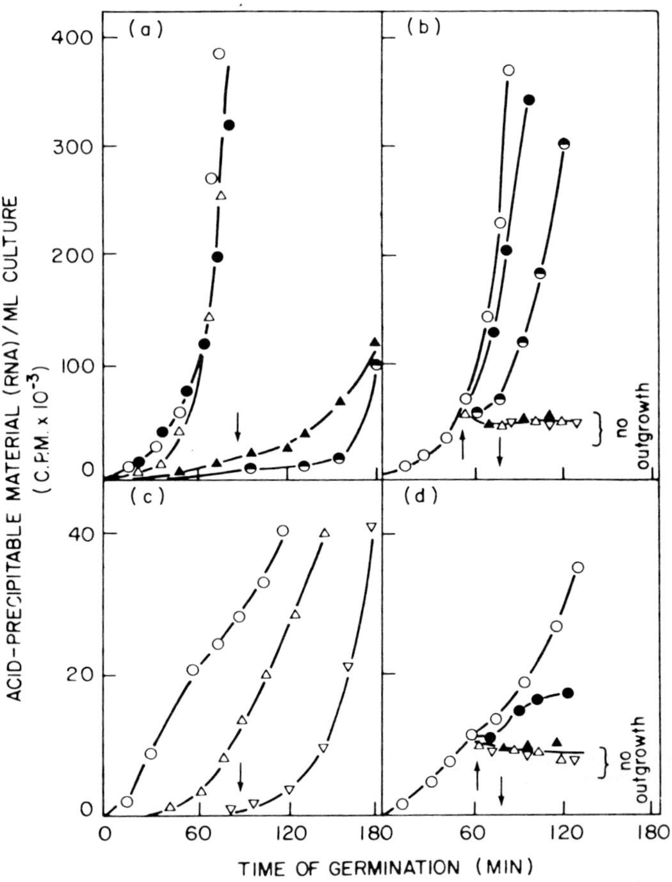

FIG. 2 *Incorporation of radioactive precursors into RNA during germination of E-1. Incubation conditions as described in Methods. ↓, time of commencement of outgrowth unless otherwise stated. ↑, addition of actinomycin D, rifampicin or GS. A: ^{14}C-uracil, ◐, 0.1 μCi/ml pulse labelling, 2 x 10^8 spores/ml; , Δ, o - continuous labelling 0.1, 0.2, 1.0 μCi/ml, 2 10^8 spores/ml; , continuous labelling 0.2 μCi/ml, 10^9 spores/ml. B: ^{14}C-uracil, 0.2 μCi/ml, 10^9 spores/ml. o, control; , , Δ - GS 1, 2 and 4-10 μg/ml; , actinomycin D (0.5 μg/ml); ∇, rifampicin (10 μg/ml). C: ^{3}H-uridine. o, Δ, ∇ - 10^9, 2 x 10^8, 10^7 spores/ml. D: ^{3}H-uridine, 10^9 spores/ml. o, control; , Δ - GS, 2 and 4-10 μg/ml; , actinomycin D (0.5 μg/ml); ∇, rifampicin (10 μg/ml). (Similar findings were made with B1-7).*

of germination similar to the GS^- mutant spores (Fig. 1, a and d) in that chromatic aggregation towards the centre of the spore core took place and this was followed by outgrowth at the normal time. It would appear that GS in some way prevents the normal conformational changes of chromatin material during germination that is a prerequisite for outgrowth.

Since in the GS^- mutant spores chromatin aggregation could be detected as early as 10 min into the germination sequence it was of interest to determine the earliest time during germination at which RNA synthesis could be detected. Using various combinations of precursor and spore concentrations, and incubation conditions, the earliest time at which we detected incorporation of RNA precursors into acid-precipitable material was 15 min (Fig. 2). Both ^{14}C-uracil and ^{3}H-uridine incorporation studies indicated this as the time of commencement of RNA synthesis. As shown in Fig. 2A, using ^{14}C-uracil, the level of radioactive precursor and the method of labelling is important to the amount of radioactive acid-precipitable material detected. Continuous labelling, which provides information about the synthesis of RNA over a period of time (rather than pulse labelling, which provides information about the rate of synthesis at a particular time) showed detectably higher levels of RNA synthesised at much earlier times. It could be that the low level of RNA synthesis occurring early on in germination is also complicated by problems of precursor entry into the germinating spore. These complications would account for the later time of commencement of RNA synthesis indicated in our previous preliminary study (9). The present study suggests that RNA synthesis commences soon after chromatin material has aggregated towards the centre of the spore core to form a nuclear body.

That the incorporation of precursors into acid-precipitable material is actually a measure of RNA synthesis is indicated by the finding that both actinomycin D and rifampicin inhibit this incorporation (Fig. 2, B and D). GS also blocks this incorporation in a like manner (Fig. 2, B and D) and wild-type spores (which contain GS) (5,8) germinated under conditions non-permissive for outgrowth (2×10^8-10^9 spores/ml) did not synthesise RNA under our experimental conditions. It should be noted that the use of GS levels (40 μg GS/10^{10} spores and less) much lower than those found in wild-type spores (5,8) did not completely inhibit RNA synthesis or outgrowth whereas GS concentration (40-100 μg GS/10^{10} spores) approaching endogenous wild-type spore levels did. It would appear that GS could inhibit RNA synthesis directly at the level of transcription.

To test whether or not CS inhibits directly the transcriptional apparatus, RNA polymerase activity was measured in toluene-permeabilised cells as described in the Methods section.

TABLE 1 *RNA polymerase activity and inhibition by peptide antibiotics. RNA polymerase from E-1 cells was purified through DEAE-cellulose and agarose and assayed as detailed in Methods. Herring Sperm DNA was obtained from the Boehringer Corporation. (Similar findings were obtained with the enzyme from wild-type Nagano cells.)*

Addition	Template	RNA Polymerase (amount)	% Inhibition
none	*B. brevis* DNA (6 μg)	9 μg	0
GS (6 μg)	DNA (6 μg)	9 μg	60
GS (20 μg)	DNA (6 μg)	9 μg	95
GS (20 μg)	DNA (24 μg)	9 μg	60
GS (20 μg)	DNA (6 μg)	27 μg	95
none	Herring Sperm DNA (20 μg)	9 μg	0
GS (20 μg)	Herring Sperm DNA (20 μg)	9 μg	89
tyrocidine (20 μg)	Herring Sperm DNA (20 μg)	9 μg	70
tyrothricin (20 μg)	Herring Sperm DNA (20 μg)	9 μg	9
linear gramicidin (20 μg)	Herring Sperm DNA (20 μg)	9 μg	-11

Substantial RNA polymerase activity could be monitored in germinating, outgrowing, growing and sporulating cells of *B. brevis* Nagano by this method. The activity was inhibited by both actinomycin D and rifampicin but was found not to be inhibited by GS. We do not fully understand, nor can we offer an explanation for, these findings. It is conceivable that the system was not sufficiently purified to allow GS to exert its effect and so further studies were carried out on purified *B. brevis* RNA polymerase and native DNA as template.

Table 1 shows that GS does, in fact, in a purified system, inhibit RNA polymerase activity. Indeed, contrary to previous studies (14), we found that, of the peptide antibiotics tested, GS was the most effective inhibitor. The ability to inhibit *in vitro* transcription decreased in the following

TABLE 2 *Complex formation between* B. brevis *Nagano DNA, peptide antibiotics and RNA polymerase. These studies were performed by the filter binding technique of Riggs* et al. *(17). 100 μg ^{3}H-DNA were incubated with the test substance in amounts indicated below (Final volume 1 ml). Complex formation was estimated by the % of the original activity retained on the filter.* B. brevis *Nagano RNA polymerase was the purified enzyme taken through DEAE-cellulose and agarose column chromatography.*

Addition	% DNA retained on filter
none	5 - 10
GS (10 μg)	65
GS (30 μg)	90
tyrocidine (10 μg)	35
tyrocidine (30 μg)	65
tyrothricin (10 μg)	20
tyrothricin (30 μg)	40
linear gramicidin (up to 50 μg)	5 - 10
B. brevis RNA polymerase (9 μg)	95

order GS > tyrocidine > tyrothricin. Linear gramicidin showed no or very little inhibitory activity. These findings were observed with RNA polymerase purified from the GS^{-} mutant E-1 and also wild-type cells. Increasing the concentration of native DNA in the system reversed the GS inhibition whereas increasing the concentration of RNA polymerase had no effect (Table 1). This implies interaction between GS and the DNA template and indeed filter binding assays (17) confirmed this interpretation (Table 2). GS at a concentration of 10 μg ml^{-1} led to retention of 60% of the DNA (100μ g ml^{-}) as a complex on the filter. Other peptide antibiotics were found to complex with *B. brevis* DNA with a lower efficiency (Table 2) analogous to their ability to inhibit *in vitro* transcription. Linear gramicidin showed no complex formation and is in agreement with reports from other laboratories (16,19). Our results are in contradiction to those of Marahiel *et al.* (4) who found neither inhibition of *in vitro* transcription nor GS-DNA complex formation in *B. brevis* ATCC 9999 and its GS^- mutants. These workers also found that their GS^- mutant spores had different levels of DPA and heat resistance than the wild-type strain (6). Our GS^- mutants show no such differences (5,8) and these findings have been confirmed by other workers (1). The discrepancy may reflect strain differences or the use of pleitropic mutants by Marahiel

From the results presented it does seem possible that the block on chromatin aggregation at stage II of germination could be caused by association of GS with the DNA. The hypothesis would be that this GS-DNA complex formation prevents transcription of the DNA by RNA polymerase which is necessary for subsequent outgrowth. Only when GS is lost from the spore can subsequent transcription, and hence outgrowth, take place. The results however are only correlative and we have previously presented a similar argument supporting the relationship between GS and respiration during germination (2). Both sets of data are not necessarily mutually exclusive since it is thought that *in vivo* respiration and the conformation and activity of DNA are both dependent on membrane organisation and it is known that HS interacts with bacterial membranes (20,21).

ACKNOWLEDGEMENTS

We thank Professor Y. Saito for supplying the organisms. This work was supported, in part, by a grant from the Science and Engineering Research Council.

REFERENCES

1. Piret, J.M. and Demain, A.L. (1981) in *Sporulation and Germination. Proc. 8th Internat. Spore Conf.* (Levinson, H.S., Sonenshein, A.L. and Tipper, D.J., eds.), pp.243-245, Am. Soc. Microbiol., Washington D.C.
2. Nandi, S., Lazaridis, I. and Seddon, B. (1981) *FEMS Microbiol Lett.* **10**, 71-75.
3. Egorov, N.S., Vyipiyach, A.N., Zharikova, G.G. and Maksimov, V.N. (1975) *Mikrobiologiya* **44**, 237-240.
4. Marahiel, M.A., Danders, W., Kraepelin, G. and Kleinkauf, H. (1982) in *Peptide Antibiotics Biosynthesis and Functions* (Kleinkauf, H. and von Döhren, H., eds.), pp.389-397, Walter de Gruyter, Berlin.
5. Lazaridis, I., Frangou-Lazaridis, M., MacCuish, F.C., Nandi, S. and Seddon, B. (1980) *FEMS Microbiol. Lett.* **7**, 229-232.

6. Marahiel, M.A., Danders, W., Krause, M. and Kleinkauf, H. (1979) *Eur. J. Biochem.* **99**, 49-55.
7. Shimura, K., Iwaki, M., Kanda, M., Hori, K., Kaju, E., Hasegawa, S. and Saito, Y. (1974) *Biochim. Biophys. Acta* **338**, 577-587.
8. Nandi, S. and Seddon, B. (1978) *Biochem. Soc. Trans.* **6**, 409-411.
9. Seddon, B. and Nandi, S. (1978) *Biochem. Soc. Trans.* **6**, 412-413.
10. Kellenberger, E., Ryter, A. and Sechaud, J. (1958) *J. Biophys. Biochem. Cytol.* **4**, 671-676.
11. Fisher, S., Rothstein, D. and Sonenshein, A.L. (1975) in *Spores VI* (Gerhardt, P., Costilow, R.N. and Sadoff, H.L., eds.), pp.226-230, Am. Soc. Microbiol., Washington D.C.
12. Rothstein, D.M., Keeler, C.L. and Sonenshein, A.L. (1976) in *RNA Polymerase* (Losick, R., and Chamberlin, M., eds.), pp.601-616, Cold Spring Harbor Laboratory.
13. Marmur, J. (1961) *J. Mol. Biol.* **3**, 208-218.
14. Sarkar, N. and Paulus, H. (1972) *Nature (London) New Biol.* **239**, 228-230.
15. Paulus, H. and Sarkar, N. (1976) in *Molecular Mechanisms in The Control of Gene Expression* (Nierlich, D.P., Rutter, W.J. and Fox, C.F., eds.), pp.117-194, Academic Press, New York.
16. Sarkar, N., Langley, D. and Paulus, H. (1977) *Proc. Natl. Acad. Sci. USA* **74**, 1478-1482.
17. Riggs, A.D., Suzuki, H. and Bourgeois, S. (1970) *J. Mol. Biol.* **48**, 67-83.
18. Rousseau, M., Flechon, J. and Hermier, J.C. (1966) *Ann. Inst. Past (Paris)* **111**, 149-160.
19. Ristow, H., Schazschneider, B., Bauer, K. and Kleinkauf, H. (1975) *Biochim. Biophys. Acta* **390**, 246-252.
20. Ostrovskii, D.N., Bulgarova, V.G., Zhukova, I.G., Kaprel'yants, A.S., Rozantzev, E.G. and Simakova, I.M. (1976) *Biokhimiya* **41**, 175-182.
21. Kaprel'yants, A.S., Nikiforov, V.V., Mironikov, A.I., Snezkova, L.G., Eremin, V.A. and Ostrovskii, D.N. (1977) *Biokhimiya* **42**, 329-337.

METAL ION CONTENT OF *STREPTOMYCES* SPORES: HIGH CALCIUM CONTENT AS A FEATURE OF *STREPTOMYCES*

J.A. SALAS, J.A. GUIJARRO and C. HARDISSON

Departamento de Microbiologia, Universidad de Oviedo, Oviedo, Spain

SUMMARY

The metal ion content of spores of five *Streptomyces* species was studied. A general feature of this study was the finding of a very high calcium content (1.2 - 2.1% of the dry weight). Spore calcium was located in the integuments fraction and more than 95% of the calcium was removed from intact spores by 10 mM EGTA. Several divalent cations (Mg^{2+}, Mn^{2+}, Zn^{2+} or Fe^{2+}) induced darkening of spores, loss of heat-resistance and release of calcium. Darkening of spores was blocked by metabolic inhibitors, whereas calcium excretion was not affected. Therefore, two different stages in the initiation of germination of *Streptomyces* spores may be differentiated. The first stage is energy-independent and is characterised by the release of calcium from spores. The second stage is energy-dependent and is accompanied by a decrease in optical density and loss of refractility and heat-resistance of spores.

INTRODUCTION

Germination of *Streptomyces* spores is initiated by divalent cations in an energy-requiring process (1,2). However, it is not yet known how divalent ions interact with the spores and are involved in the subsequent release of dormancy. In *S. viridochromogenes*, it has been suggested that calcium ions initiate spore germination by acting externally to the cytoplasmic membrane (2). Here, we have studied the metal ion

FUNDAMENTAL AND APPLIED ASPECTS OF BACTERIAL SPORES

content of the spores of various *Streptomyces* species and report that a high calcium content is a general feature of the spores examined. Furthermore, our data indicate that most of the spore calcium is located outside the membrane and is rapidly released into the medium as one of the earliest events in the germination of *Streptomyces antibioticus* spores.

MATERIALS AND METHODS

Microorganisms and Culture Conditions

Several *Streptomyces* species were used in this work: *S. antibioticus* ATCC 11891, *S. griseus* ATCC 11429, *S. scabies* CMI 99049, *S. viridochromogenes* ATCC 14290 and *S. aureofaciens* ETH 13387. Conditions for sporulation, and for harvesting and germination of spores were as previously described (1,3).

Metal Ion Measurements

The metal ion content of spores was determined by atomic absorption spectrophotometry with a Perkin Elmer model 372 atomic absorption spectrophotometer. Analytical wavelengths (nm) were: Ca, 422.7; Mg, 285.2; Na, 589; K, 767.5; Fe, 248.3; Cu, 324.7; Zn, 213.9; Co, 240.7. Plasticware was used throughout these experiments after removal of ion contamination by successive washings in Radiacwash containing EDTA for 12 h, in 1N HCl for 12 h, and finally extensive washings in double distilled water. For measurements of the metal ion content, spores were dried at 160°C for 5 h and then ashed at 500°C for 6 h. The ash was dissolved in 100 µl of concentrated HCl and diluted to 0.1N HCl. Lanthanum chloride (1%, w/v) was added to the samples prior to analysis for calcium to free the analysis from interference by phosphates.

Labelling of Spore Calcium during Sporulation

After 7-9 days at 28°C, spores were harvested from GAE solid medium plates supplemented with 0.1 µCi $^{45}CaCl_2$ per ml. The radioactivity present in spore suspensions was measured, showing values between 3000 and 6000 c.p.m. per 10^8 spores. Here, we will refer to these as ^{45}Ca-spores.

Calcium Release Studies

Determinations of calcium release from spores were made by measuring the ^{45}Ca in the supernatant after incubation of ^{45}Ca-spores under different conditions. Spore suspensions (5 ml; 3.10^8 spores/ml; 9000-18000 c.p.m./ml) were incubated in distilled water at 35°C in a water bath with shaking. The experiment was initiated by adding small volumes (10-100 μl) of different compounds and, at subsequent times, samples (500 μl) were removed into centrifuge tubes. The samples were centrifuged at 12000 x g for 30 s in an Eppendorf minifuge and 400 μl of the supernatant was counted for radioactivity. Total radioactivity initially present in the spores was measured by counting samples of the spore suspension taken immediately before starting the experiment. Values for radioactivity in the supernatants of the sample at zero time were negligible (less than 3% of the total) and were always discounted from the time samples.

Heat Resistance Studies

The loss of heat resistance during germination of spores was determined as follows: spores were incubated at 35°C in various media and, at different time intervals, samples were withdrawn and treated at 55°C for 30 min. After cooling in ice water, the samples were diluted in distilled water and spores surviving the treatment enumerated by viable counts.

TABLE 1 *Metal ion content of* Streptomyces *spores*[a]

	% Dry weight						
	Na	K	Mg	Ca	Fe	Zn	Co or Cu
S. aureofaciens	0.06	0.88	0.27	1.21	0.05	0.03	<0.01
S. griseus	0.02	1.06	0.20	1.10	0.01	0.04	<0.01
S. scabies	0.02	1.21	0.13	1.36	0.05	0.09	<0.01
S. antibioticus	0.16	0.80	0.40	2.10	0.05	0.09	<0.01
S. viridochromogenes	0.04	0.92	0.54	1.52	0.04	0.06	<0.01

[a]The metal ion content was determined by atomic absorption spectrophotometry as described in Material and Methods. Each value is the average of 3 independent determinations. All the values obtained were in a range ± 10% of the average.

RESULTS

Metal Ion Content of Spores and Changes during Germination

Spores of different *Streptomyces* species showed a high potassium content: 0.80 - 1.21% of the dry weight, and a very high calcium content: 1.10 - 2.10% of the dry weight (Table 1). Lower levels of Na, Mg, Fe and Zn were found whereas the content of Co and Cu was negligible (less than 0.01% of the dry weight).

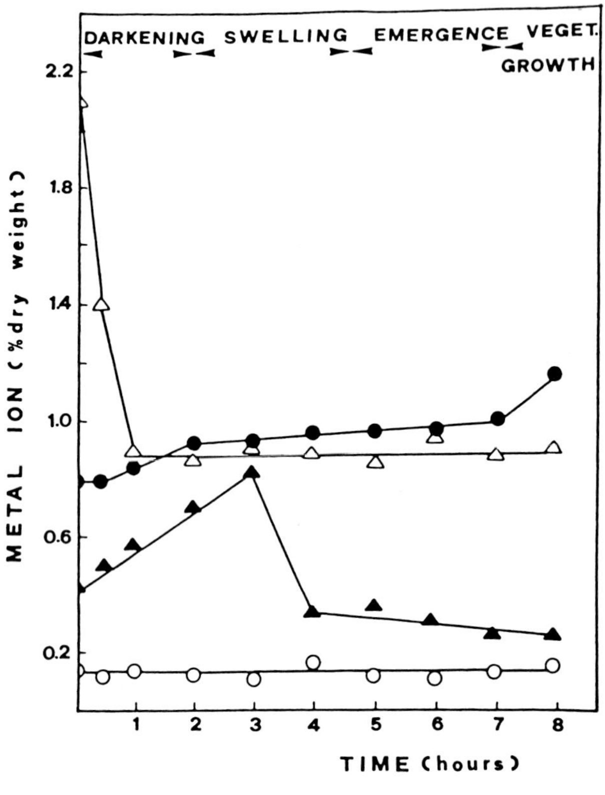

FIG. 1 *Changes in metal ion content of* Streptomyces antibioticus *spores during germination. Spores were incubated at 35°C in a minimal synthetic medium containing the 4 metal ions studied and, at different times, samples removed and washed 4 times in distilled water. The metal ion content of spores was then analysed by atomic absorption spectrotometry. Each value is the average of three independent determinations. (o) Na, (•) K, (Δ) Ca and (▲) Mg.*

We also studied the content of four metal ions (Ca, Mg, Na and K) during germination of *S. antibioticus* spores in a minimal synthetic medium containing the four ions cited (Fig. 1). The Na content remained constant during the germination process, while the K content increased slightly. However, the Mg level of spores progressively increased during the darkening process and at the beginning of spore swelling. Later, coinciding with the presence of the highest proportion of swollen spores, the Mg level decreased to values slightly lower than those found in the dormant spores and remained constant throughout germ tube emergence. The calcium content of spores decreased by about 55-65% in the first hour of germination and then remained constant during the remainder of the germination process.

Changes in Spore Calcium Content as Influenced by Environmental Conditions

The finding of a very high calcium content in dormant spores of several *Streptomyces* species prompted us to study the influence of different conditions on the calcium content and the relative binding state of the ion to spore components. The calcium content of dormant spores increased when calcium was added to the sporulation medium. *S. antibioticus* spores harvested from GAE solid medium plates supplemented with 0.2 mM and 1 mM $CaCl_2$, had calcium levels of about 2.3% and 3.1% of the dry weight respectively. The calcium content remained constant during extensive washings of the spores in distilled water. Treatment of spores with 10 mM EGTA removed more than 95% of the calcium. However, calcium deficient spores (EGTA-treated spores) did not recover their initial calcium content upon incubation in 40 mM $CaCl_2$. Furthermore, this newly bound calcium was removed simply by washing the spores with distilled water. These experiments suggest either that a permeability barrier makes the calcium binding sites in the spore inaccessible to externally added calcium or that removal of calcium by EGTA modifies, in some way, the calcium binding sites and thus prevents the reinsertion of exogenously added calcium. In addition, these experiments prove that the calcium content of spores cannot be increased by incubation with calcium once the spores are mature.

Location of Calcium in Spores

In order to locate calcium in dormant spores, ^{45}Ca-spores were broken and the integuments fraction sedimented by centrifugation. Nearly all the radioactive calcium (80-85%) was found to be associated with the integuments, and 95% of this

^{45}Ca remained firmly bound during repeated washings with distilled water. It could be argued that disintegration of the spores might have liberated calcium from its original site (not in the integuments) and that redistribution of the element then occurred with consequent binding to the integuments. However, the following experiments allowed us to rule out this possibility.

(1) Non-radioactive spores were broken before and after removal of calcium with EGTA. Then, ^{45}Ca was added and incubation carried out at room temperature for 30 min. The integument fraction of both EGTA-treated and non-treated spores were then obtained and the radioactivity associated with this fraction determined. Under these conditions, assuming that the

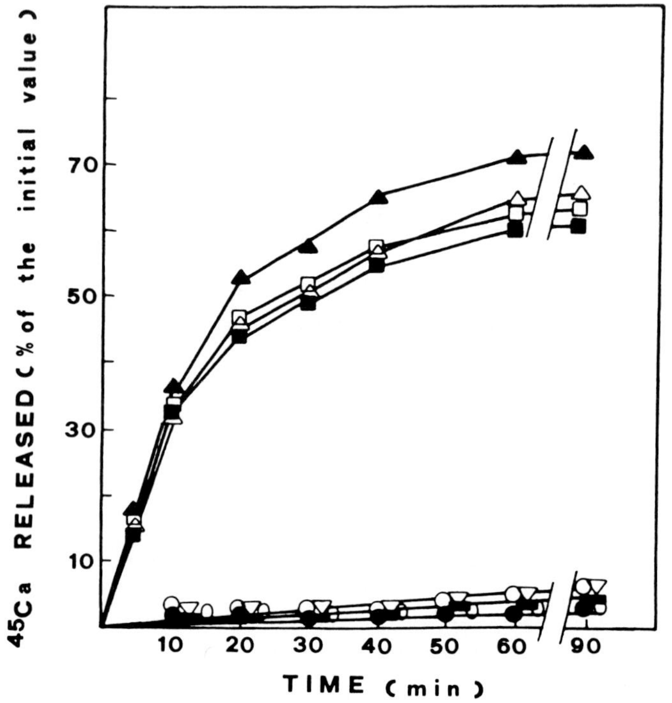

FIG. 2 *Calcium release during germination of* Streptomyces antibioticus *spores.* ^{45}Ca*-spores were incubated at 35°C in several nutritional conditions. At different times of incubation, samples (0.5 ml) were withdrawn and centrifuged at 12000* x *g for 30 s. Radioactivity present in 400 μl samples of the supernatant were then counted. (o) Ca ions. (●) Distilled water. (Δ) Zn or Fe ions. (▲) Minimal synthetic medium. (□) Mg ions. (■) Mn ions. (▬) Na or K ions. (∇) Ba ions. (⊙) Li ions.*

results cited above depended upon redistribution of calcium following spore breakage, more radioactive calcium should have been recovered bound to the integuments of EGTA-treated spores than of non-treated spores. However, this prediction was not borne out by the results. The amounts of calcium bound to both preparations of integuments were almost identical (32 and 39% of the total added calcium for EGTA-treated and non-treated spores, respectively). Furthermore, in both cases, 80% of the calcium was removed simply by washing the integument fraction with distilled water.

(2) ^{45}Ca-spores were broken in the presence and in the absence of an excess (10 mM) of non-labelled calcium and the integument fraction isolated. If, during spore breakage, redistribution of ^{45}Ca were to occur, this would be accompanied by dilution of the ion in the presence of excess calcium with a consequent reduction in the level of ^{45}Ca recovered in the integument fraction. In fact (data not given), similar levels of ^{45}Ca were found in the integuments under both sets of conditions. These experiments strongly suggest that calcium in *S. antibioticus* spores is firmly bound to some structural component(s) of the integuments, and therefore that the redistribution hypothesis may be discarded.

Calcium Release during Germination

Spore germination in a minimal synthetic medium was accompanied by a progressive loss of ^{45}Ca (Fig. 2). This release started within 5 min of the addition of spores to the germination medium and proceeded at a constant rate during the first 10 min of germination (about 35% of the radioactive calcium was excreted at this time). Then, during the following 50 min, the rate of calcium release diminished so that a maximum of about 65-75% of the calcium was lost after 60 min. A similar pattern of calcium release was found upon incubation of ^{45}Ca spores in the presence of divalent cations (Mg, Fe, Mn or Zn) (Fig. 2). In contrast, incubation of ^{45}Ca-spores in distilled water, glucose, asparagine, glucose plus asparagine, monovalent cations (Na, K or Li) or some divalent cations (Ca or Ba) did not stimulate excretion of significant amounts of calcium. In other experiments (data not given), the release of calcium from ^{45}Ca spores during germination was not affected by the presence of different metabolic inhibitors such as sodium azide, atabrine, 2,4-dinitrophenol, CCCP, DCCD or potassium cyanide (all at 1 mM final concentration).

Loss of Heat Resistance During Spore Germination

Dormant spores of *S. antibioticus* were resistant to heat-shock

TABLE 2 *Relationship between calcium release and loss of heat-resistance in* Streptomyces antibioticus *spores*

Additions	% ^{45}Ca released	% loss of viability
None (Control)	2	0
1% glucose and/or 0.2% asparagine	3	0
1 mM Li, Na, K, Ca or Ba ions	2-5	1-5
1 mM Mg ions	63	70
1 mM Mn ions	60	64
1 mM Zn ions	64	68
1 mM Fe ions	65	72
Minimal synthetic medium	70	73

Ca spores were incubated at 35°C in distilled water with the additions mentioned in the table. After 90 min of incubation, samples were removed for determination of radioactive calcium and assays of heat-resistance (55°C for 30 min) as described in Materials and Methods.

for 30 min at 55°C. However, most of the spores lost their viability when submitted to this treatment after 90 min of incubation in the minimal synthetic medium, by which time 60-70% of the spores were phase dark. Accordingly, we used the loss of heat resistance as a parameter to measure germination. When the heat resistance of spores was followed under the nutritional conditions employed in the studies of calcium release, a clear correlation was observed between the latter effect and the appearance of heat sensitive spores (Table 2). It was also observed (data not given) that the response of EGTA-treated spores to the heat treatment was similar to that of non-treated spores.

DISCUSSION

Descriptions of the metal ion composition of *Streptomyces* spores are scarce. In *S. viridochromogenes*, the spore content of Mg, Ca and K ions was reported to be 0.17, 0.28 and 2.0% of the dry weight, respectively (2). In this communication, we present a more complete study of the content of eight metal ions in the spores of five *Streptomyces* species. Each of the ions was present in amounts typical of those usually found in bacterial endospores (4). Significantly, a very high calcium content (1.10 - 2.10% of the dry weight) was found in the spores of all five species. This level is comparable only with that encountered in bacterial endospores since prokaryotic organisms tend to maintain low cytoplasmic levels of calcium. In *S. antibioticus* spores, this divalent cation seems to be bound to some structural component(s) of the integuments. Thus, the location of calcium in *S. antibioticus* spores is clearly different from that in bacterial endospores which is located in the core (5). Accumulation of calcium in *Streptomyces* spores seems to occur preferentially, although not exclusively, during the sporulation process (data not shown), although we do not know to which component(s) of the spore integuments it binds.

Many potential anionic sites in the walls of gram-positive bacteria are available to sequester cations. These include the phosphodiester groups of teichoic acid, the free carboxyl groups of peptidoglycan and the sugar hydroxyl groups of both wall polymers. Few studies have been made of the *Streptomyces* spore wall, hence its composition and structure are not well known. Accordingly, we cannot speculate concerning the component(s) which bind calcium in *Streptomyces* spores, except to note the likely presence of peptidoglycan and teichoic acid and the fact that aspartic acid is present in the spore wall in quantity (6). Moreover, regardless of its binding site, no function can be assigned to *Streptomyces* spore calcium at this time. Certainly, since EGTA-treated spores were as resistant to heat as were non-treated spores, any role for calcium in such resistance cannot be proposed.

A release of 65-75% of the spore calcium was observed during germination. Incubation of spores in the presence of some divalent cations also induced excretion of calcium and simultaneous loss of spore heat resistance. Furthermore, darkening of spores occurred during the period in which calcium was excreted and heat sensitive spores appeared (in the presence of either Mg, Fe, Mn or Zn ions). Consequently, it could be suggested that one of the earliest events occurring during initiation of germination of *S. antibioticus* spores is the release of calcium from spore integuments into the medium. This would be a consequence of the interaction of the initiator cation and some spore receptor(s). Darkening of spores is an energy-requiring process (1,2), but calcium release was not affected by different metabolic inhibitors. In the presence of these agents, neither changes in optical density nor loss of spore refractility was observed (1). Therefore, the use of these inhibitors permits us to differentiate two different stages in the darkening process. The first stage is energy-independent and is characterised by the release of calcium from the spores. During the second stage, which requires energy, the spores lose refractility and heat resistance and this is accompanied by a decrease in their optical density. However, the mere removal of calcium by EGTA is not sufficient to cause the spores to proceed to the second stage, since they remain dormant as monitored by their refractility and heat resistance.

In *S. viridochromogenes*, it has been proposed that the site of calcium-initiated spore germination is external to the cytoplasmic membrane (2). We do not know where the calcium-receptor is in *S. antibioticus* spores. This remains to be elucidated together with other points such as the nature of the receptor and the type of interaction involved. Nor is it clear how this early event initiates spore germination and how it is connected to the other changes which subsequently occur in the spore. Further experiments now in progress in our laboratory will focus their attention upon these questions.

ACKNOWLEDGEMENTS

This work was supported, in part, by a grant from the Comisión Asesora par el Desarrollo de la Investigación Cientifica y Técnica, Spain.

REFERENCES

1. Hardisson, C., Manzanal, M.B., Salas, J.A. and Suarez, J.E. (1978) *J. Gen. Microbiol.* **105**, 203-214.
2. Eaton, D. and Ensign, J.C. (1980) *J. Bacteriol.* **143**, 377-382.
3. Hardisson, C., Salas, J.A., Guijarro, J.A. and Suarez, J.E. (1980) *FEMS Microbiology Letters* **7**, 233-235.
4. Murrell, W.G. (1969) in *The Bacterial Spore* (gould, G.W. and Hurst, A., eds.), pp.215-273. Academic Press, New York.
5. Johnstone, K., Ellar, D.J. and Appleton, T.C. (1980) *FEMS Microbiology Letters* **7**, 97-101.
6. De Jong, P.J. and McCoy, E. (1966) *Can. J. Microbiol.* **12**, 985-994.

Applied Aspects of Spores

TOXIGENIC SPORE FORMING BACTERIA

P.D. WALKER

Bacteriology R & D, Wellcome Research Laboratoties, Langley Court, Beckenham, Kent, UK

SUMMARY

In a number of toxigenic spore-forming bacteria it can be demonstrated that components of the bacterial spore or metabolites produced as part of the sporulation process are toxic for certain species of animals and man.

The bacterial spore provides a means by which pathogenic organisms can survive in unfavourable environments until such time as conditions are more favourable when germination, bacterial multiplication and toxin production occur.

INTRODUCTION

It would be inappropriate in the context of this symposium to present an exhaustive review of toxigenic spore-forming bacteria. Instead, this paper will concentrate on a number of examples of toxigenic spore-forming bacteria which illustrate the relationship between spore formation and toxin production and the bacterial spore as an essential component of the pathogenic process.

The family, Bacillaceae, is divisible into two genera; the genus *Bacillus* and the genus *Clostridium*. While the genus *Bacillus* has a small number of frank human pathogens, e.g. *B. anthracis*, the cause of anthrax in man and animals and to a lesser extent *Bacillus cereus* which has been implicated in human food poisoning, the genus *Clostridium* contains a large number of species pathogenic for man and animals. As far as disease is concerned, it is probably one of the most economically important groups.

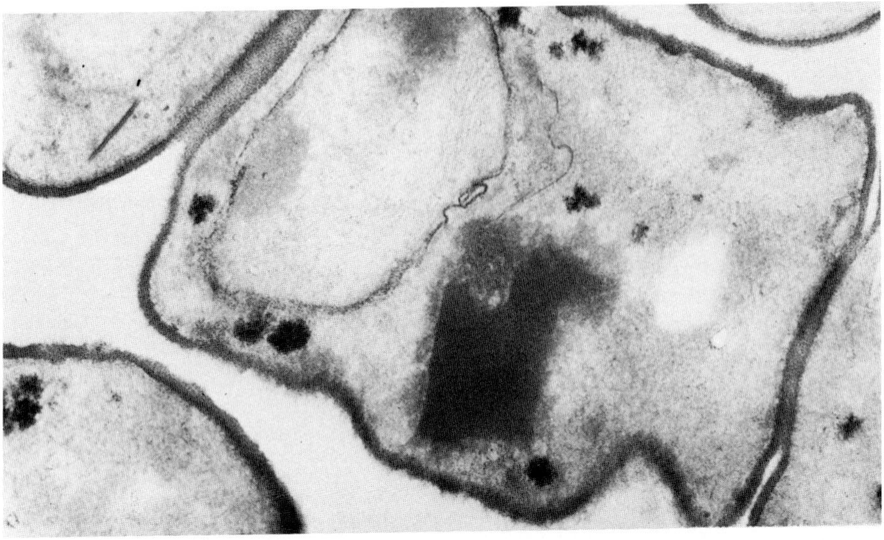

FIG. 1 *Ultrathin section of sporulating cell of* Bacillus thuringiensis *showing developing crystal inclusion associated with the exosporium.* x 82,500

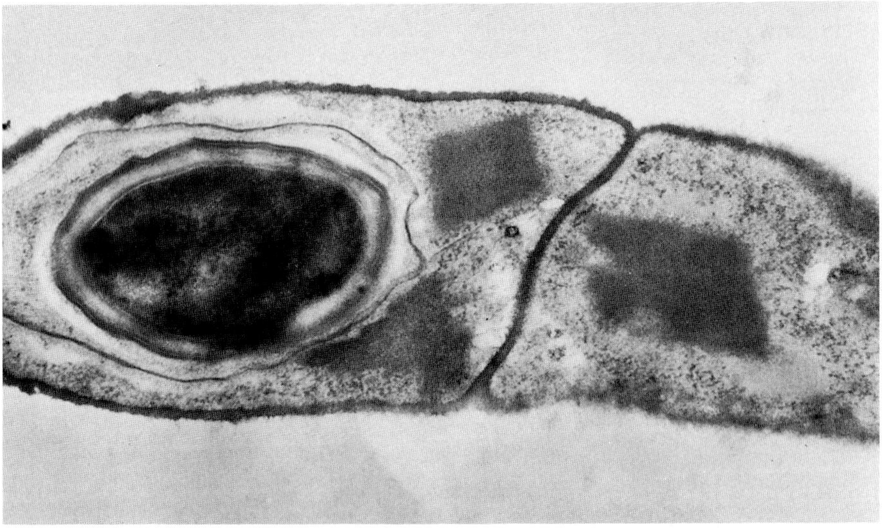

FIG. 2 *Ultrathin section of mature spore of* Bacillus thuringiensis *showing crystal inclusions within the cytoplasm.* x 61,500

RELATIONSHIP BETWEEN SPORE FORMATION AND TOXIN PRODUCTION

The relationship between spore formation and toxin production is well characterised in two species, *Bacillus thuringiensis*, an organism pathogenic for lepidopterous larvae (1,2) and *Clostridium perfringens* type A, one of the causative agents of cases of human food poisoning (3).

Sporulation in *Bacillus thuringiensis* is accompanied by the formation of a crystalline inclusion body (Figs. 1 and 2) associated with the developing exosporium (4-6). Upon maturation of the spore, spores and crystals are released from the disintegrating bacilli and can be separated by physical procedures. It can be demonstrated that there is a structural similarity between the developing inclusion and layers of the bacterial spore coat (7). Preparations of sporulating cells subjected to freeze etching reveal a structural similarity between the crystal protein and layers of the spore coat (Fig. 3). The antigenic similarity between these structures can be demonstrated by staining mixtures of spores and crystals and partially disintegrated spores with ferritin-labelled antibody to crystal protein. The crystals are stained with ferritin-labelled antibody to the crystal protein but the outer surface of the whole spore is unstained (Fig. 4). On the other hand, following partial disintegration of the spore, it can be seen that ferritin labelled antibody to crystal protein stains a layer of the spore coat (Fig. 5). It has been postulated that an excess of spore coat component material produced during sporulation escapes through the pits which can be demonstrated on the surface of the exosporium of *Bacillus thuringiensis* spores (Fig. 6) and become assembled on a template associated with the membrane (7).

A somewhat different phenomenon is observed during sporulation of *Clostridium perfringens* type A. As early as forespore formation sections of sporulating cells of *Cl. perfringens* type A stained with an antiserum to purified enterotoxin using the PAP technique (8) show accumulation of enterotoxin in vacuoles in the cytoplasm of the cell (Fig. 7). These are also observed initially in the cytoplasm of the developing forespore and continue to accumulate in the cytoplasm of the sporulating cell as the spore matures (9). However, unlike *Bacillus thuringiensis* no staining of any of the spore structures is observed (Fig. 8). Crystalline structures similar to those seen in *Bacillus thuringiensis* have been observed in sporulating cells of *Cl. perfringens* type A (10) but these were not observed in the strain used for these experiments. Unlike the case of *Bacillus thuringiensis* where the crystal protein appears to be a component of the spore itself, the enterotoxin of *Cl. perfringens* type A appears to be a sporulation associated product.

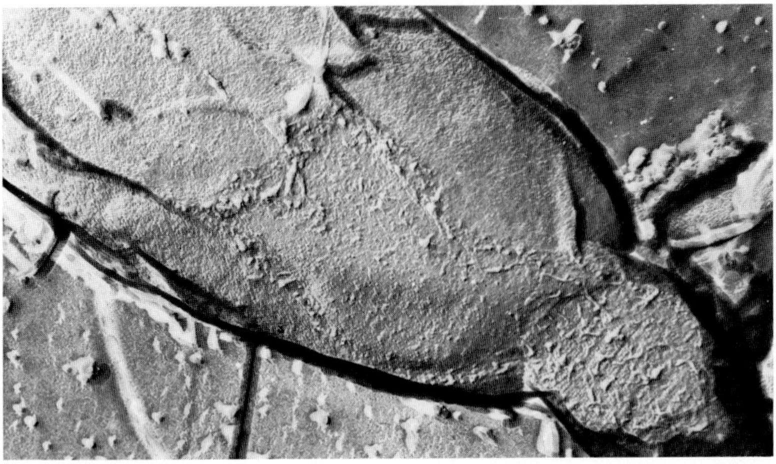

FIG. 3 *Freeze etch preparation of sporulating cell of* Bacillus thuringiensis *showing cell and crystal (bottom right). Note the structural similarity between the crystal and the inner spore coat.* x 67,100

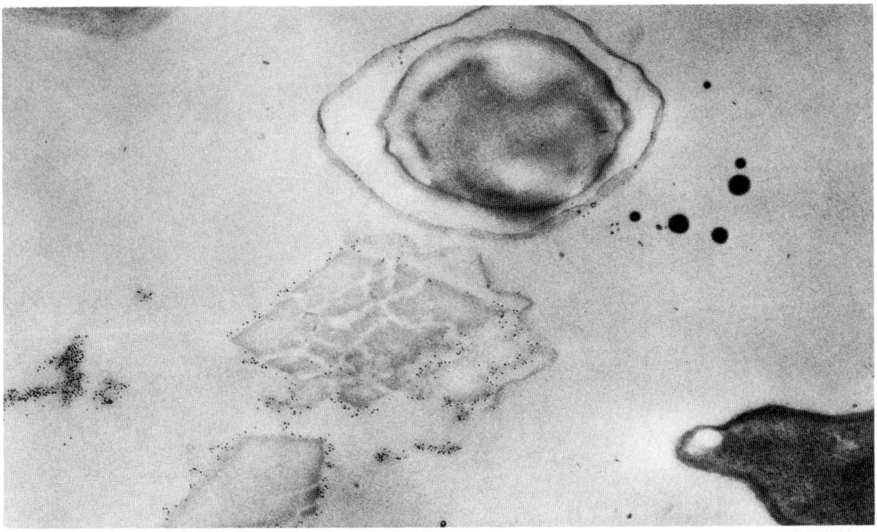

FIG. 4 *Ultrathin section of mixture of spore and crystal of* Bacillus thuringiensis *stained with ferritin-labelled anti-serum to the crystalline protein. Note deposits of ferritin on the crystal but not on the spore.* x 45,750

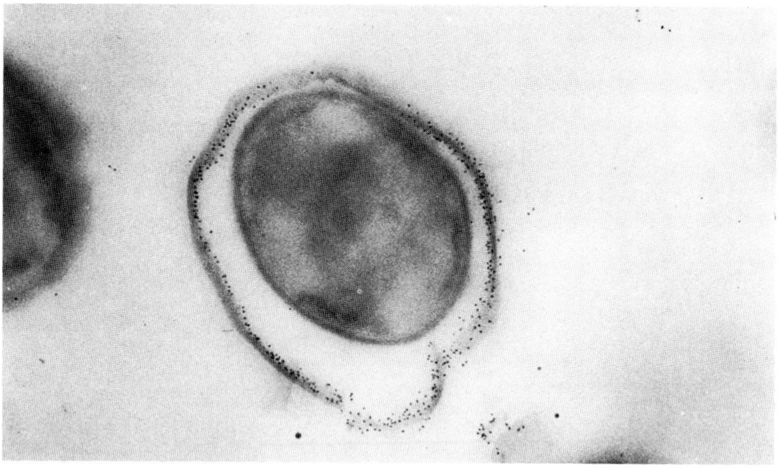

FIG. 5 *Ultrathin section of partially disintegrating spore of* Bacillus thuringiensis *stained with ferritin-labelled antibody to the crystal protein. Note staining of the inner spore coat with ferritin particles.* x 82,500

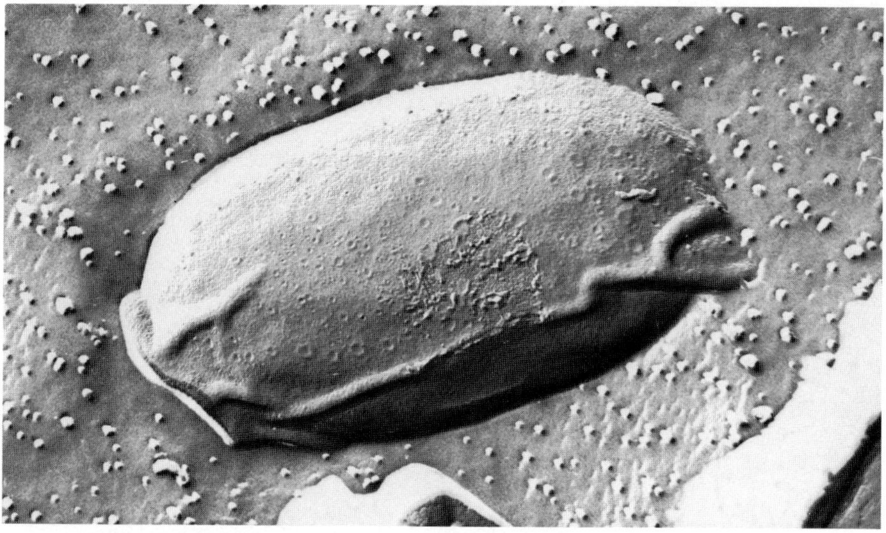

FIG. 6 *Freeze etch preparation of spore* Bacillus thuringiensis. *Note pits on the surface of the exosporium.* x 52,500

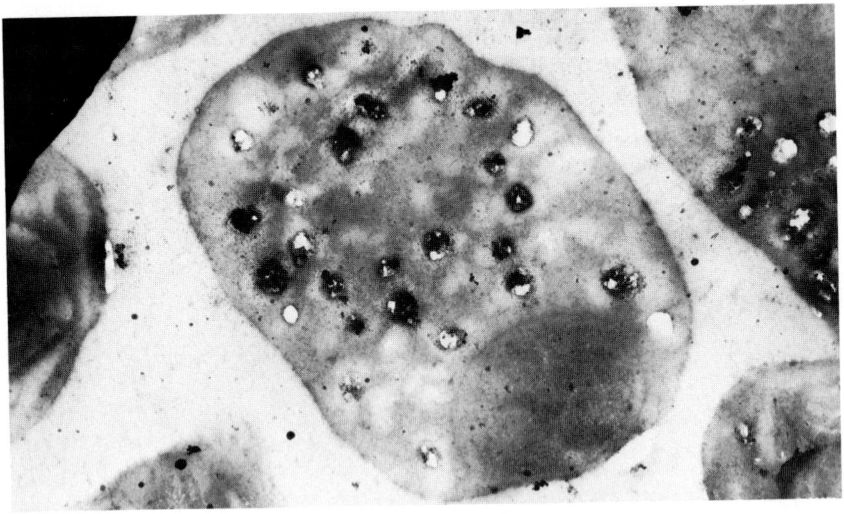

FIG. 7 *Ultrathin section of sporulating cell of* Cl. perfringens *type A stained with unlabelled antiserum to purified enterotoxin and peroxidase antiperoxidase complex. Note staining of cytoplasmic vacuoles.* x 18,750

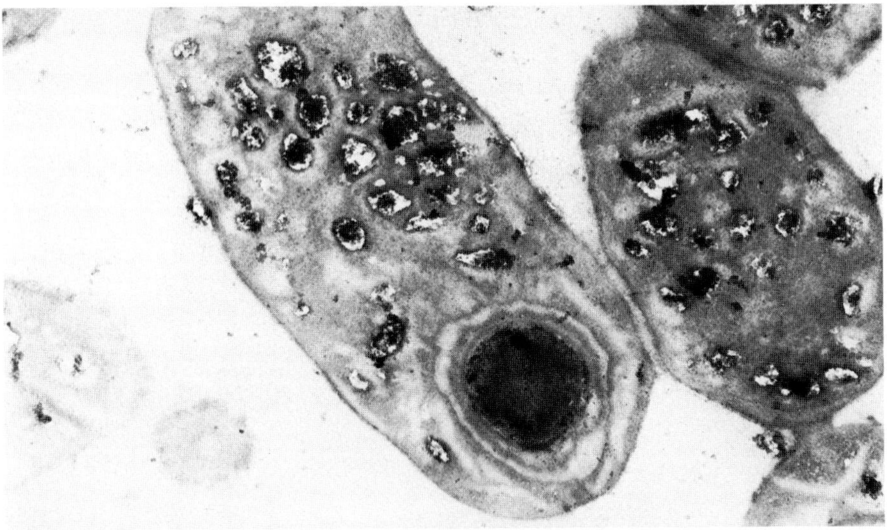

FIG. 8 *Ultrathin section of sporulating cell of* Cl. perfringens *type A with a mature spore stained with unlabelled antiserum to purified enterotoxin and peroxidase antiperoxidase complex. Note staining of cytoplasmic vacuoles.* x 18,750

THE ROLE OF SPORES IN PATHOGENICITY

The bacterial spore provides a means by which pathogenic organisms can survive in unfavourable environments but once conditions are favourable can germinate resulting in bacterial multiplication and toxin production.

Wound Infections

Two wound infections commonly occurring in humans are infections due to *Cl. tetani* and *Cl. perfringens/Cl. oedematiens* type A. Contamination of superficial wounds with spores of *Cl. tetani* can lead to limited bacterial multiplication and toxin production which, in view of the extreme toxicity of tetanus toxin leads to clinical cases of tetanus in man and animals. On the other hand, gas gangrene produced by strains of *Cl. perfringens* type A and *Cl. oedematiens* type A usually involves contamination of deep wounds such as occurs during severe injuries following accidents or war wounds. Experiments carried out in sheep have shown conclusively that spores of *C. oedematiens* type A impregnated on cloth in the immediate vicinity of the leg are sucked into wounds caused by high velocity missiles and that these subsequently germinate resulting in bacterial multiplication and death (11).

Enteric Disease

Cl. perfringens. Although species of *Cl. perfringens* frequently cause gas gangrene and food poisoning, it can be seen from Table 1 that many of the diseases associated with this organism are enterotoxaemias in man and animals. More recently enterotoxaemias due to *Cl. difficile* have also been described (12,13).

Enterotoxaemias can occur in either the neonatal or adult animal. In the neonatal animal infection takes place before the development of a normal flora with the result that the pathogen multiplies free of constraint. On the other hand, in the adult animal the presence of a resident bacterial flora confers significant protection on the host and it is only when changes occur in the flora and the normal controlling mechanisms are interfered with that multiplication of pathogens introduced either externally or present as part of the normal flora can occur.

In the case of *Cl. perfringens* type C infection of neonatal calves, lambs and piglets, a high level of contamination in the environment leads to colonisation of the small intestine. The organism adheres to the intestinal villi and produces necrotic toxins (Figs. 9 and 10) which destroy the villi

TABLE 1 *Diseases caused by* Clostridium perfringens *(welchii)*

Type A	(1)	Gas Gangrene, man and animals Intestinal commensal, man and animals Putrefactive processes, soil etc. (United States)
	(2)	Food poisoning (Britain)
Type B	(1)	Lamb Dysentery Enterotoxaemia, foals (Britain)
Type C	(1)	Enterotoxaemia ("struck") sheep (England)
	(2)	Enterotoxaemia, calves, lambs, (England)
	(3)	Entertoxaemia, piglets (Britain)
	(4)	Necrotic enteritis, man (formerly type F) (Germany)
	(5)	Necrotic enteritis, man (Papua)
Type D		Enterotoxaemia, sheep, lambs, goats, bovines and possibly man (Australia)
Type E		Sheep and cattle, pathogenicity doubtful Britain

surface (14). In the adult animal such enterotoxaemias usually follow changes in diet which result in the creation of a new environment in which pathogens can multiply free of normal controlling mechanisms.

In the case of *Cl. perfringens* type D infection, the consumption by sheep of large quantities of grass in the Spring leads to high carbohydrate levels and intestinal stasis results from gas production by organisms present in the gut (15). Under these circumstances the normal peristaltic movement is interfered with and pathogens can attach to the intestinal villi and multiply in a similar manner to the neonatal animal.

In humans a similar disease follows a change from a simple vegetable diet of sweet potato to pork (16,17). Intermittent ingestion of large quantities of pork by the young adult population in Papua New Guinea (Fig. 11) follows pig feasting which occurs on various ceremonial occasions. Pigs are butchered

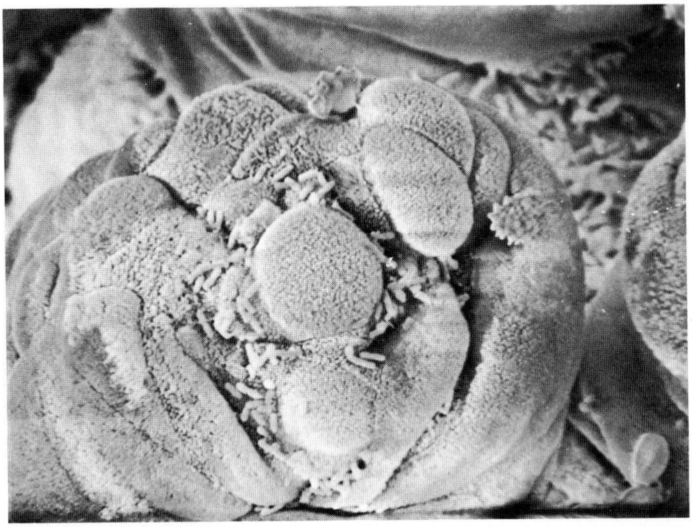

FIG. 9 *Scanning electron micrograph of villus of a piglet 4 h after infection with* Cl. perfringens *type C. Note adherence of the organisms to the villus surface.* x 2,500

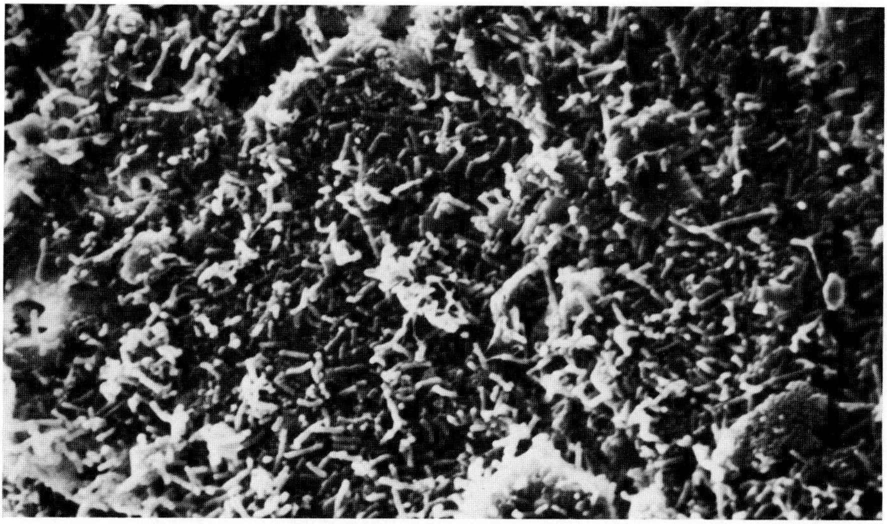

FIG. 10 *Scanning electron micrograph of piglet dying from piglet enteritis caused by* Cl. perfringens *type C. Note complete destruction of intestinal villi.* x 1,280

FIG. 11 *Young adult girl from the native population in the highlands of Papua New Guinea.*

FIG. 12 *Killing of pigs at ceremonial pig feast.*

FIG. 13 *Pig meat butchered ready for insertion into the cooking oven seen in the background.*

FIG. 14 *Cooking oven comprising heating stones prior to loading.*

FIG. 15 *Loaded oven filled with meat and earth. The long hollow tube is for insertion of water to give a "pressure cooker" effect.*

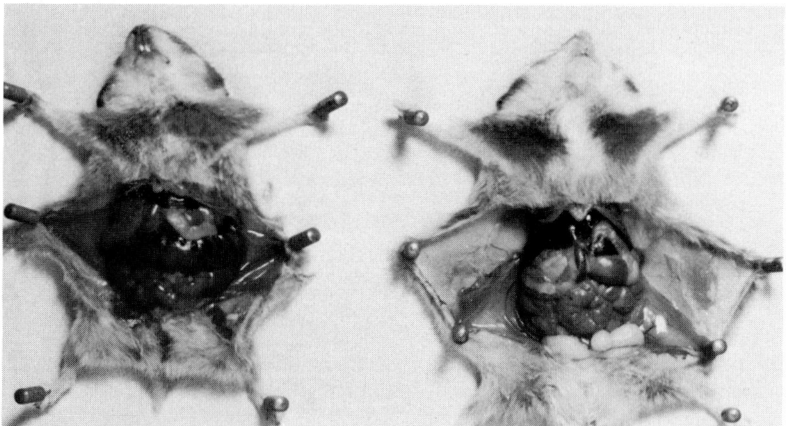

FIG. 16 *Post mortem appearance of the intestines of a hamster infected with* Cl. difficile *following Clindomycin treatment (left) compared to a normal hamster (right).*

(Fig. 12) and the meat cooked in ovens dug in the ground (Figs. 13-15). Samples of soil from the ovens show the presence of *Cl. perfringens* type C organisms. As a result of ingestion of pork a rich dietary medium is provided in the small intestine and resident *Cl. perfringens* type C organisms or organisms introduced with the meat can multiply rapidly. The resulting production of toxins causes interference with normal peristalsis and attachment of the organism and destruction of the tissues occurs. Trypsin levels in the population would normally be sufficient to destroy the toxins produced. However, trypsin levels in the intestine of the native population are low due to consumption of a low protein diet and trypsin inhibitors present in sweet potato, the staple diet, and also secreted by ascaris worms with which infestation is common (18,19).

Cl. difficile. More recently enterocolitis due to *Cl. difficile* has been observed in groups of patients following antibiotic treatment with antibiotics such as Lincomycin and Clindomycin which interfere with the normal anaerobic flora of the inestine. Under these circumstances, *Cl. difficile* can multiply free of normal intestinal constraints producing toxins causing enterocolitis. The effects of the toxins can be seen in hamsters in which the disease has been induced by Clindomycin treatment and dosing with *Cl. difficile* (20,21). Figures 16 and 17 show the effect on the caecum of normal and

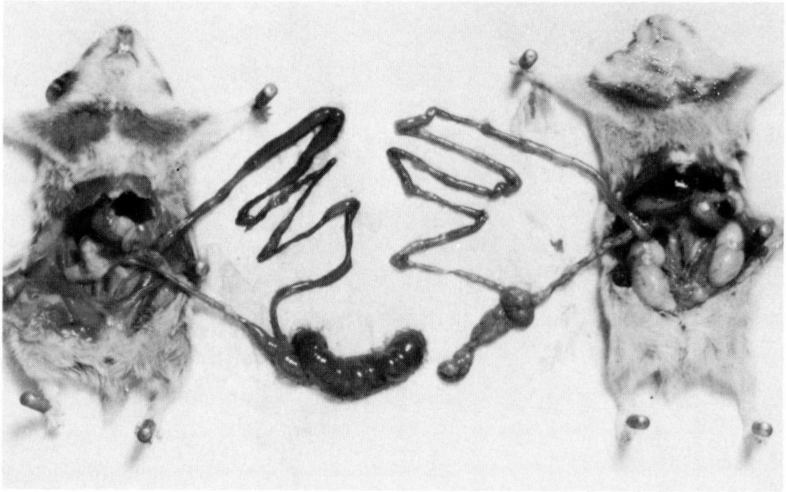

FIG. 17 *Comparison of the caecum of a hamster infected with* Cl. difficile *following Clindomycin treatment (left) compared to a normal hamster (right).*

FIG. 18 *Scanning electromicrograph of the caecal surface of a normal hamster.* x 2,500

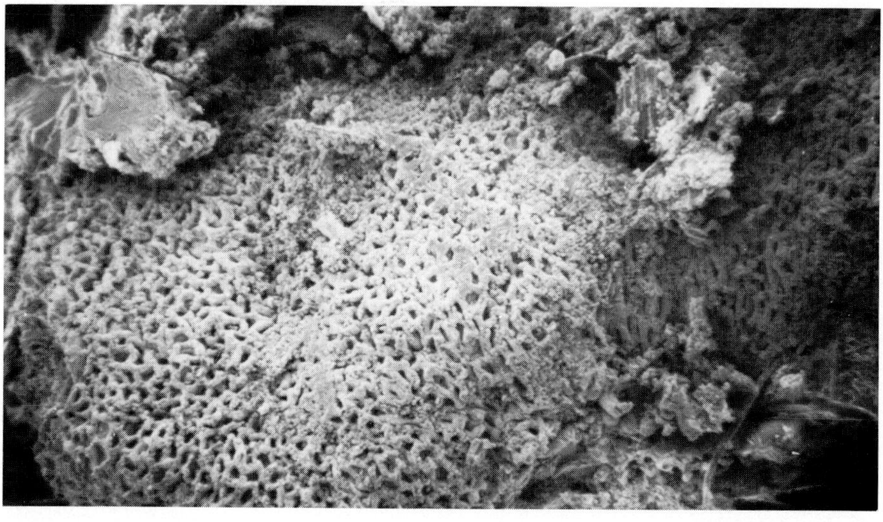

FIG. 19 *Scanning electron micrograph of the appearance of the caecal surface of a hamster infected with* Cl. difficile. x 2,500

infected hamsters. The denuding of the epithelium of the caecum by the toxins is shown by scanning electron micrographs of the tissue (Figs. 18 and 19).

Clostridium botulinum. Botulism is a neuroparalytic disease affecting both man and animals and is the most dangerous form of bacterial food poisoning. It is caused by ingestion of pre-formed toxin in food. The foods most seriously implicated are home-bottled vegetables, meat, fish, home-cured salted and smoked ham and various raw, smoked and fermented fish products. Seven distinct types of *Cl. botulinum* are recognised, of which A, B, E and F are responsible for botulism in man. All types produce toxin in food and with some non-proteolytic strains of type B, E and F this can occur at temperatures as low as 5 C (22,23).

After absorption into the circulatory system the toxin acts by preventing the pre-synaptic release of acetyl choline which results in the failure of nerve impulses to be transmitted across nerve fibre junctions.

More recently the ingestion of *Cl. botulinum* spores and subsequent intra-intestinal growth with production of toxin *in vivo* has been recognised as a form of botulism occurring in young infants. Symptoms include lethargy and reduced suck reflexes. As descending paralysis becomes generalised the baby is "floppy", particularly noted by lack of head control. Flaccidity of the airway may cause respiratory arrest. The most common types of *Cl. botulinum* associated with these cases are A and B, although type F has recently been implicated in New Mexico. *Cl. botulinum* spores have been isolated from products fed to infants prior to illness (24).

Clostridium septicum. C. septicum is responsible for an infection of the lamb's stomach known as "braxy". In countries where frosts may be severe and prolonged the disease is thought to develop through the ingestion of grass from frozen pastures, the frozen grass resulting in irritation of the stomach lining and infection with *Cl. septicum*. The animal dies from a profound toxaemia and bacteraemia. The organism is also responsible for a disease of sheep in Britain known as "bloody guts". This is typified by the presence of large quantities of blood in the abomasum (25). Large numbers of organisms can be seen in the haemolysed fluid using fluorescent staining and sections of the stomach wall show the presence of invading organisms.

Clostridium oedematiens type B. Although not strictly an intestinal pathogen, *Cl. oedematiens* type B causes Black Disease, an infection of the liver in sheep. The disease in sheep is associated with liver fluke. Eggs from mature flukes are shed in the faeces onto the pasture where they undergo various maturation phases including a stage in the garden snail before again being consumed by sheep. Final maturation takes place in the liver causing tracts of dead tissue. Spores of *Cl. oedematiens* present in these areas germinate under such conditions, the organisms producing lethal toxins causing death of the sheep. Sudden death in cattle and pigs associated with *Cl. oedematiens* may be precipitated by causes other than fluke. Many of the animals affected were on intensive feeding systems and possibly local poisoning of the liver may be the "triggering factor" (16).

In summary, the bacterial spore, in addition to being an agent whereby the organism can survive in adverse circumstances and is able to germinate and reproduce when the circumstances become favourable, also has components which are toxic for certain species. These are either components of the spore itself or metabolites arising from spore-related events which take place. It is these factors which make the spore forming bacteria important in the pathogenic context.

ACKNOWLEDGEMENTS

Figures 11 to 15 are reproduced with the kind permission of Professor T.G.C. Murrell and Dr G. Lawrence, and Figures 16-19 by Dr D. Fernie.

REFERENCES

1. Hannay, C.L. and Fitz-James, P.C. (1955) *Can. J. Microbiol.* **1**, 694-710.
2. Angus, T.A. (1956) *Can. J. Microbiol.* **2**, 416-425.
3. Hobbs, B.C., Smith, M.G., Oakley, C.L., Warrack, G.H. and Cruickshank, J.C. (1953) *J. Hyg. (Cam)* **51**, 75-101.
4. Fitz-James, P.C. (1962) *Proc. V Int. Cong. Elect. Microscop.* **2**, RR10.
5. Somerville, H.J. and James, C.R. (1970) *J. Bact.* **102**, 580-583.
6. Somerville, H.J. (1971) *Europ. J. Biochem.* **18**, 226-237.
7. Short, J.A., Walker, P.D., Thomson, R.O. and Somerville, H.J. (1974) *J. Gen. Microbiol.* **84**, 261-276.
8. Sternberger, L.A., Hardy, P.H. Jun., Cuculis, J.J. and Meyer, H.G. (1970) *J. Histochem. Cytochem.* **18**, 315-333.
9. Roper, G., Short, J.A. and Walker, P.D. (1976) in *Spores 1976* (Gould, G.W., Dring, G.J., Ellar, D., Wolf, J. and Barker, A.N., eds.), pp.279-296, Academic Press, London.
10. Duncan, C.L., King, G.J. and Frieben, W.R. (1973) *J. Bact.* **114**, 845-859.

11. Boyd, N.A., Walker, P.D. and Thomson, R.O. (1972) *J. Med. Microbiol.* **5**, 459-465.
12. Larson, H.E. (1979) *J. Infect.* **1**, 221-222.
13. George, R.H., Symonds, J.M., Dimock, F., Brown, J.D., Arabi, Y., Shinagawa, N., Keighley, M.R.B., Alexander Williams, J. and Burdon, J.W. (1978) *Brit. Med. J.* **1**, 695.
14. Arbuckle, J.R. (1972) *J. Path.* **106**, 65-72.
15. Bullen, J.J. and Scarisbrick, R. (1957) *J. Path. Bact.* **73**, 495-509.
16. Murrell, T.C.G., Egerton, J.R., Rampling, A., Samuels, J. and Walker, P.D. (1966) *J. Hyg. (Camb.)* **64**, 375-396.
17. Murrell, T.C.G., Egerton, J.R., Roth, L., Samuels, J. and Walker, P.D. (1966) *Lancet* **1**, 217-222.
18. Lawrence, G. and Walker, P.D. (1976) *Lancet* **1**, 125.
19. Lawrence, G. and Cooke, R. (1981) *Brit. J. Exptl. Path.* **61**, 261-271.
20. Allo, M., Silva, J., Fekety, R., Rifkin, G.D. and Waskin, H. (1978) *Gastroenterol.* **76**, 351-355.
21. Bartlett, J.G., Onderdonk, A.B., Asneros, R.L. and Kaspar, D.L. (1977) *J. Infect. Dis.* **136**, 701-705.
22. Ball, A.P. and Farrell, I.D. (1979) *J. Infect.* **1**, 121-125.
23. Gilbert, R.J. and Willis, A.T. (1980) *Comm. Med.* **2**, 25-27.
24. Arnon, S.S. (1980) *Ann. Rev. Med.* **31**, 541-560.
25. Sterne, M. and Batty, I. (1975) *Pathogenic Clostridia*, Butterworth, London and Boston.
26. Batty, I., Buntain, D. and Walker, P.D. (1964) *Vet. Rec.* **76**, 115-117.

MODIFICATION OF RESISTANCE AND DORMANCY

G.W. GOULD

*Unilever Research Laboratory, Colworth House,
Sharnbrook, Bedford MK44 1LQ, U.K.*

SUMMARY

The resistance of endospores can be raised or lowered over many orders of magnitude by a variety of treatments. For example: by exposure to extreme pH values; cation exchange treatment; oxidising and reducing agents; surfactants; heavy metals; ionizing radiation; ionizing and non-ionizing radiation in the presence of "sensitizers"; raised hydrostatic pressure; raised osmotic pressure; lowered equilibrium relative humidity or water activity. Some of these treatments alter the intrinsic resistance of spores, whilst others damage their germination mechanisms so as to induce extreme dormancy and bring about an apparent reduction in resistance whilst leaving most protoplast components undamaged and the spore thus potentially viable. So far, only a few of these procedures have been utilized for the development of improved practical spore control procedures.

INTRODUCTION

For many years the resistance and dormancy of bacterial endospores were considered to be relatively immutable properties. Almost all attempts to discover additives or procedures that might usefully reduce the heat resistance of spores, for example screening of the effects of about 600 substances by Michener *et al.* (1) in 1959, were disappointing. More recently, however, there have been reported increasing numbers of procedures that substantially alter the resistance, the apparent resistance, or the dormancy of spores, and this new information is important for two major reasons.

TABLE 1 *Modification of resistance and dormancy*

Treatments	Major effects	Possible mechanisms of action
Low pH	'Reversible' reduction in heat resistance. 'Reversible' increase in dormancy.	Slow protonation of something (peptidoglycan?) Inactivation of part of germination mechanism.
Cation exchange	Reversible recovery of lost heat resistance.	Replacement of H^+ at some site.
Chelates	Reduction in heat resistance via the initiation of germination.	Rise in free Ca^{++} concentration in the protoplast.
Oxidising and reducing agents	Increase in sensitivity to enzymes and to heat in the presence of multivalent cations.	Increased leakiness of coat.
High pH (with reducing agents)	Increase in sensitivity to a number of chemicals, multivalent cations and enzymes. Increase in dormancy.	Increased leakiness of coat. Removal of lytic enzyme.
Surfactants	Reduction in heat resistance via lysis during outgrowth unless osmotically stabilized.	Interference with membrane stability.
Alkylamines	Initiation of germination.	Ca^{++} replacement.
Lysozyme	Lysis, or initiation of germination, of leaky-coated spores.	Hydrolysis of peptidoglycan (in all spores)

Ionizing radiation.	Reduction in resistance to heat, low a_w and low pH	Hydrolysis or decarboxylation of spore polymers.
Ionising radiation plus "sensitizers".	Reduction in apparent radiation resistance via prevention of germination.	Inactivation of part of germination mechanism by free radicals.
Ultraviolet radiation plus H_2O_2	Strong synergism.	Inactivation by (hydroxyl?) free radicals.
Heavy metals	Increase in sensitivity of metal-permeant spores to H_2O_2.	Intracellular generation of hydroxyl free radicals via Fenton Reaction?
Ultrasonication	Reduction in heat resistance	Structural change?
Hydrostatic pressure.	Reduction in heat and radiation resistance via initiation of germination.	Electrostriction, disturbance of Donnan equilibrium. Hydration of protoplast.
Inert gases	Reversible induction of dormancy.	Interference with membrane function; "narcosis".
Raised osmotic pressure; lowered a_w or ERH.	Increase in heat resistance.	Removal of water from protoplast.

Firstly, the new information may help us to learn more about mechanisms of resistance and dormancy in general. Secondly, the ability to confidently alter the apparent resistance of spores could lead to valuable improved practical methods for the control of spores in industry, for instance in disinfection, sterilization and preservation.

This paper therefore reviews some of the most studied factors that influence resistance and dormancy, and comments on these two aspects: how the information helps our understanding of mechanisms and how we might use the information to move towards more practically useful spore control procedures. The major factors and comments considered are summarized in Table 1.

TREATMENTS MODIFYING RESISTANCE

Low pH Values and Cation Exchanges

Heating spores at low pH values, or pretreating them at low pH, may decrease their dormancy (2) but decreases their heat resistance and, provided that the pH shift is not too great, it does so reversibly in that, on raising the pH, resistance is partially or completely restored at a rate that depends on the nature of the ions present. The studies of Alderton *et al.* have clearly shown that this phenomenon influences dry as well as wet heat resistance and occurs with all spores so far studied (3,4,5) including the toxinogenic food poisoning clostridia (6,7).

Following the early work of Alderton proposals were made for the use of the phenomenon in preservation procedures that would reduce the heat treatment necessary for the sterilization, for instance in foods, by a preliminary acidification followed by neutralization prior to (reduced) thermal processing.

These acid treatments removed a fraction of up to 50% or so of the spores' calcium and other cations (8), however, Marquis *et al.* (9) described experiments in which spores of a particular strain of *Bacillus megaterium* lost essentially all of their cations in acid and yet remained viable.

There are two distinct possible mechanisms that have been suggested to explain the observed heat sensitization. First, the protonation of some structural components of the spore, for instance cortex peptidoglycan, reducing its contractile (10), expansive (3,11), osmotic (12), reverse-osmotic (13) or otherwise pressure exerting functions (14,15) so as to allow the enclosed protoplast to partly hydrate or increase in water activity, and thereby lose resistance. Second, the low pH, or the consequent removal of cations, could have a more specific effect, as indicated by the observation (16) that spores of

Clostridium perfringens which had been acid-treated, then heated to inactivate them, could be recovered if lysozyme was added to the medium used to count them. Evidently the acid plus heat treatment only inactivated the germination mechanism rather than denaturing the whole of the protoplast contents, and lysozyme, by directly hydrolysing peptidoglycan in the cortex, "by-passed" the inactivated system.

Re-equilibration with cations reimposes resistance on acid-treated spores, and the effectiveness of calcium in this role is most often quoted, however, with regard to possible mechanisms it is important to remember that a variety of salts of monovalent and divalent cations will restore resistance with various degrees of effectiveness (4).

Treatment with Chelates

Calcium dipicolinate (or DPA) has long been known to reduce spore resistance by initiating germination (17), and possible mechanisms of action have been proposed (see 18) based on structural effects and on interference with, and mobilization of, endogenous calcium dipicolinate.

The possibility that a normal function of CaDPA within the spore is to act as a calcium buffer, maintaining a constant, but low, intracellular level of soluble calcium (19) has raised the possibility that exogenous CaDPA, entering the spore as the soluble ion pair may, whilst it remains in solution, initiate germination by raising, temporarily, the intracellular soluble calcium level. Calcium can activate some spores (see Douthit and Preston, this volume), and the fact that Ca^{2+}, with the correct adjuncts, initiated germination of spores of *Cl. perfringens* led Ando (20) to suggest that in many forms of germination an increase in membrane permeability to, or transport of, Ca^{2+} may be a key event. If this is so it would bring dormancy control mechanisms in spores much closer to the better-understood mechanisms that involve triggering *via* change in intracellular free Ca^{2+} typical of eukaryotic cells (43,44,45,46).

Oxidising and Reducing Agents

A number of oxidising and reducing agents will increase the sensitivity of spores to enzymes like lysozyme, to heating in the presence of multivalent inorganic and organic cations and to some chemical agents, but without influencing resistance to heat directly. With fair certainty the effects result from breakage of disulphide bonds in protein components of the spore coat and consequent increases in coat permeability. Evidently a normal function of the coat is therefore to protect the underlying structures from hydrolysis by adventitious enzymes and perhaps also from interference by multivalent cations and, by inference, maintenance of the normal cation status of the spore is important.

From the practical viewpoint, unfortunately, procedures discovered so far that make coats leaky and increase the sensitivity of spores, whilst valuable in laboratory studies, have been too severe to allow the development of usefully improved practical spore control procedures.

High pH

The effectiveness of reducing agents on spores is greatly potentiated by some concurrent treatments like incubation in high concentrations of urea or in surfactants like sodium dodecyl sulphate, and by some sequential treatments including treatment at high pH which removes alkali-soluble coat proteins and thus further increases coat permeability. If severe, such alkali treatments will reduce apparent viability, possibly by removing lytic enzymes that operate during germination (21).

Surfactants

A number of surface-active chemicals decrease the apparent heat resistance of spores in the sense that heated spores are much more surfactant sensitive than unheated ones (22). The same is true for some other chemicals, sodium chloride and nitrite having received most attention because of their importance in food preservation (23). It has been reasonably suggested that the surfactants act by interfering with membrane stability (24), and the only anti-spore antibiotic with a food preservation use, nisin, may have a similar site of action.

The surfactant alkyl primary amines reduce resistance in a different way, by actually initiating germination. Their mode of action has not been elucidated, but it has been suggested (25) that they act by replacing calcium from some site within the spore. Germination may then follow the unblocking of the site or, analogously with the possible action of calcium dipicolinate discussed above, by causing a temporary rise in the concentration of intracellular Ca^{2+}.

Lysozyme

Lysozyme is capable of reducing the resistance of all spores so far studied, once their coats have been permeabilized, whatever the lysozyme-sensitivity of the vegetative form, because peptidoglycan in the outer part of the cortex seems to be always lysozyme-sensitive. The mode of action is, of course, well known and the enzyme is very effective. Practically, the phenomenon of lysozyme-initiated germination has been important in indicating that some supposedly relatively heat-sensitive spores, like *Cl. botulinum* type E, are really much more resistant than realized hitherto, and it is just their germination mechanism that is inactivated by heat, and the inactivated mechanism is then by-passable by the use of lysozyme (26).

Ionizing Radiation

Application of ionizing radiation for food preservation is likely to steadily grow now that legislative and toxicological barriers are falling. It is well documented that ionizing radiation sensitizes spores to heat (but not vice versa), it has been suggested through the chain scission, or decarboxylation of peptidoglycan and other spore polymers (27,28). Sensitization may well involve a partial rehydration of the spore protoplast, since suspension of sensitized spores in solutions of high osmolarity reimposes resistance on them (29).

Radiation plus Sensitizers

When so-called sensitizers are present, the effectiveness of ionizing irradiation against spores increases greatly. It was originally thought that such sensitizers as iodoacetamide acted on vegetative cells by inhibiting the repair of radiation-induced damage to DNA. However, it was later shown that radiation plus iodoacetamide gave rise to short-lived free radicals and it was these that killed the cells (30). The same is true for spores. Irradiation plus iodoacetamide, iodate, iodide, or bromide, for example, kill by a completely different mechanism to irradiation alone. This is most obvious from an examination of the stage of growth affected. Spores irradiated alone, then placed in a suitable medium, initiate germination normally, or even faster than usual, then develop as far as the stage of outgrowth at which replication of DNA normally commences. In contrast, spores irradiated plus sensitizers are inhibited much earlier and are unable to even initiate germination.

Regarding the possibilities of practical use of sensitizers, the major problem is that the free radicals that react so quickly with spores also react quickly with almost everything else so that in a complex substrate the synergism is greatly suppressed.

Ultraviolet Irradiation plus Hydrogen Peroxide

The strong synergism of ultraviolet irradiation with hydrogen peroxide in killing spores (32) probably results from formation of hydroxyl radicals, which are about the most reactive and damaging of the oxygen-derived radicals. Indeed, .OH, generated from H_2O_2 + Cu^{2+} or H_2O_2 + Fe^{2+} by the Fenton reaction will almost completely dissolve spores (33) and sensitize them to heat (34).

Practically, the ultraviolet - H_2O_2 synergism is of real potential for the decontamination of surfaces and packaging materials.

Metals

The increased effectiveness of H_2O_2 on spores in the presence of metal ions like Cu^{2+} has been shown to correlate with permeation of the metal into the spore protoplast and may thus represent site-directed generation of the toxic .OH species (34).

Probably many more practical uses could be made of hydroxyl radicals, generated by other means.

Ultrasonication

Whilst ultrasonication will readily remove exosporia from those spores that have one, spores are otherwise very resistant indeed. However, ultrasonication can evidently decrease their heat resistance by mechanisms so far unknown, but presumably mediated by some sort of structural change (Sanz *et al.* this volume). The effect is interesting from the practical point of view, because combinations of purely physical processes are particularly attractive, if sufficiently effective, since they may require no use of additives or even changes in formulation, for example, of foods.

High Pressure

High hydrostatic pressure sensitizes spores to heat and to ionizing radiation. In the mid-pressure range it causes the initiation of germination and very big "kills" can be obtained (35,36). Its mode of action has been suggested to be via electrostriction, causing an increase in the ionization of spore components and disturbing a Donnan equilibrium in such a way as to allow the protoplast to hydrate (36,37).

The impediment to the wider use of high pressure as a spore control procedure has been that spores of some species are very much more resistant to pressure than others, and the pressures needed to sufficiently kill mixed populations of spores, for instance in foods, are therefore too high to be practical.

Inert Gases

Under pressure, inert gases prevent the initiation of germination (or induce dormancy) reversibly, in the same order that they induce narcosis, so that it has been reasonably suggested that their mechanism of action operates at the level of a spore membrane, which must therefore be important in the normal germination reaction (38).

Raised Osmotic Pressure

Most of the procedures summarized above bring about a reduction in resistance, either directly or through the imposition or breaking of dormancy. However, the most well-known treatments that alter heat resistance are those that raise it by causing a loss of water, for instance through drying, equilibration at low equilibrium relative humidities or at low water activities in solutions of high osmolality containing solutes that do not easily permeate the spore (39,40).

Such increases in resistance are characteristic for cells of most types: vegetative and spore, prokaryote and eukaryote, and the magnitude of the effect can be very great. For instance, the heat resistance of *Salmonella typhimurium* suspended in non-penetrant solutes rose over 700-fold (41). With spores of the DPA-negative mutant of *B. cereus* T, shrinkage of the cell and "buckling" of the coat can be seen in freeze-etch electron micrographs of spores in high concentrations of sucrose (42).

CONCLUSIONS

In contrast to the situation not so many years ago, the resistance and dormancy of bacterial endospores now seem to be much less immutable properties, but rather shallow, and influenced by a large variety of treatments.

One can identify key "targets" for manipulation of resistance and dormancy. For instance: the spore coats, and their important function in restricting permeability; the cortex, the integrity and functionality of which are certainly essential parts of the resistance mechanism; the cation-exchange sites within the spore; the free radical-sensitive sites.

On the practical side, however, although there is now a formidable list of resistance-modifying procedures available, we must admit that developments of improved spore control procedures based on the new observations have been disappointingly few.

REFERENCES

1. Michener, H.D., Thompson, P.A. and Lewis, J.C. (1959) *Appl. Microbiol.* **7**, 166-173.
2. Keynan, A., Evenchik, Z., Halvorson, H.O. and Hastings, J.W. (1964) *J. Bact.* **88**, 313-318.
3. Alderton, G. and Snell, N. (1963) *Biochem. Biophys. Res. Commun.* **10**, 139-143.
4. Alderton, G., Thompson, P.A. and Snell, N. (1964) *Science, N.Y.* **143**, 141-143.
5. Alderton, G. and Snell, N. (1969) *Science, N.Y.* **163**, 1212-1213.
6. Alderton, G., Ito, K.A. and Chen, J.K. (1976) *Appl. Environ. Microbiol.* **31**, 491-498.
7. Alderton, G., Chen, J.K. and Ito, K.A. (1980) *Appl. Environ. Microbiol.* **40**, 511-515.
8. Rode, L.J. and Foster, J.W. (1966) *J. Bact.* **91**, 1582-1588.
9. Marquis, R.E., Carstensen, E.L., Child, S.Z. and Bender, G.R. (1981) in *Sporulation and Germination* (Levinson, H.S., Sonnenshein, A.L. and Tipper, D.J., eds.), pp.266-268, Amer. Soc. Microbiol., Washington D.C.
10. Lewis, J.C., Snell, N.S. and Burr, H.K. (1960) *Science, N.Y.* **132**, 544-545.
11. Gould, G.W. and Dring, G.J. (1975) in *Spores VI* P., Costilow, R.N. and Sadoff, H.L., eds.), pp.541-546, Amer. Soc. Microbiol., Washington D.C.
12. Gould, G.W. and Dring, G.J. (1975) *Nature (Lond.)* **258**, 402-405.
13. Algie, J.E. (1980) *Curr. Microbiol.* **3**, 287-290.
14. Warth, A.D. (1978) *Adv. Microbial Physiol.* **7**, 1-45.
15. Warth, A.D. (1980) *J. Bact.* **143**, 27-34.
16. Ando, Y. and Tsuzuki, T. (1982) *J. Appl. Bact.* **54**, 197-202.
17. Riemann, H. and Ordal, Z.J. (1961) *Science N.Y.* **133**, 1703-1704.
18. Lewis, J.C. (1969) in *The Bacterial Spore* (Gould, G.W. and Hurst, A., eds.), pp.301-358, Academic Press, London.
19. Gould, G.W. and Dring, G.J. (1974) *Adv. Microbial Physiol.* **11**, 137-164.
20. Ando, Y. (1981) in *Sporulation and Germination* (Levinson, H.S., Sonnenshein, A.L. and Tipper, D.J., eds.), pp.240-242, American Society for Microbiology, Washington D.C.
21. Brown, C.W. (1977) in *Microbiology - 1977* (Schlessinger, D., ed.), pp.75-84, American Society for Microbiology, Washington D.C.
22. Flowers, R.S. and Adams, D.M. (1976) *J. Bact.* **125**, 429-434.

23. Roberts, T.A., Gibson, A.M. and Robinson, A. (1981) *J. Fd. Technol.* **16**, 337-355.
24. Chumney, R.K. and Adams, D.M. (1980) *J. Appl. Bact.* **49**, 55-63.
25. Ellar, D.J., Eaton, M.W. and Posgate, J. (1974) *Trans. Biochem. Soc.* **2**, 947-948.
26. Cassier, M. and Sebald, M. (1969) *Ann. Inst. Pasteur, Paris* **117**, 312,324.
27. Stegeman, H., Mossel, D.A.A. and Pilnick, W. (1977) in *Spore Research 1976* (Barker, A.N., Wolf, J., Ellar, D.J., Dring, G.J. and Gould, G.W., eds.), pp.565-587, Academic Press, London.
28. Gombas, D.E. and Gomez, R.F. (1978) *Appl. Environ. Microbiol.* **36**, 403-407.
29. Gomez, R.F., Gombas, D.E. and Herrero, A. (1980) *Appl. Environ. Microbiol.* **39**, 525-529.
30. Dewey, D.L. and Michael, B.D. (1965) *Biochem. Biophys. Res. Commun.* **21**, 392-396.
31. Gould, G.W. (1970) *J. Gen. Microbiol.* **64**, 301-309.
32. Bayliss, C.E. and Waites, W.M. (1969) *J. Appl. Bact.* **47**, 263-269.
33. King, W.L. and Gould, G.W. (1969) *J. Appl. Bact.* **32**, 481-490.
34. Waites, W.M., Bayliss, C.E., King, N.R. and Davies, A.M.C. (1979) *J. Gen. Microbiol.* **112**, 225-233.
35. Clouston, J.G. and Wills, P.A. (1970) *J. Bact.* **103**, 140-143.
36. Gould, G.W. and Sale, A.J.H. (1970) *J. Gen. Microbiol.* **60**, 335-346.
37. Murrell, W.G. and Wills, P.A. (1977) *J. Bact.* **129**, 1272-1280.
38. Enfors, S.O. and Molin, N. (1975) in *Spores VI* (Gerhardt, P., Costilow, R.N. and Sadoff, H.L., eds.), pp.506-512, Amer. Soc. Microbiol., Washington D.C.
39. Murrell, W.G. and Scott, W.J. (1966) *J. Gen. Microbiol.* **43**, 411-425.
40. Harnulv, B.G. and Snygg, B.G. (1972) *J. Appl. Bact.* **35**, 615-624.
41. Corry, J.E.L. (1974) *J. Appl. Bact.* **37**, 31-43.
42. Bothipaksa, K. and Busta, F.F. (1978) *Appl. Environ. Microbiol.* **35**, 800-808.
43. Dring, G.J. and Gould, G.W. (1975) in *Spores VI* (Gerhardt, P., Costilow, R.N. and Sadoff, H.L., eds.), pp.488-494, Amer. Soc. Microbiol., Washington D.C.

44. Dring, G.J. and Gould, G.W. (1977) in *Spores Research 1976 2* (Barker, A.N., Wolf, J., Ellar, D.J., Dring, G.J. and Gould, G.W., eds.), pp.771-791, Academic Press, London.
45. Dring, G.J. and Gould, G.W. (1981) *Spore Newsletter* **7** 130-131.
46. Gould, G.W. (1978) in *Spores VII* (Chambliss, G. and Vary, J.C., eds.), pp.21-26, Amer. Soc. Microbiol., Washington D.C.

INACTIVATION OF SPORES WITH CHEMICAL AGENTS

W.M. WAITES

Agricultural Research Council Food Research Institute, Colney Lane, Norwich NR4 7UA, U.K.

SUMMARY

Chemicals do not inactivate bacterial spores rapidly or efficiently at ambient temperatures but may provide the only practical method of sterilising thermolabile materials. Chemicals with sporicidal activity include ethylene oxide, glutaraldehyde, formaldehyde, the halogens, peracetic acid, HCl vapour and hydrogen peroxide. Other chemicals, including those which kill vegetative cells, are inactive against spores. Combinations of chemicals or of chemicals and physical agents have therefore been used together in attempts to improve sporicidal activity. Sporicidal activity depends on the bacterial species, medium and method of spore production and the conditions and time of storage before testing as well as the conditions during the test. The basis of resistance to chemicals is not understood although resistance to chlorine, octanol and ethylene oxide may be produced by the intact spore coat, since spores with defective coats have a markedly reduced resistance. Spores without dipicolinic acid also have increased sensitivity to chemicals such as phenol although the absence of dipicolinic acid may introduce other changes into the mature spore. Spores with extreme heat resistance are not especially resistant to chemicals suggesting that the dehydration of the protoplast, which is important in conferring heat resistance on spores, is less important in making spores resistant to chemicals.

INTRODUCTION

Chemicals do not kill bacterial spores rapidly or efficiently at ambient temperatures. Nevertheless, to avoid damage to thermolabile materials treatment with chemicals may be the only practical alternative to heat sterilisation. Spores are much more resistant than other forms of life so that their complete inactivation will result in sterilisation. The resistance of spores varies depending on the chemical, the bacterial species and strain, the medium and method of spore production and even on the conditions and time of storage before testing as well as during the actual test (1).

As spores are resistant to individual chemicals, attempts have been made to improve their inactivation by using combinations both of chemicals and of chemicals and physical agents (2). Such combinations offer the possibility of reducing the severity of treatment and allowing the sterilisation of labile materials. However, many chemicals, such as mercurials, quanternary ammonium compounds, phenolics and alcohols, will kill vegetative bacteria but will not kill spores under the conditions and concentrations at which they are generally used although alcohols, for example, inhibit germination (3) and must be removed before examining viability to prevent confusion of sporostasis with sporicidal activity. In view of the differences between the methods of different authors in testing sporicidal activity the resistances to individual chemicals described in Table 1 should only be viewed as approximations.

RESISTANCE TO CHEMICALS

Ethylene Oxide

Gaseous sterilisation by ethylene oxide including its sporicidal action has been reviewed (4). Although spores are only about 10 times more resistant to ethylene oxide than vegetative cells (5,6) sterilisation is slow, normally taking at least 4 h and sometimes even 18 h. Spores are less resistant to mixtures of ethylene oxide and methyl formate than to either gas alone (7). Germinated spores are intermediate in resistance between dormant spores and the corresponding vegetative cells (8) although drying markedly increases resistance (9). Spores killed by treatment with ethylene oxide were able to germinate but not outgrow (10).

Glutaraldehyde

The microbicidal, including sporicidal, properties of glutaraldehyde have been reviewed by Gorman *et al.* (11) and Scott

TABLE 1 *Resistance of spores to chemicals*

Chemical	Organism	Kill	Time (h)	Concentration (% w/v)	Temperature (°C)	Reference
Peracetic acid	*B. subtilis*	10^3	0.1	4	20	27
HCl vapour	*B. subtilis*	10^3	0.08	31*	20	34a
Ethylene oxide	*B. subtilis*	10^3	0.7	0.07	40	46
Hydrogen peroxide	*B. subtilis var. globigii*	10^3	0.17	25.8	24	29
Hypochlorous acid	*B. subtilis*	10^3	3.0	0.01^+	10	68
Glutaraldehyde	*B. pumilis*	10^3	0.5	2.0	37	13
Formaldehyde	*B. subtilis var. niger*	10^3	1.5	1.0	40	21

*0.25 ml in a 300 ml bottle

$^+$Free chlorine

and Gorman (12). Low concentrations of glutaraldehyde (0.1% w/v) inhibit germination and must be neutralised to prevent sporostasis. High concentrations (2% w/v) are rapidly sporicidal (13) and a 3 h contact period is sufficient to give a six-fold decimal reduction in viable count of spores of *Bacillus subtilis* (14). A number of other chemicals have been combined with glutaraldehyde in order to increase its sporicidal activity. These include cationic surfactants (15), mixtures of non-ionic ethoxylates of isomeric linear alcohols (16), a highly ionizable salt of a monovalent or divalent cation exchangeable with calcium (17) and small quantities of anionic and non-ionic surfactants (18). In addition, Sierra and Boucher (19) found that ultrasound reduced the time for 1% aqueous glutaraldehyde to completely inactivate spores of *B. subtilis* from 10 to 4 min, although ultrasonically treated spores were as resistant as untreated spores.

Formaldehyde

Early claims of rapid sporicidal activity were shown to be caused by sporostasis (20), due to reversible inhibition of spore germination (21). Spicher and Peters (22) have also reported that subsequent heating will reactivate spores apparently killed by formaldehyde. Sterilisation procedures take 1 to 2 h although mixtures of formaldehyde and glutaraldehyde are ten times as sporicidal as either chemical alone (2).

Halogens

The sporicidal efficiency of halogens increases in the order bromine : iodine : chlorine (23). Chlorine (as hypochlorous acid, chloramines or chlorine dioxide) is only slowly sporicidal although Death and Coates (24) have shown that buffered methanol/sodium hypochlorite is more rapidly sporicidal than buffered sodium hypochlorite alone. Hypochlorite (at 100 ppm available chlorine) took 5 min to produce a 5-fold decimal reduction in spores of *B. subtilis* while addition of methanol (1% v/v) produced this kill in 3 min. Survivor curves are not logarithmic and often have a marked lag (25). Mild treatment with chlorine may activate spores so that germination is faster (26) or a greater proportion of spores in the population form colonies (27).

Peracetic acid

Breakdown of peracetic acid produces acetic acid, hydrogen peroxide, water and oxygen and leaves no residues. Peracetic acid is rapidly sporicidal, 0.1% (v/v) at 20 C producing 0.1% survivors of spores of *B. subtilis* in 15 min. Kills are exponential, often after a lag until tailing occurs at about a 4-fold decimal reduction although mild treatment may increase the percentage of the population able to form colonies (27). Used as a vapour, 80% relative humidity is optimal, spores on porous surfaces being less resistant than those on glass surfaces (28).

Hydrogen Peroxide

Hydrogen peroxide is slowly sporicidal at ambient temperatures, 25.8% taking 11 min to kill 99.99% of spores of *B. subtilis* at 24°C but only 30 s at 76°C (29). After an initial lag, kills may be logarithmic but tailing occurs at about a 4-fold decimal reduction (30). Less severe treatment may increase the number of spores able to form colonies (27). Ultra-violet light increases the destruction of spores of *B. subtilis* by hydrogen peroxide by as much as 4,000-fold (31), probably by catalysing the production of free hydroxyl radicals (32), and since spores of *B. licheniformis*, *B. cereus*, *B. megaterium*, *B. pumilis*, *B. stearothermophilus* and *Clostridium sporogenes* are also rapidly inactivated by simultaneous treatment with both agents (33), the combination may form a useful and novel method of sterilisation (34).

HCl Vapour

Gaseous HCl is rapidly sporicidal (Table 1; 34a).

MECHANISMS OF RESISTANCE

Differences in Resistance

Spores of different species exhibit different resistances to the same chemical but spores of the same population also differ in the resistance to different chemicals, suggesting that the mechanism of resistance cannot be the same. Bayliss *et al.* (27) examined the destruction of spores of *B. subtilis* produced on five different media and found that the resistance to one chemical was not related to that of another (Table 2). For example, those spores most resistant to hydrogen peroxide and chlorine were among the least resistant to peracetic acid while those least resistant to peracetic acid and chlorine were relatively much more resistant to hydrogen peroxide. In addition, heat resistance was not related to chemical resistance and those spores most resistant to hydrogen peroxide at 25°C were among the least resistant at 90°C. Toledo *et al.*

TABLE 2 *Resistance of spores of* B. subtilis *produced on different media* (27)

Sporulation medium	Survivors (%) after treatment with				
	H O (10% w/v at 25°C)	H O (1.5% w/v at 90°C)	Chlorine (200 ppm at 0°C)	Peracetic acid (0.1% at 20°C)	Heat (90°C 20 min)
Potato agar	4.0	0.029	0.040	105	200
Nutrient agar	73	3.0	0.0035	0.02	60
Supplemented nutrient agar	83	0.016	0.020	0.02	65
Bacillus spore agar	120	0.23	0.10	0.02	7.5
Glucose salts medium	10	2.0	0.0046	9.0	30

(29) also found that the relative resistances of different strains of *B. subtilis* to hydrogen peroxide varied with temperature. It is apparent, therefore, that the extent of inactivation of spores by a chemical at high temperatures cannot be predicted from results obtained at lower temperatures.

Changes in Resistance During the Cell Cycle

As indicated in the paragraph on ethylene oxide, dormant spores are more resistant to chemicals than germinated spores. In general, resistance is lost immediately on germination but is gained slowly during spore formation. However, even under optimum conditions, sporulation requires 7 h while germination is much quicker, often taking place in an individual spore in less than 1 min. During sporulation of *B. subtilis* Balassa *et al.* (35) have shown that resistances to toluene, benzene, octanol, butanol and chloroform are the first to appear followed by those to ethanol, pyridine, phenol, trichloroacetic acid and to heating at 80°C for 10 min. Appearance of resistance to toluene coincided with the beginning of cortex formation and to pyridine with the beginning of coat formation, suggesting that the spore coats are responsible for resistance to the second group (36). Such circumstantial evidence is not conclusive but, since disruption leads to loss of resistance in spore components, the structure of the dormant spore is of obvious importance in resistance.

Resistance and Structure (Table 3)

The bacterial spore consists of exosporium, protein coat, outer membrane, cortex which is of peptidoglycan (similar but not identical in structure to that of the wall of the vegetative cell), an inner membrane and the protoplast which contains DNA, RNA and most of the spore enzymes, in addition to several components unique to spores. There is mounting evidence to support the suggestion that the heat resistance of spores is due to the dehydration of the spore protoplast but less to relate this to resistance to chemicals. The action of chemicals must be different to that of heat since they must pass through the outer layers of the spore so that their effectiveness will be reduced by permeability barriers. As with vegetative cells (37), membrane damage is likely to result in failure of the spore to maintain its energy potential after germination, particularly since energy-rich compounds undergo rapid changes early in germination (38,39).

TABLE 3 *Resistance and structure*

Spore component producing resistance	Resistance	Reference
Coat	Chlorine	25,45
	Ethylene oxide	46
	Octanol	48, 49
	Lysozyme	40, 41
	Di- and tri-valent cations at elevated temperatures	42
	Hydrogen peroxide	44
Cortex (and the dehydrated protoplast)	Octanol	56
	Xylene	56
Dipicolinic acid (and Ca^{2+})	Phenol	35
	Pyridine	35
	Trichloroacetic acid	35

Permeability

The coat protects spores against the lytic enzyme, lysozyme, since spores with defective coats (arising from mutation or extraction with chemicals) will undergo germination — like changes in the presence of lysozyme (40,41). The coat also protects spores against the combined effects of di- and tri-valent cations and heat, since coat-defective spores are rapidly killed by mild heat in the presence of high concentrations of Mg^{2+} or Ca^{2+} (42). Gould and Dring (42) suggested that multivalent cations cross-link the carboxyl groups of the cortical peptidoglycan and cause it to contract, allowing partial rehydration and loss of heat resistance within the protoplast. Warth (43) has suggested, as an alternative, that heating with multivalent cations may damage other spore components. In addition, intact coat protects spores against some chemicals. For example, spores of *C. bifermentans*, *B. cereus* and *B. subtilis* with intact coats were up to 10 -fold as resistant to chlorine as those pre-treated to reduce disulphide bonds in their coats (25). The resistance of spores of

C. bifermentans with intact coats to hydrogen peroxide was also about 500-fold greater than that of spores with altered coats (44) but spores of *B. cereus* made sensitive to lysozyme were only slightly less resistant (45), suggesting that the coat was unimportant in resistance. Treatment which would be likely to change the coat also reduced the resistance of spores of *B. subtilis* to ethylene oxide but resistance was regained on storage (46). Since lysozyme resistance is also regained on storage, probably as a result of oxidation of reduced disulphide bonds in the coats (47), this is added evidence that intact coats protect against ethylene oxide.

Aronson and Fitz-James (48) found that spores of *B. cereus* were lysozyme and octanol sensitive after chemical removal of coat protein and a mutant of *B. cereus* whose spores were lysozyme-sensitive also had reduced resistance to octanol. In a further study, Stelma *et al.* (49) using temperature-sensitive mutants of *B. cereus*, suggested that octanol resistance was produced by the outer coat layers.

However, intact coats do not provide resistance to all chemicals. In addition to defective coats not decreasing the resistance of spores of *B. cereus* to hydrogen peroxide, sensitivity to sodium hydroxide or ethanol did not parallel sensitivity to lysozyme (45). Thomas and Russell (13) also found that the intact coat of spores of *B. pumilus* also provided little protection against glutaraldehyde although McErlean *et al.* (50) showed that coat removal from spores of *B. subtilis* reduced their resistance to glutaraldehyde, to an iodophor and to a mixture of hypochlorite/methanol by 8-, 8- and 12-fold respectively.

It is apparent that an intact coat will protect spores against some chemicals, such as chlorine, octanol and ethylene oxide. With other chemicals, like hydrogen peroxide, the intact coat protects spores of some species more than others but since those strains in which the coat is more important are generally less resistant (compare results in (44) with those in (45)), then it is possible that the more resistant strains have an additional protective mechanism.

Other workers (51) have suggested that the outermost layer of the spore, the exosporium, acts as a permeability barrier to hydrogen peroxide. In the absence of techniques to remove exosporia without damaging the rest of the spore, this suggestion is difficult to evaluate but in at least one species the exosporium has a pore and cannot prevent the passage of chemicals (52).

Dehydration

Although some work has cast doubt on the suggestion that the spore protoplast is dehydrated relative to the rest of the spore (53), it is generally accepted that during sporulation the protoplast contracts (43,54) and loses water, perhaps as a result of expansion of the cortex (43,55), and that such changes are at least partly responsible for the extreme heat resistance of the dormant spore. As described earlier, Sousa *et al.* (36) found that the appearance of the cortex and coat during spore formation coincided with the development of resistance to a variety of chemicals. In addition, Imae and Strominger (56) used mutants of *B. sphaericus* defective in synthesis of meso-diaminopimelate to examine heat and chemical resistance. They found that muramic lactam in the cortex increased linearly with an increase in the concentration of meso-diaminopimelic acid in sporulation medium. Heat resistance did not occur until 90% of the maximum cortex concentration was present but spores became resistant to both octanol and xylene when 25% of the maximum cortex concentration was produced. Thus a complete cortex, perhaps by contracting and dehydrating the protoplast, is important in conferring resistance to some chemicals. However, an incomplete cortex may prevent development of other spore components and spores without cortices were irregularly shaped (57), suggesting that the coats could also have been defective.

Cations

Incubation of spores at acid pH results in removal of Ca^{2+} and other cations and such spores have reduced heat resistance which can be restored after the uptake of either monovalent or divalent metals, although Ca^{2+} and Mn^{2+} produce the most heat resistant spores (58,59). Such changes did not alter the resistance of spores to glutaraldehyde, to an iodophor or a mixture of hypochlorite/methanol (50). However, Balassa *et al.* (35) found that spores of a mutant of *B. subtilis* which did not synthesise dipicolinic acid during sporulation were less resistant to phenol, to pyridine and to trichloroacetic acid than those of the parent. The mutant took up dipicolinic acid added to the sporulation medium and then formed spores which were resistant to phenol, pyridine and trichloroacetic acid. Absence of dipicolinic acid in the mature spore may have led to other changes, however, since dipicolinic acid is involved in Ca^{2+} uptake (60) and, in the absence of dipicolinic acid, the spores of the mutant contained less Ca^{2+} than those of the parent (35). Nevetheless Balassa's work suggests that the presence of dipicolinic acid in spore protoplasts plays a part in making spores resistant to some chemicals.

Small Molecular Weight Proteins within the Protoplast

Setlow (61) has isolated unique low molecular weight basic proteins which comprise about 15% of the protein of dormant spores of both *Bacillus* (62) and *Clostridium* (63). These proteins accumulate during sporulation at the time when the developing spore becomes resistant to ultraviolet light and are degraded during germination when resistance to ultraviolet light is lost (64). Irradiation of intact dormant spores results in covalent cross-linking of the proteins to other macromolecules, including spore DNA (65). The mutant of *B. sphaericus*, unable to make spore cortex because of a block of diaminopimelic acid biosynthesis (56), accumulated and maintained levels of the proteins similar to those of the parent and also acquired resistance to ultraviolet light to the same extent as the parent, suggesting that the proteins, but not the cortex, were required for the acquisition of resistance to ultraviolet light (66). When chemicals, such as hydrogen peroxide, damage DNA (67) cross-links with these proteins may protect but in the absence of mutants unable to accumulate the proteins, positive proof is lacking.

Conclusions

The mechanisms which produce chemically resistant spores are not understood. An intact spore coat provides resistance to some chemicals and reduced levels of DPA and cortical peptidoglycan reduce resistance to others presumably by altering conditions within the protoplast. In addition, chemicals might be expected to damage spore membranes, resulting in failure to maintain energy potential after germination. The mechanisms which make spores resistant to heat may be important in determining their resistance to chemicals but other factors must also be involved in particular since heat resistant spores may have a low resistance to chemicals. To obtain further evidence, changes in specific spore components, produced either by mutation or by altered sporulation conditions, must be used to initiate differences in resistance. Past attempts to change resistance have resulted in multiplicity of structural and chemical changes, suggesting that the bacterial spore is too complex for changes in one component not to affect those in others.

REFERENCES

1. Waites, W.M. (1982) in *Principles and Practice of Disinfection, Preservation and Sterilisation* (Russell, A.D., Hugo, W.B. and Ayliffe, G.A.J., eds.), pp.207-220, Oxford, Blackwell.

2. Waites, W.M. and Bayliss, C.E. (1982) in *Revival of Injured Microbes* (Russell, A.D. and Andrew, M.H.E., eds.), Society for Applied Bacteriology, Loughborough, 1982.
3. Yasuda-Yasaki, Y., Namika-Kanie, S. and Hachisuka, Y. (1978) *J. Bacteriol.* **136**, 484-490.
4. Christensen, E.A. and Kristensen, H. (1982) in *Principles and Practice of Disinfection, Preservation and Sterilisation* (Russell, A.D., Hugo, W.B. and Ayliffe, G.A.J., eds.), pp.548-568, Oxford, Blackwell.
5. Phillips, C.R. (1949) *American J. Hyg.* **50**, 280-289.
6. Znamirowski, R., McDonald, S. and Roy, T.E. (1960) *Canad. Med. Assoc.* **83**, 1004-1006.
7. Dadd, A.H., McCormick, K.E. and Daley, G.M. (1977) *J. Appl. Bacteriol.* **43**, XVIII-XIX.
8. Church, B.D., Halvorson, H., Ramsey, D.S. and Hartman, R.S. (1956) *J. Bacteriol.* **72**, 242-247.
9. Kaye, S. and Phillips, C.R. (1949) *American J. Hyg.* **50**, 296-306.
10. Dadd, A.H. and Cockshott, J. (1982) *Society for Applied Bacteriology*, Loughborough.
11. Gorman, S.P., Scott, E.M. and Russell, A.D. (1980) *J. Appl. Bacteriol.* **48**, 161-190.
12. Scott, E.M. and Gorman, S.P. (1982) in *Disinfection in the Hospital and Home - Current Topics*. Society for Applied Bacteriology, Loughborough, 1982.
13. Thomas, S. and Russell, A.D. (1974) *J. Appl. Bacteriol.* **37**, 83-92.
14. Forsyth, M.P. (1975) *Develop. Indust. Microbiol.* **16**, 37-47.
15. Stonehill, A.A. (1966) U.S. Patent No. 3,282,775.
16. Boucher, R.M.G. (1971) U.S. Patent Application No. 155,233.
17. Cowan, S.M. (1976) British Patent No. 1,443,786.
18. Shattner, R.I. (1978) U.S. Patent No. 4,103,001.
19. Sierra, G. and Boucher, R.M.G. (1971) *J. Appl. Microbiol.* **22**, 160-161.
20. Klarmann, E.G. (1956) *American J. Pharm.* **128**, 4-18.
21. Trujillo, R. and David, T.J. (1972) *Appl. Microbiol.* **23**, 618-622.
22. Spicher, G. and Peters, J. (1981) *Zentralb, Bakteriol. Mikrobiol. Hyg. Abt.* **174**, 133-150.
23. Marks, H.C. and Strandskov, F.B. (1950) *Ann. N.Y. Acad. Sci.* **53**, 163-171.
24. Death, J.E. and Coates, D. (1979) *J. Clin. Pathol.* **32**, 148-153.
25. Wyatt, L.R. and Waites, W.M. (1975) *J. Gen. Microbiol.* **89**, 327-334.
26. Wyatt, L.R. and Waites, W.M. (1973) *J. Gen. Microbiol.* **78**, 383-385.

27. Bayliss, C.E., Waites, W.M. and King, N.R. (1981) *J. Appl. Bacteriol.* **50**, 379-390.
28. Portner, D.M. and Hoffman, R.K. (1968) *Appl. Microbiol.* **16**, 1782-1785.
29. Toledo, R.T., Escher, F.E. and Ayres, J.C. (1973) *Appl. Microbiol.* **26**, 592-597.
30. Cerf, O. and Metro, F. (1977) *J. Appl. Bacteriol.* **42**, 405-415.
31. Bayliss, C.E. and Waites, W.M. (1979) *FEMS Microbiol. Letts.* **5**, 331-333.
32. Symons, M.C.R. (1960) in *Peroxide Reaction Mechanisms* (Edwards, J.O., ed.), pp.137-151, New York, Interscience.
33. Bayliss, C.E. and Waites, W.M. (1979) *J. Appl. Bacteriol.* **47**, 263-269.
34. Peel, J.L. and Waites, W.M. (1981) U.S. Patent No. 2,063,070A.
34a Tuynenberg Muys, G., Van Rhee, and Lelieveld, H.L.M. (1978)*J. Appl. Bacteriol.* **45**, 213-217.
35 Balassa, G., Milhaud, P., Raulet, E., Silva, M.T. and Sousa, J.C.F. (1979) *J. Gen. Microbiol.* **110**, 365-379.
36. Sousa, J.C.F., Silva, M.T. and Balassa, G. (1978) *Ann. Microbiol.* **129B**, 377-390.
37. Dawes, E.A. (1982) in *Revival of Injured Microbes* (Russell, A.D. and Andrew, M.H.E., eds.), Society for Applied Bacteriology, Loughborough, 1982.
38. Setlow, P. and Kornberg, A. (1969) *J. Bacteriol.* **100**, 1155-1160.
39. Hausenbauer, J.M., Waites, W.M. and Setlow, P. (1977) *J. Bacteriol.* **129**, 1148-1150.
40. Suzuki, Y. and Rode, L.J. (1969) *J. Bacteriol.* **98**, 238-245.
41. Gould, G.W. and Hitchins, A.D. (1963) *J. Gen. Microbiol.* **33**, 413-423.
42. Gould, G.W. (1978) in *Spores VII* (Chambliss, G. and Vary, J.C., eds.), pp.21-26, American Society for Microbiology, Washington D.C.
43. Warth, A.D. (1978) in *Adv. Microbial Physiology* (Rose, A.H. and Morris, J.G., eds.), Vol. 17, pp.1-45.
44. Bayliss, C.E. and Waites, W.M. (1976) *J. Gen. Microbiol.* **96**, 401-407.
45. Waites, W.M. and Bayliss, C.E. (1979) *J. Appl. Biochem.* **1**, 71-76.
46. Marletta, J. and Stumbo, C.R. (1970) *J. Fd. Sci.* **35**, 627-631.
47. King, W.L. and Gould, G.W. (1969) *J. Appl. Bacteriol.* **32**, 481-490.
48. Aronson, A. and Fitz-James, P.C. (1976) *Bacteriol. Rev.* **40**, 360-402.

49. Stelma, G.N., Aronson, A.I. and Fitz-James, P. (1978) *J. Bacteriol.* **134**, 1157-1170.
50. McErlean, E.P., Gorman, S.P. and Scott, E.M. (1980) *J. Pharm. and Pharmac.* **32**, 32p.
51. Ahmed, F.I.K. and Russell, C. (1975) *J. Appl. Bacteriol.* **39**, 31-40.
52. Mackey, B.M. and Morris, J.G. (1972) *J. Gen. Microbiol.* **73**, 325-338.
53. Bradbury, J.H. (1981) *Spore Newsletter VII*, 26-27.
54. Murrell, W.G. (1981) *Spore Newsletter VII*, 75-77.
55. Gould, G.W. and Dring, G.J. (1975) *Nature* **258**, 402-405.
56. Imae, Y. and Strominger, J.L. (1976) *J. Bacteriol.* **126**, 907-913.
57. Imae, Y., Strominger, M.B. and Strominger, J.L. (1976) *J. Bacteriol.* **127**, 1568-1570.
58. Alderton, G. and Snell, N. (1963) *Biochem. Biophys. Res. Commun.* **10**, 139-143.
59. Marquis, R.E., Carstenson, E.L., Bender, G.R. and Child, S.Z. (1981) *Spore Newsletter VII*, 34-35.
60. Ellar, D.J. (1978) in *Relations between Structure and Function in the Prokaryotic Cell* (Stanier, R.Y., Rogers, H.J. and Ward, B.J., eds.), pp.295-325, Society for General Microbiology, Cambridge.
61. Setlow, P. (1983). This volume.
62. Yuan, K., Johnson, W.C., Tipper, D.J. and Setlow, P. (1981) *J. Bacteriol.* **146**, 965-971.
63. Setlow, P. and Waites, W.M. (1976) *J. Bacteriol.* **127**, 1015-1017.
64. Setlow, P. (1975) *J. Biol. Chem.* **250**, 8159-8167.
65. Setlow, B. and Setlow, P. (1979) *J. Bacteriol.* **139**, 486-494.
66. Setlow, B., Hawes Hacket, R. and Setlow, P. (1982) *J. Bacteriol.* **149**, 494-498.
67. Bayliss, C.E., Shah, J. and Waites, W.M. (1982) *FEMS Microbiology Letts.* **13**, 147-149.
68. Dye, M. and Mead, G.C. (1972) *J. Fd. Technol.* **7**, 173-181.

INCREASED SENSITIVITY OF SURVIVING BACTERIAL SPORES IN IRRADIATED SPICES

J. FARKAS and É. ANDRÁSSY

International Facility for Food Irradiation Technology, Wageningen, The Netherlands and Central Food Research Institute, Budapest, Hungary.

SUMMARY

The water activity of black pepper did not influence the radiation resistance of its bacterial flora. The survivors of 3 kGy gamma radiation were more heat sensitive than the microflora of the untreated spice. The radiation treatment, the post-irradiation application of heat treatment, reduced pH and addition of sodium chloride or nitrite to the plating media synergistically reduced the colony counts. The sensitizing effect of irradiation was maintained during the 6-month post-irradiation storage.

Contrary to irradiation, the effect of ethylene oxide was considerably influenced by the water activity of the spice. Ethylene oxide sensitized the survivors in a much less degree than radiation.

INTRODUCTION

Both the producing countries and the food industries, which use spices as ingredients of various food products, face the problem of the high microbial counts of spices, especially the contamination with moulds and heat resistant bacterial spores.

In general, black pepper, turmeric, paprika, chili and thyme are the most highly contaminated spices. Not infrequently, their aerobic plate count may exceed the 10 million per gram level (1).

The majority of bacterial flora of spices consists of aerobic spore-froming bacteria (2), many of which are in the spore state (3), which, in one study were, in order of frequency, *B. subtilis*, *B. licheniformis*, *B. megaterium*, *B. pumilis*, *B. firmus* and *B. brevis* (4). Relatively high incidences of *Bacillus cereus* and *Clostridium perfringens* were also reported (5,6). Thermophilic bacteria usually constitute 0.1 - 10% of the bacterial population (3,7).

In many cases, contaminated spices were responsible for the spoilage of canned meats or caused defective sausage products (8-10). The destruction of the thermoduric bacterial spores introduced to the product by seasonings often require a severe heat treatment, which ensures microbial stability only at the cost of substantial reduction of nutritional and sensory quality of the spiced product. This accounts for the attempts to reduce the viable cell counts in spices by an appropriate decontamination treatment.

Fumigation with ethylene oxide is used commercially in several countries to kill insect pests in seasonings and to reduce the viable count. This chemical treatment has, however, disadvantages. The wrapping or packaging material of items to be fumigated must be permeable to ethylene oxide and moisture (11). Fumigation is a time consuming batch-procedure, appears to be beset with problems of uniformity of sterilization, requires a complex process-monitoring (12) and is inseparable from the danger of chemical residues. The absorptive-bound ethylene oxide remains fairly high after gassing even after repeated evacuation (13). Although the absorptive ethylene oxide residues decrease continuously during storage, this is frequently due to further chemical reactions rather than to additional loss of the gas.

In addition to ethylene glycol, ethylenechlorohydrin may be formed during the ethylene oxide fumigation. Ethylene chlorohydrin is more persistent than ethylene oxide and it is frequently found in ethylene oxide-treated spices in concentrations exceeding even the 1000 mg/kg level (14-16). Several studies indicate that ethylene oxide and ethylene chlorohydrin are mutagens and they are suspected of causing other chronic or delayed toxic effects (17-19). Fumigants represent a significant occupational health hazard. Due to these problems the future of fumigation is very much in question.

As an alternative, treatment of spices by ionising radiations is suitable for the decontamination of many spices and other dry food materials (20). Irradiation of spices is a technically and economically feasible physical process. The irradiation procedure is direct, simple, requires no additives and is highly efficient. A radiation dose of 4 to 8 kGy proved to be sufficient in most instances to reduce the viable cell counts to the level required in the food industries. The sensory properties of spices are not influenced at these radiation doses. The radiation treatment results in less loss of essential oils than treatment with ethylene oxide (21).

Since the cost of the radiation treatment and whatever chemical changes induced by irradiation are practically proportional with the radiation dose, it is worthwhile to study the properties of the surviving microbial population in irradiated spices, in order to establish a realistic dose requirement and avoid unnecessarily high doses with a view to the use of the irradiated spices by food processors.

Earlier studies in our laboratories revealed that the surviving microflora of irradiated spices is more sensitive to the environmental conditions and to subsequent food processing treatments than that of untreated spices (22,23). The present study was performed to see the radiation sensitization of the spore flora of black pepper to heat, reduced water activity and/or reduced pH, and to determine whether this sensitization would depend on the water activity of the spice during irradiation. For comparison, similar studies were conducted also with ethylene oxide-treated samples. Further, how long the sensitizing effect of irradiation is maintained during the post-irradiation storage of black pepper under various conditions was studied. The pH- and a_w-values, the concentration of curing salts and storage temperatures were selected as to represent the variations of these parameters under practical conditions in the meat industry.

For the evaluation of the effects and interactions of treatments and media conditions the logarithms of cell counts were evaluated by analysis of variance.

METHODS

Duplicate samples of ground black pepper were equilibrated over H_2SO_4 solutions of appropriate a_w-s at 10°C and 25°C to a_w = 0.25, 0.50 and 0.75, respectively. The required a_w values of the samples were checked by a NOVASINA instrument. The equilibrium moisture-content proved to be 8.5, 11.0 and 15.0%.

After the water-activity equilibration, one of each sample was irradiated by a self-shielded ^{60}Co-gamma irradiator of type RH-γ-30 at room temperature under aerobic conditions. The radiation dose and the dose-rate have been 3 kGy and 5.7 $kGy.h^{-1}$, respectively. The other samples with the same water activity level have been kept untreated as controls. Subsequent to the radiation treatment, both the untreated and irradiated samples were stored under the same conditions as for the water-activity equilibration. The microbiological studies were performed directly after the irradiation and after 1, 2, 4 and 6 months of post-irradiation storage.

For the microbiological studies 10% suspensions were prepared from the black pepper sample by a dilution-liquid containing 0.1% bacteriological peptone (OXOID) and 0.9% NaCl. The suspensions were heated under continuous stirring at 90°C for 20 min or kept at 20°C for the same period as unheated controls for the total plate counts. For the heat treatment the spice was suspended when the dilution-liquid already reached the 90°C temperature. Decimal dilution series were prepared from the unheated and the heat-treated suspensions and the number of colony forming units were established by triplicate plating of selected dilution levels with plate counting agar (OXOID) adjusted to NaCl concentrations of 0.5 and 2.5%, and pH values of 6.0 and 5.0, respectively. Colonies were counted after 72 h of incubation at 30°C.

In an additional experiment the effect of 50 ppm mitrite was studied on the colony-forming capacity of the untreated and irradiated and/or heat-treated microflora of black pepper in the same way as just described, except that the place countings were performed only at the beginning and after 6 months of storage.

The effect of ethylene oxide treatment on the survival and heat resistance of the aerobic microflora of black pepper was studied with the same plating media as in other experiments but without further storage of the fumigated samples. The fumigation has been performed in a DEGESCH-type industrial fumigation chamber at the Azeged Paprika Processing Company. An EO-concentration of 600 $g.m^{-3}$ was applied for 6 h at 22°C.

RESULTS AND DISCUSSION

The total aerobic plate count of untreated samples exceeded the 10^6/g level and the number of cells surviving the 20 min heat treatment at 90°C amounted to more than 50% of the total plate count. This shows again that the majority of the bacterial flora consisted of aerobic bacterial spores.

According to the statistical analysis of the results obtained with the untreated and the irradiated samples, the water activity of the black pepper did not influence either the total aerobic plate counts, or the number of cells surviving the 90°C for 20 min. This is in agreement with other observations (24,25) that no significant effect of water activity was observed on the radiation resistance of the microflora in spices. Härnulv and Snygg (16) reported that spores of *B. subtilis* as well as *B. stearothermophilus*, when irradiated in water vapour under anoxic conditions showed a slight increase in radiation resistance with decreasing a_w from 1.0 - 0.0. However, this increase was also practically negligible, when one considers the a_w range of our spice

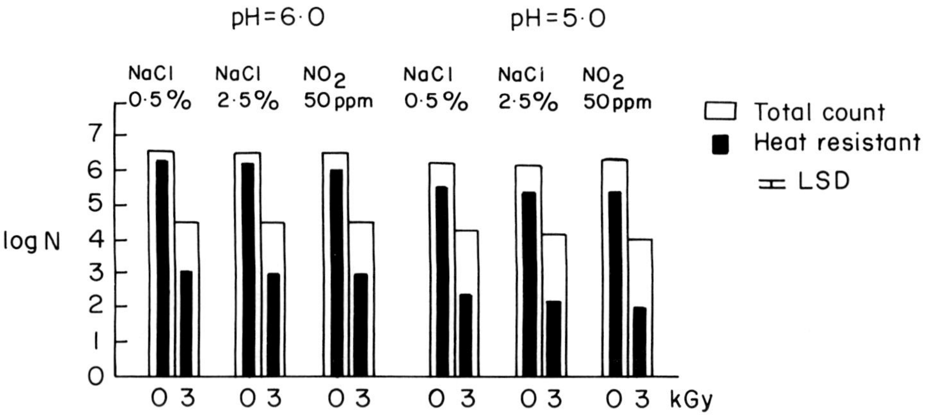

FIG. 1 *The effect of irradiation on the total viable count of black pepper and the heat resistant fraction of the survivors as affected by the pH, salt content and nitrite content of the plating media. LSD: least significant difference, p = 0.05.*

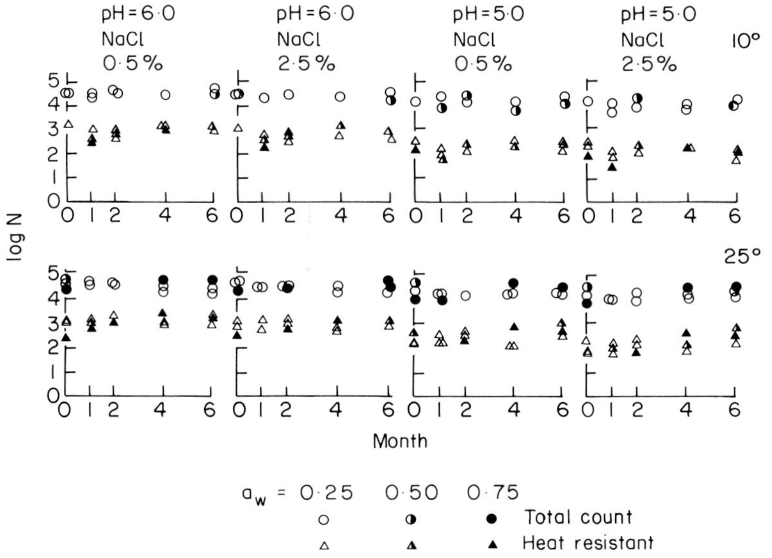

FIG. 2 *The logarithmic total counts of irradiated black pepper samples and those of colony-forming units surviving a heat treatment at 90°C for 20 min, as a function of the water activity and storage time of the spice at 25 and 10°C, respectively, and the pH and salt concentration of the plating media.*

factors calculated under the assumption of additive effects. These ratios were called as synergy factors, and such synergy factors are shown in Fig. 4.

Since the fumigation at a_w = 0.25 resulted in the same magnitude of cell count reduction using the control medium as the 3 kGy irradiation, for a direct comparison of the alternative decontamination treatments the synergy factors related to fumigation are shown in the case of a_w = 0.25. It can be seen from Fig. 4, that the radiation treatment and the alterations of the plating medium had only additive effects under our experimental conditions (i.e. the synergy factors were very close to 1.0). The radiation treatment strongly sensitized the surviving microflora to the subsequent heat treatment and this synergistic effect was even more pronounced when the survivors were plated in a medium with decreased pH especially when the water activity was also decreased, or nitrite was added to the plating medium. The fumigation treatment at the same overall cell-count-reducing effect as that of the radiation treatment heat-sensitized the survivors in a much lesser degree than the radiation treatment. The fumigation-related synergy factors showed less increase with reduced pH or in the presence of added nitrite, than the irradiation-related synergy factors. However, the (apparent) efficiency of the ethylene oxide treatment was also considerably increased when the survivors were counted in a medium of simultaneously reduced pH and water activity.

Our observations are in agreement with the literature on the existence of injury in bacterial cells and spores after exposure to one or more environmental stresses, and on the influence of injury on apparent survival.

Our earlier studies showed already that radiation damaged and heat injured spores are more dependent on pH and temperature of incubation and on curing salts in the recovery medium (29,35-37). Increased sensitization of spores to sodium chloride was observed with combinations of heat and gamma radiation (29,38). Most sensitive were spores irradiated first, then heated.

Others have also demonstrated that both aerobic and anaerobic spores surviving bactericidal treatments have an increased sensitivity to incubation temperature and recovery medium-pH (39,40) or to the presence of curing agents (NaCl, KNO_3, $NaNO_2$) in the recovery medium (41-44).

Since the residual viable counts in our irradiated spice samples did not increase significantly during a six-month period, either at 10°C or at 25°C storage temperature, and the increased sensitivity of the residual spore flora in the spice samples was maintained, it can be concluded that no significant repair of the damage induced during the aerobic irradiation conditions was occurring during the post-irradiation storage. This is in harmony with Roberts' studies (38), who has been unable to demonstrate reproducibly any increase in count of irradiated clostridial spores during post-irradiation storage. It is believed that DNA breaks produced in presence of O_2 cannot be rejoined (45).

The differences in the sensitizing effects of irradiation and fumigation, as shown by their different "synergy profiles" (Fig. 4) make it probable that the nature of injury of bacterial spores should be different after radiation than after fumigation. It is known from the literature that irradiated and ethylene oxide treated anaerobic spores behaved in an exactly opposite way regarding the effect of incubation temperature on the recovery (39). Although it is reasonable to consider that multiple mechanisms may result in the manifestation of spore injury in all instances, damage of different components may be critical with different damaging agents (46,47).

Comparing the radiation treatment with fumigation, one can also conclude that the efficiency of irradiation is much less dependent on the moisture conditions than the efficiency of ethylene oxide. The residual microflora of irradiated black pepper would be more sensitive to antimicrobial factors prevailing during a subsequent use of the spice than that of fumigated black pepper.

The sensitization of the surviving microflora by radiation of the black pepper, and probably other dry ingredients, seems to be a permanent feature and does not diminish upon storage.

REFERENCES

1. Pivnick, H. (1980) in *Microbial Ecology of Foods (ICMSF)*, Vol. 2, pp.731-751, Academic Press, New York.
2. Elter, B. and Scharner, E. (1968) *Fleisch* **22**, 219-220.
3. Fábri, I., Deák, T., Nagel, V., Szabad, J., Tabajdi-Pintér, V. and Zalavári, Zs. (1982) Bacterial spore profiles in red paprika and black pepper. This Symposium.
4. Goto, A., Yamazaki, K. and Oka, M. (1971) *Food Irradiation Japan* **6**, (1) 35-42.
5. Powers, E.M., Lawyer, R. and Masuoka, Y. (1975) *J. Milk Fd. Technol.* **38**, 683-687.
6. Powers, E.M., Latt, T.G. and Brown, T. (1976) *J. Milk Fd. Technol.* **39**, 668-670.
7. Mattada, R.R., Thangamani, R.S. and Ramanathan, L.A. (1974) *Indian J. Microbiol.* **14**, 139-141.
8. Mossel, D.A.A. (1955) *Ann. Inst. Past. Lille* **7**, 171-186.
9. Julseth, R.M. and Deibel, R.H. (1974) *J. Milk Fd. Technol.* **37**, 414-419.

10. Palumbo, S.A., Rivenburgh, A.I., Smith, J.L. and Kissinger, J.C. (1975) *J. Appl. Bacteriol.* **38**, 99-105.
11. Kereluk, K. and Lloyd, R.S. (1969) *The J. Hosp. Res.* **7**, 8-75.
12. Frohnsdorff, R.S.M. (1981) *Revue IRE* **5**, (2), 7-17.
13. Kröller, E. (1966) *Deutsche-Lebensm.-Rundsch.* **102**, 227-234.
14. Wesley, F., Rourke, B. and Darbischire, O. (1965) *J. Fd. Sci.* **30**, 1037-1042.
15. Stijve, T., Kalsbach, R. and Eyring, G. (1976) *Trav. Chim. Aliment. Hyg.* **67**, 403-428.
16. Gustafsson, K.H. (1981) *Vår Föda* **33**, 15-21.
17. Kucerová, M., Zhurkov, V.S., Polivková, Z. and Ivanova, J.E. (1977) *Mutation Res.* **48**, 355-360.
18. Environmental Protection Agency (1978) *Federal Register* **43**, 3802-3815.
19. Hogstedt, C., Malmquist, N. and Wadman, B. (1979) *J. Amer. Med. Ass.* **241**, 1132-1133.
20. Farkas, J. (1982) in *Preservation of Food by Ionizing Radiations* (Josephson, E.S. and Peterson, M.S., eds.), Vol. 3, CRC Press, Inc., Bota Raton (in press).
21. Vajdi, M. and Pereira, R.R. (1973) *J. Fd. Sci.* **38**, 893-895.
22. Farkas, J., Beczner, J. and Incze, K. (1973) in *Radiation Preservation of Food*, pp.389-402, International Atomic Energy Agency, Vienna.
23. Kiss, I. and Farkas, J. (1981) in *Combination Processes in Food Irradiation*, pp.107-122, International Atomic Energy Agency, Vienna.
24. Török, G. and Farkas, J. (1961) *Communications of the Central Food Research Institute*, Budapest, (3) 1-7.
25. Anon. (1982) Progress in food irradiation: The Netherlands. *Food Irradiation Information*, No.12 (In press).
26. Härnulv, B.G. and Snygg, B.G. (1973) *J. Appl. Bact.* **36**, 677-682.
27. Morgan, B.H. and Reed, J.M. (1954) *Food Res.* **19**, 357-366.
28. Farkas, J. (1963) A study into the radiation damage of bacterial spores (In Hungarian) Thesis, University of Politechnics, Budapest.
29. Farkas, J. and Roberts, T.A. (1976) *Acta Alimentaria* **5**, 289-302.
30. Kempe, L.L. (1955) *Appl. Microbiol.* **3**, 346-359.
31. Ernst, R.R. and Shull, J.J. (1962) *Appl. Microbiol.* **10**, 342-344.
32. Kelsey, J.C. (1967) *J. Appl. Bacteriol.* **30**, 92-100.
33. Mayr, G.E. and Suhr, H. (1973) in *Proceedings of the Conference on Spices* held at the London School of Pharmacy,

London, 10-14 April, 1972, pp.210-207, Tropical Products Institute, London.
34. Murányi, E. and Nagy, J. (1975) *Konserv- és Paprikaipar* No.6, 218-221.
35. Farkas, J., Kiss, I. and Andrássy, É. (1967) in *Radio-sterilization of Medical Products*, pp.343-354, IAEA, Vienna.
36. Farkas, J. (1970) in *Spore Group Meeting*, SIK Göteborg, June 16-18, 1970, p.3, SIK:s Service-Seris Nr.321, SIK, Göteborg.
37. Kiss, I., Rhee, C.O., Greca, N., Roberts, T.A. and Farkas, J. (1978) *Appl. Environm. Microbiol.* **35**, 533-539.
38. Roberts, T.A. (1970) *J. Appl. Bact.* **33**, 74-94.
39. Futter, B.V. (1968) The detection and viability of anaerobic spores surviving bactericidal influences. Ph.D. thesis, Council for National Academic Awards. Ref.: Roberts (38).
40. Futter, B.V. and Richardson, G. (1970) *J. Appl. Bact.* **33**, 321-330.
41. Roberts, T.A., Ditchett, P.J. and Ingram, M. (1965) *J. Appl. Bact.* **28**, 336-348.
42. Roberts, T.A. and Ingram, M. (1966) *J. Fd. Technol.* **1**, 147-163.
43. Briggs, A. and Yazdany, S. (1970) *J. Appl. Bact.* **33**, 621-632.
44. Chumney, R.K. and Adams, D.M. (1980) *J. Appl. Bact.* **79**, 55-63.
45. Grecz, N., Waitr, C., Durban, E., Kang, T. and Farkas, J. (1978) *J. Food Processing Preserv.* **2**, 315-337.
46. Hurst, A. (1977) *Canad. J. Microbiol.* **23**, 935-944.
47. Foegeding, P.M. and Busta, F.F. (1981) *J. Fd. Protection* **44**, 776-786.

HEAT INACTIVATION AND INJURY OF *CLOSTRIDIUM BOTULINUM* SPORES IN SAUSAGE MIXTURES

F.-K. LÜCKE

Federal Centre for Meat Research, D-8650 Kulmbach, Federal Republic of Germany

SUMMARY

Decimal reduction times of *Clostridium botulinum* spores at 100 and 105°C were longer in liver sausage mixture (an emulsion prepared from pre-cooked ingredients) than in Brühwurst mixture (a dispersion made from raw ingredients only). This difference was most distinct when survivors were recovered in inhibitory media. Nitrite did not prevent development of the surviving spores in liver sausage, but was markedly inhibitory in Brühwurst. The results suggested that the fat in liver sausage mixture protected the spores from heat injury, and that nitrite was "inactivated" from iron liberated from liver.

INTRODUCTION

The behaviour of *Clostridium botulinum* in meat products has been studied in several laboratories in order to evaluate whether changes in the manufacture and storage of meat products may result in an increased risk of botulism. Most of these studies dealt with model systems (1-3), bacon, frankfurters or related products (see (4) for review). From these studies, the widely accepted opinion emerged that omission of nitrite from meat products leads to an increased risk of botulism.

We studied the survival and development of *C. botulinum* spores in meat products common in the Federal Republic of Germany and concentrated on those actually involved in cases of botulism. Such products include large raw hams (5) and

Kochwurst (heat-processed sausage made from pre-cooked ingredients, e.g. liver sausage and blood sausage), but not Brühwurst (heat-processed sausage made only from raw ingredients, e.g. frankfurters, bologna). This paper presents some results of an investigation designed to ascertain why Kochwurst is more often involved in cases of botulism than Bruhwurst. It will be shown that this difference can be partially explained in terms of differences in the rates of heat inactivation and heat injury of *C. botulinum* spores.

METHODS

Spore Inocula, Growth Media and Sausage Mixtures

Spores of the proteolytic *C. botulinum* strains 52A, 77A, 33A, 12885A, 36A, 41B, 53B, 213B, Lamanna B and ATCC 794B (all strains provided by T.A. Roberts, Langford, UK) were produced in trypticase-peptone-thioglycolate medium (6), washed three times with water and stored frozen until required. They were enumerated after heat shock (80°C, 10 min) in egg-yolk agar (7) by a pour-plate procedure; plates were incubated in anaerobic jars for 2-3 days at 30°C.

Cooked Meat Medium (CMM) was prepared by adding 1 g of dried beef heart particles (Lab-M Ltd., London, UK) to 10 ml Standard I-nutrient broth (Merck, Darmstadt, West Germany) containing 1 g glucose per litre. After inoculation with sausage samples (5 g per 20 ml), the pH was 6.3 - 6.5. To prepare Cooked Meat Medium with salt (CMMS), sodium chloride was added to give a final concentration of 3.5% (w/v), and the pH was adjusted to 6.0 with lactic acid; the water activity a_w of this medium was approximately 0.973. Both CMM and CMMS were used either fresh or freshly steamed and overlaid with vaseline-paraffin mixture (1:1) after inoculation.

The methods to prepare sausage mixtures are outlined in Fig. 1. Cutters of 2 or 8 litre working volume were used. For the TDT experiments, Brühwurst mixture was prepared from 40% lean beef, 40% pork belly and 20% ice. To each kg was added 18 g nitrite curing salt (corresponding to 80 mg $NaNO_2$), 0.24 g sodium ascorbate, 3 g sodium diphosphate and 2.5 g Brühwurst spice mixture. For all other experiments with Brühwurst, the mixture was prepared from 55% lean beef, 25% pork back fat and 20% ice; 18 g NaCl, variable amounts of sodium nitrite (0 or 83 mg) and 2.5 g Brühwurst spice mixture were added to each kg. All liver sausage mixtures were prepared with variable amounts of pork liver (12.5 and 25%), pork jowls (25, 37.5 and 50%) and lean pork shoulder (25, 37.5, 50 and 62.5%). To each kg mixture was added 16 or 20 g sodium chloride, variable amounts of sodium nitrite (0, 67 and

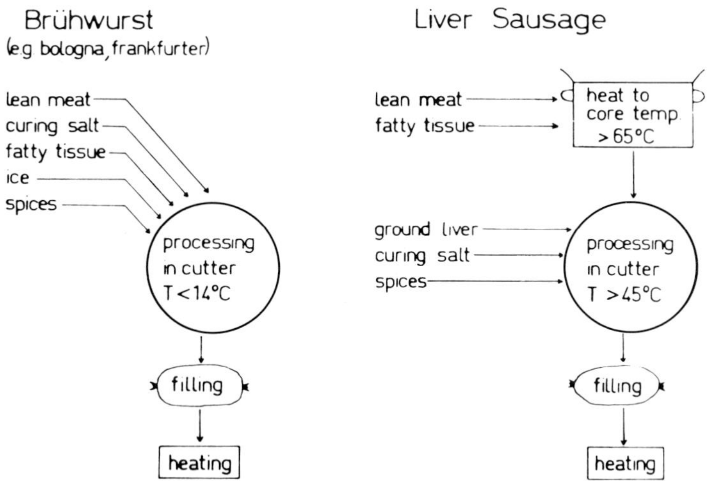

FIG. 1 *Simplified processing scheme for Brühwurst and liver sausage.*

83 mg/kg), 3 g liver sausage spice mixture and 20 g milk proteins.

Determination of Decimal Reduction Times by a TDT Method

Serum vials of 20 mm outer diameter were filled with sausage mixtures inoculated with approximately 50,000 spores of *C. botulinum* strain 213B per g. Each vial contained 5.5 g of mixture and not more than 1 ml of headspace. The vials were stoppered and sealed with aluminium caps. For each of a number of heating times, 21 vials with inoculated mixture and 9 vials with uninoculated mixture were heated in a glycerol bath of 100 or 105°C and subsequently chilled in ice water. Then, equal numbers of vials were randomly assigned to three groups. the contents of each vial of groups I and II were aseptically transferred to screw-cap bottles each containing 20 ml of CMM and CMMS, respectively. To determine survival of the spores using the sausage mixture as a recovery medium, the vials of group III were incubated as such. After four weeks at 30°C, the growth of *C. botulinum* in the cultures was examined by the method given later. From the numbers of bottles or vials showing growth, the heating time necessary to prevent development of *C. botulinum* in 50% of the replicas (ET_{50}) was estimated graphically assuming a log-normal dose-response curve (8). Decimal reduction times (D_{100} and D_{105} values) were calculated using the formula $D = (ET_{50} - t_o)/\log N_o - \log 0.693)$, where

N_o is the "initial" number of spores per vial at t_o = 10 or 7 min after insertion of the vials into the heating baths of 100 and 105°C, respectively. Both of these heat treatments correspond to a process exposing the spores to 100°C for 3 min. During heating and cooling, the core temperature in some vials was measured using a copper-constantan thermocouple. ET_{50} values were then corrected for heating and cooling lags. N_o was determined both by colony count in egg-yolk agar and by a 5-tube MPN technique employing the same media (CMM, CMMS, sausage mixture) that were used for recovery.

Probability of Spore Development after Heat Processing

Sausage mixtures were inoculated with decimal dilutions of a composite of *C. botulinum* spores (adjusted to contain equal numbers of 10 proteolytic strains) to give 5 to 50,000 spores/g). Screw-cap bottles (approximately 18 per inoculum level + 10 uninoculated controls) of 28 mm outer diameter were filled with the mixtures. Each bottle contained approximately 30 g of mixture and not more than 4 ml of headspace. After being sealed with rubber liners and screw caps, the bottles were heated by total immersion in boiling water (99 ± 0.5°C). In most experiments a core F_o value of 0.34 was obtained (as calculated from thermo-electrically measured core temperatures)*. The bottles were incubated at 21 ± 1°C for 90 days. At regular intervals, they were visually examined for gas formation and proteolysis, and representative samples were tested for botulinum toxin. After incubation, all remaining bottles were examined for growth of *C. botulinum* (see below). From the number of positive bottles, the most probable number (MPN) of *C. bolutilum* spores capable of developing in the sausage was calculated using a programmable pocket calculator (9). The total number of survivors was determined immediately after heat processing using a 3-tube MPN technique and CMM as recovery medium.

Analytical Methods

Gas formation and proteolysis in inoculated media but not in the uninoculated controls were taken as evidence for growth of *C. botulinum*. Material from all samples showing no or equivocal signs of clostridial spoilage was directly streaked onto

---

*F_o denotes the equivalent, in minutes at 121.1°C, of all heat capable to destroy spores. F_o is calculated assuming a z-value of 10°C, where z denotes the increase of temperature necessary to diminish the decimal reduction time by a factor of 10.

egg-yolk agar. If no typical *C. botulinum* colonies (surrounded by an opalescent layer) appeared after 2-3 days' incubation, the sample was tested for botulinum toxin. This was done by homogenizing it in a 1 : 9 mixture with 30 mM sodium phosphate buffer, pH 6.0, containing 0.2% (w/v) gelatine, and then injecting 0.5 ml of the centrifuged extract into each of two white mice. The mice were observed over a period of four days for signs of botulism.

An electrolyte hygrometer supplied by Novasina AG, Zürich, Switzerland, was employed to measure the water activity (a_w) of the sausages (10). The pH was measured electrometrically with an Ingold No.404 combination electrode. Sodium chloride, nitrite, nitrate and fat content was analyzed according to official methods (11).

RESULTS

In the first series of experiments, decimal reduction times for spores of *C. botulinum* 213B were calculated after determination of thermal death times of approximately 50,000 spores/g in Brühwurst and liver sausage (25% liver). D values at 100°C (D_{100}) were higher in liver sausage, especially if suboptimal recovery media (such as CMMS or the sausage) were used (Table 1). Incubation of the spores in the same sausage in which they were heated gave D_{100} values of the spores that were more than three times longer in liver sausage than in Brühwurst. Similar results were also obtained at a heating temperature of 105°C although D_{105} values were between 24 and 30% of those determined at 100°C, and indicated z-values between 8.1 and 9.5°C. This evidence suggests that liver sausage protects the spores much better against heat injury than Bruhwurst does.

In further experiments we used a spore composite of 10 proteolytic *C. botulinum* strains as inoculum and determined D_{100} values by counting the spores surviving a defined heat treatment (F_o between 0.27 and 0.35). Again, inactivation (measured as an inability to grow in CMM) was significantly slower in liver sausage than in Brühwurst (Table 2). D_{100} values were slightly lower than those given in Table 1, which could be due to the different inoculum and to a "tailing" of survivor curves often observed in heat resistance studies. The heat resistance of the spores in liver sausage appears not to be related to the fat and liver content of the product (Table 2). In addition, a reduction of its water activity to 0.97 did not increase the heat resistance determined with CMM or CMMS as recovery medium. Further, the spores were inactivated with similar rates in nitrite-free and nitrite-containing sausage mixtures. Since no discernible differences between these rates were found, both sets of data were pooled

growing in the sausage by the number of spores inoculated and by the MPN of total survivors estimated in CMM after heat processing, respectively. After being heat treated in liver sausage mixture, more than 5% of the survivors retained the capability to grow in this product. In one experiment this percentage was higher than 40%; however, due to the inherent lack of precision when counting very few survivors, both values are subject to some experimental error. Nevertheless, it may be seen that addition of nitrite did not diminish the probability of growth in liver sausage. In contrast, the addition of 80 mg/kg sodium nitrite to Brühwurst diminished the probability of growth in this product by a factor of 260. If P_i rather than P_s is considered, it can be seen that the chances of a *C. botulinum* spore present in the raw material to both survive F_o = 0.34 and form toxin during subsequent unrefrigerated storage are more than 10^4 times higher in nitrite-containing liver sausage than in nitrite-containing Brühwurst. With liver sausage, we sometimes observed a delay of one to three weeks until onset of toxin formation; this, however, is of limited practical significance because canned sausages are usually stored much longer than that.

DISCUSSION

Most cured meat products, especially liver sausage and blood sausage, would become unacceptable to the consumer if they received a "botulinum cook", i.e. a heat treatment to F_o = 2.5. Therefore, shelf-stable cured meat products such as canned luncheon meat or sausages in hermetically-sealed containers are usually heated only to F_o values between 0.2 and 0.8. Such processes will reduce the viable count of heat-resistant *C. botulinum* spores by only one- to four-log cycles. In most of these heat-processed meat products, the water activity and the pH are not sufficiently low to prevent growth of *C. botulinum*. The satisfactory microbiological stability and safety of these products, if produced on a commercial scale, must therefore be due to a low spore level in the raw material and to a sensitization of the surviving spores to the curing salts. Various authors (12-16) have shown that if the recovery media contain similar amounts of sodium chloride or sodium nitrite as meat products do, the apparent rate of heat destruction of clostridial spores is accelerated. This effect became significant at heating temperatures above 85°C (16). Similar observations have also been made with luncheon meat and comparable cured meat products (17-19). When we heated *C. botulinum* spores in Brühwurst, we also found a markedly reduced ability of the spores to tolerate even the low concentrations of NaCl and $NaNO_2$ present in our mixture. However, after being heated in sausage, a greater proportion of the survivors retained their ability to grow in the mixture during subsequent incubation at 21 or 30°C. Further, the spores were killed more slowly in liver sausage than in Brühwurst although the effect was small. This may explain why Lerche (20) did not observe an increased heat resistance of *C. sporogenes* spores in liver sausage.

Variations in the water activity and pH values of our sausages were too small to account for the different rates of heat inactivation and injury observed. Likewise, the liver and fat content of the sausages had little if any influence on the heat resistance of the spores (Table 2). One could also rule out the possibility that liver sausage favours repair of injured spores: if so, recovery rates in CMMS would have been similar after heating in liver sausage and Brühwurst. Therefore, we suggest that the degree of protection of the spores is related to the structure of the sausage mixture. During the processing of Brühwurst mixture, the ingredients are not pre-heated, and the mixture can be considered as a dispersion, i.e. a protein gel network around fat particles (21). In the manufacture of Kochwurst (liver sausage and blood sausage), fatty tissues are pre-heated so that the fat is molten and an oil-in-water emulsion is obtained (22). Thus, spores may be in closer contact to fat which is known to protect them from heat inactivation (20,23), possibly because the water activity of fat is low during heating (24). Preliminary data from this laboratory (H. Hechelmann, unpublished observations) indicate that blood sausage protects the spores from inactivation in a similar manner.

The addition of nitrite (83 mg $NaNO_2$/kg) did not diminish the probability of growth of *C. botulinum* in liver sausage. In principle, the same applies to the non-proteolytic *C. botulinum* strains (25). Hauschild *et al.* (26) studied growth of *C. botulinum* in liver sausage comparable to ours, and found that addition of 100 mg/kg $NaNO_2$ or less reduced the risk of toxin formation only if the product contained about 5% salt in water. By extending heating temperature and time, the inhibitory effect of the added nitrite diminished in parallel to the residual nitrite content in the sausage, being practically absent after a heat process of 90 min at 80°C. This suggests that upon heating sausage mixtures containing liver, increased amounts of iron are liberated from iron storage proteins (e.g. ferritin) of liver and become available to "inactivate" the nitrite by formation of ferric nitrosyl complexes or by catalysing the autoxidation of nitrite to nitrate. Pork liver contains 200-300 mg/kg iron, which is 10 to 15 times more than in muscle (27). Accordingly, we only found between 5 and 15 mg/kg $NaNO_2$ but up to 70 mg/kg KNO_3 in our liver sausages.

A striking result of the present study is that canned, cured liver sausage with water activity of 0.98 is four orders of magnitude more likely to become toxic than Brühwurst with the same a_w, manufactured with the same amount of nitrite. In practice, the relative risk may be even higher since Kochwurst tends to contain more clostridial spores (from pork skins, blood or liver) than Brühwurst. Indeed, of the cases of botulism in West Germany due to the consumption of sausages (probably in the order of 10 per year), almost all could be traced back to liver sausages or blood sausages. To the author's knowledge, all of the sausages involved have been canned and stored in the home or in small butcher's shops where heat treatment and refrigeration are poorly controlled. Shelf-stable Kochwurst produced on a commercial scale has a good safety record, although not infrequently manufactured without any nitrite addition. This could be due to a lower water activity of commercial products (in a survey, we mostly found a_w values around 0.97) and to a more advanced heat processing technique but it also depends on a low content of clostridial spores in the raw material.

ACKNOWLEDGEMENTS

I wish to thank Inge Hûbner for skilful technical assistance and Professor L. Leistner for critically reading the manuscript.

REFERENCES

1. Christiansen, L.N., Johnston, R.W., Kautter, D.A., Howard, J.W. and Aunan, W.J. (1973) *Appl. Microb.* **25**, 357-362.
2. Rhodes, A.C. and Jarvis, B. (1976) *J. Fd. Technol.* **11**, 13-23.
3. Roberts, T.A., Gibson, A.M. and Robinson, A. (1981) *J. Fd. Technol.* **16**, 239-266.
4. Sofos, J.N., Busta, F.F. and Allen, C.E. (1979) *J. Fd. Protect.* **42**, 739-770.
5. Lücke, F.-K., Hechelmann, H. and Leistner, L. (1982) *Fleischwirtschaft* **62**, 203-206.
6. Schmidt, C.F. and Nank, W.K. (1960) *Food Research* **25**, 231-327.
7. Hauschild, A.H.W. and Hilsheimer, R. (1977) *Can. J. Micro. biol.* **23**, 829-832.
8. Meynell, G.G. and Meynell, E. (1965) *Theory and Practice in Experimental Bacteriology*, pp.179-181, Cambridge University Press.
9. Koch, A.L. (1981) in *Manual of Methods for General Bacteriology* (Gerhardt, P., ed.), pp.188-192, American

Society of Microbiology, Washington D.C.
10. Rödel, W., Krispien, K. and Leistner, L. (1979) *Fleischwirtschaft* **59**, 831-836.
11. Anonymous (1980) in *Amtliche Sammlung von Untersuchungsverfahren* nach § 35 LMBG (Bundesgesundheitsamt, ed.), part L.06.00.
12. Roberts, T.A. and Ingram, M. (1966) *J. Fd. Technol.* **1**, 147-163.
13. Duncan, C.L. and Foster, E.M. (1968) *Appl. Microbiol.* **16**, 401-405.
14. Pivnick, H. and Thacker, C. (1970) *Can. Inst. Food Technol. J.* **3**, 70-75.
15. Ingram, M. and Roberts, T.A. (1971) *J. Fd. Technol.* **6**, 21-28.
16. Jarvis, B., Rhodes, A.C., King, S.E. and Patel, M. (1976) *J. Fd. Technol.* **11**, 41-50.
17. Silliker, J.H., Greenberg, R.A. and Schack, W.R. (1958) *Fd. Technol.* **12**, 551-554.
18. Riemann, H. (1963) *Fd. Technol.* **17**, 38-42, 45, 46, 49.
19. Pivnick, H., Barnett, H.W., Nordin, H.R. and Rubin, L.J. (1969) *Can. Inst. Food Technol. J.* **2**, 141-148.
20. Lerche, M. (1941) *Fleischwirtschaft* **21**, (18), 1-7.
21. Honikel, K.O. (1982) *Fleischwirtschaft* **62**, 16, 19-22.
22. Fischer, A. and Killeit, U. (1980) *Z. Lebensm.-Technol.-Verfahrenstechnik* **31**, 1-7.
23. Molin, N. and Snygg, B.-G. (1967) *Appl. Microbiol.* **15**, 1422-1426.
24. Senhaji, A.F. and Loncin, M. (1977) *J. Fd. Technol.* **12**, 203-215.
25. Lücke, F.-K., Hechelmann, H. and Leistner, L. (1981) in *Psychotrophic Microorgsniams in Spoilage and Pathogenicity* (Roberts, T.A., Hobbs, G., Christian, J.H.B. and Skovgaard, N., eds.), pp.491-497, Academic Press, London.
26. Hauschild, A.H.W., Hilsheimer, R., Jarvis, G. and Raymond, D.P. (1981) *J. Fd. Protect.* **45**, 500-506.
27. Anynomous (1981) in *Food Composition and Nutrition Tables 1981/82* (Deutsche Forschungsanstalt fürLebensmittelchemie, ed.), 2md Ed., pp.288, 336, 370, Wissenschaftl. Verlagsgesellschaft Stuttgart.

THE HEAT RESISTANCE OF ASCOSPORES OF *KLUYVEROMYCES* SPP. ISOLATED FROM SPOILED HEAT-PROCESSED FRUIT PRODUCTS

HENRIETTE M.C. PUT

Carl. C. Conway Laboratories,
Thomassen & Drijver-Verblifa N.V.
P.O. Box 103, 7400 GB Deventer, The Netherlands

SUMMARY

The heat resistance (D- and z-values) of nine *Kluyveromyces* strains (5 species: *K. bulgaricus; K. cicerisporus; K. fragilis; K. lodderi; K. veronae*), was investigated by a previously described method (Put *et al.* 1977). $D_{60°C}$ values vary between ½-50 min; z-values vary between 4.0 - 7.0°C. Heat resistance of *Kluyveromyces* is less dependent on the species than on the strain.

Kluyveromyces ascospores are protected against heat inactvation by sucrose, fruit sugars and to a lesser extent by NaCl. More fundamental research is required to understand the mechanism of yeast ascospore heat resistance. Since statistical data and laboratory models for studying yeast ascospore heat resistance are not yet available, the practical implications are at present neither predictable nor calculable.

INTRODUCTION

Kluyveromyces spp. form one or more ascospores in fragile asci which rupture readily at maturity. Ascospores tend to agglutinate, are kidney shaped, weakly acid-fast and hence difficult to recognize in routine examinations. The inter-breeding relationship in *Kluyveromyces* spp. is relatively high (Johannsen, 1980; Van der Walt, 1980; Van der Walt and Johannsen, 1978 and 1979). Soil, the external surface of plant leaves (phyllosphere), fruits, flowers and some insects are a natural habitat for some *Kluyveromyces* spp. (Davenport, 1976a,b).

Except in lactase production and ethanol fermentation from cheese whey by *K. fragilis* and *K. lactis*, *Kluyveromyces* species are not used in industrial fermentation processes. Both species have been isolated frequently from a variety of (lactose containing) dairy products (Phaff and Starmer, 1980; Suriyaracheh and Fleet, 1981; Janssens *et al.* 1983).

In laboratory investigations, carried out by us, *Kluyveromyces* spp. have more frequently been associated with spoilage of heat processed soft drinks and fruit products than reported by other authors (Sand, 1976; Baird-Parker and Kooiman, 1980; Röcken *et al.* 1981). We have therefore now studied the heat resistance of *Kluyveromyces* ascospores.

MATERIALS AND METHODS

Organisms, Isolation, Identification and Storage

Nine *Kluyveromyces* strains were studied (Table 1). Three strains were obtained from culture collections; the remaining six strains were isolated from spoiled heat processed canned fruit products in sugar syrup or raw materials for soft drinks production.

TABLE 1 *Types and origin of* Kluyveromyces *strains tested*

Kluyveromyces species	Strain code*		Source
K. bulgaricus	120	III	apple sauce
K. bulgaricus	291	III	raspberries
K. cicerisporus	277	III	strawberries
K. fragilis	273	I	CBS 6397
K. lodderi	276	I	CBS 2757
K. veronae	237	II	orange squash
K. veronae	272	II	soft drink plant
K. veronae	274	I	Groningen 524
K. veronae	289	III	mirabelles

*Our own yeast culture collection number; *I*, international culture collections: strains isolated from fresh fruit, fruit orchard or fruit-tree leaves; *II*, raw materials for soft drink production; *III*, spoiled canned heat processed fruit product in sugar syrup.

Methods and techniques used for isolation, identification and storage of the *Kluyveromyces* strains were as previously described for *Saccharomyces* spp. (Kreger-van Rij, 1969, 1980; Lodder, 1970; Put *et al.* 1976). In relation to the thermoduric character of *Kluyveromyces* spp. higher incubation temperatures were sometimes applied (33-35°C). In addition temperature ranges for growth (0-45°C) were tested (Phaff *et al.* 1978; Davenport, 1980).

Preparation of Vegetative Cells and Ascospores

Pre-enrichment medium. From the stock culture, mycophil broth (BBL) pH 7.0 was inoculated and incubated static at 25°C for 16-24 h. Then the yeast culture was harvested in saline solution (0.9% w/v NaCl) centrifuged, resuspended in saline solution and reinoculated in a medium for vegetative growth or ascosporulation.

Vegetative cells. These cells were grown in BBL mycophil broth to which 1,000 µg/g of erythromycin ethylsuccinate was added (Eli Lilly Benelux, Bus 1, Brussel 1.000, Belgium) to inhibit ascosporulation (Puglisi and Zennaro, 1971; Put and De Jong, 1982b). Flasks were shaken for 16-20 h at 25°C ± 1 . The yeast was harvested by centrifugation, resuspended in saline (0.9% w/v NaCl) or in the buffered heating substrate (see below).

Vegetative cell counts were made microscopically in a counting chamber, as well as by the plate count method on mycophil agar (BBL) pH 7.0 incubated at 25°C for up to 4 days.

Ascospores. Ascospores were obtained by a heavy inoculation of the washed pre-sporulation culture on: Difco potato dextrose agar (PDA, pH 5.5); Difco malt extract agar (MA, pH 4.7); vegetable juice agar (VA, pH 5.7); BBL mycophil agar (MyA, pH 7.0); Kleyn acetate agar (KA, pH 7), incubated for 2-15 days at 25°C (Lodder, 1970). Then the sporing cultures were harvested in saline solution, washed twice by centrifugation in a Sorvall RC-2B centrifuge at 10°C ± 1° for 10 min and 10,000 rev/min (13,000 g), resuspended in saline solution and stored at 5°C ± 1°C.

Total counts of asci, ascospores and vegetative cells were carried microscopically in a Bürker Türk counting chamber (W. Schreck, Hofheim, BRD). Discrimination of ascospores and vegetative cells by acid-fast staining of a thin smear according to the method of Ziehl-Neelsen modified by Lindegren (Savarese, 1974) showed a lower reproducibility of ascospore counts then obtained by the counting chamber method and phase-contrast microscopy.

For staining of ascospores the following modification of the Lindegren method was used. A thin smear of a sporulating culture was prepared on a clean glass slide, dried in hot air of 45-50°C, the slide was then flooded with Ziehl-Neelsen's carbol-fuchsin and stained for 2 min at 90°C whereafter the stain was removed and the slide flooded with an acetic acid solution of 15% w/v for 10 s for decolorization, rinsed lightly in tap water, air dried, and examined microscopically. *Kluyveromyces* ascospores stain rose/red, vegetative cells stain weak rose. *Kluyveromyces bulgaricus* ascospores are less acid fast than *K. cicerisporus*. Freshly prepared ascospore suspensions are more acid fast than spores in older suspensions, stored at 5°C. This phenomenon is related to the weak acid fastness of *Kluyveromyces* ascospores, which lose their stain (Miller and Kingsley, 1961). Measurements of turbidity and enzyme activity are unsuitable as ascospore counting methods (Von Bärwald, 1972; Bestic and Arnold, 1976).

Determination of Heat Resistance

For determination of heat resistance, amounts (1.0 ml) of the buffered yeast cell suspensions were pipetted into Thermal Death Time (TDT) tubes. The tubes were closed with rubber stoppers and clamped in specially designed racks (Put and Wybinga, 1963). The thermal treatment was carried out in a temperature-regulated 100-litre water-bath. The temperature of the water was controlled to within 0.05°C, recorded electronically. The heating lag was calculated from the heat penetration curve according to the method of Sognefest and Benjamin (1944). Each temperature/time combination was tested 12-fold. Vegetative cells were heated at 52.5°, 55°, 57.5° and 60°C. Ascospores were heated at 57.5°, 60°, 62.5°, 65°, 66° and 67.5°C.

For studying the influence of a_w (water activity) on the heat resistance, spore suspensions were equilibrated by centrifuging the ascospore suspension at 5°C and resuspending in a buffered (pH 4.5) solution of sucrose, strawberry juice or NaCl of a known a_w (0.995; 0.975; 0.950), followed by storage at 5°C for 16 h, after which the initial spore number was assessed by plate count on Mycophil Agar (BBL) incubated for 10 days at 25°C. At the same time the heat resistance was tested. A Sina equihydroscope (Sina AG, Zurich) was used for a_w measurements (Leistner and Rödel, 1975; Gal, 1975).

After the heat treatment, the TDT tubes were cooled immediately in running tap water, thereafter to 10 out of 12 tubes of each series 4.5 ml of Mycophil Broth (BBL) was added (final pH 5.1). Incubation was at 25°C for 1 month.

In tubes showing growth at the highest temperature/time combination the purity of the surviving culture was assessed. The remaining two tubes of each series were used for the assessment of the number of cells or ascospores surviving heat treatment by poured-plate counts into Mycophil Agar (pH 7.0).

Calculation of Heat Resistance

The calculations of heat resistance, TDR curves were plotted as semi-log graphs. The end-point TDT curves were plotted by the same method as were the phantom TDT curves.

The best straight line drawn through the D points, the "phantom" TDT curve, gives the slope but not the position in relation to a complete destruction for a particular concentration of vegetative cells or ascospores. The z-value represents the amount of heat (°C) required for the TDT curve to traverse one-log cycle, to bring about a ten-fold change in the TDT or D-value (Schmidt, 1957).

To calculate the D_{10} values the end-point method of Schmidt (1950, 1957) was applied, using the formula of Ball (1923, 1928): $D = U/\log a - \log b$, where D = decimal reduction time or death rate constant at T (°C), this means that the time (min) required to obtain a 90% reduction in surviving cell number, is equivalent to the time (min) for the TDT curve to transfer one-log cycle; U = heating time (min); a = initial number of vegetative cells or ascospores; and b = most probable number of surviving cells or ascospores (Stumbo, 1948, 1965).

Transmission Electron Micoscope Studies

Populations of young-, mature, germinating, and heat-inactivated spores of *Kluyveromyces bulgaricus* 120 were washed three times in saline solution followed by centrifugation. Washed cells were fixed with 1.5% w/v $KMnO_4$ solution for 20 min at room temperature. The fixed material was then washed, dehydrated with ethanol and during dehydration poststained with a saturated solution of uranyl acetate in 100% ethanol for 1 h. The cells were embedded in Epon resin and cut with glass knives on an ultramicrotome. The sections were viewed and photographed in a Philips EM microscope (Kreger-van Rij and Veenhuis, 1976).

RESULTS

$D_{60°}$ Values of *Kluyveromyces* Ascospores

$D_{60°}$ values of ascospores of nine *Kluyveromyces* strains (five species) vary between 0.5 and 40 min (Table 2). The ascospore

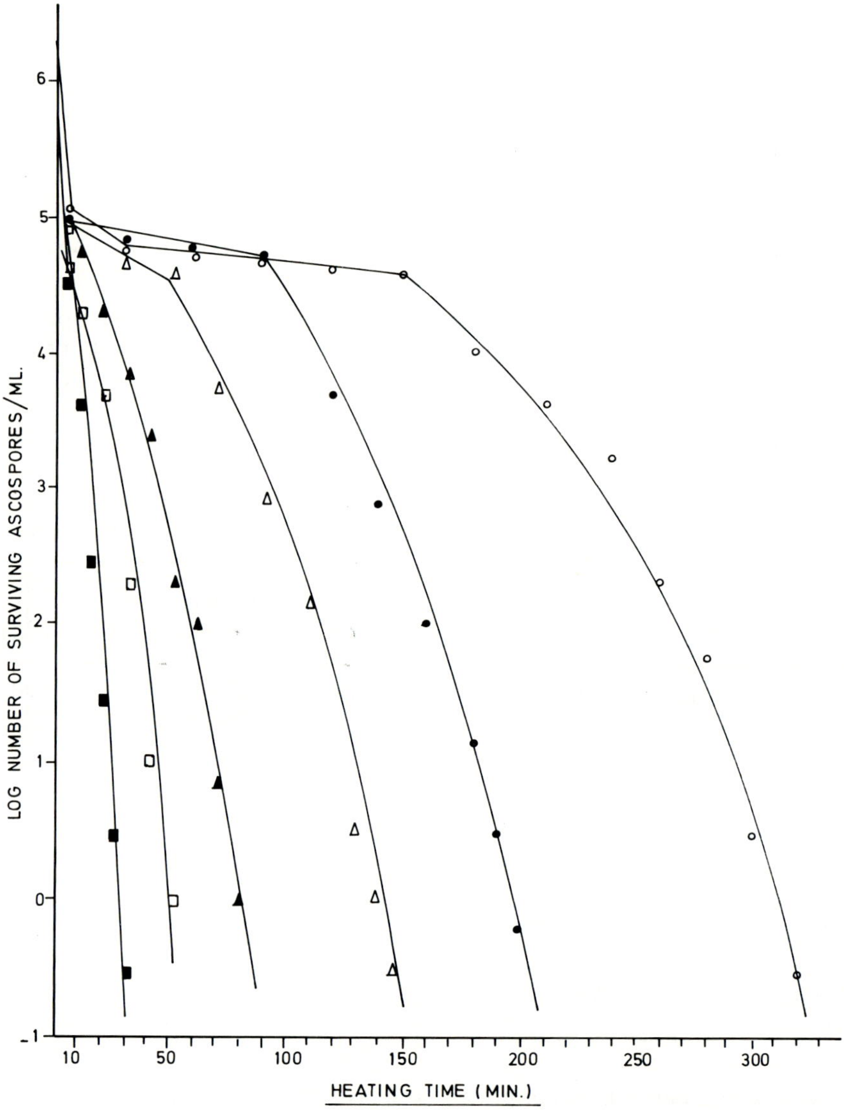

FIG. 2a *Thermal death rate curves of* K. bulgaricus *120 in 1/20 M phosphate-citrate buffer of pH 4.5. Initial number of ascospores 5 x* 10^5 *per ml. Heating temp. (°C):* o, *60°;* ●, *61°;* ▲ (open), *62°;* ▲, *63°;* □, *64°;* ■, *65°. Ascospores: veg. cells = 1 : 1. (Put and de Jong, 1980).*

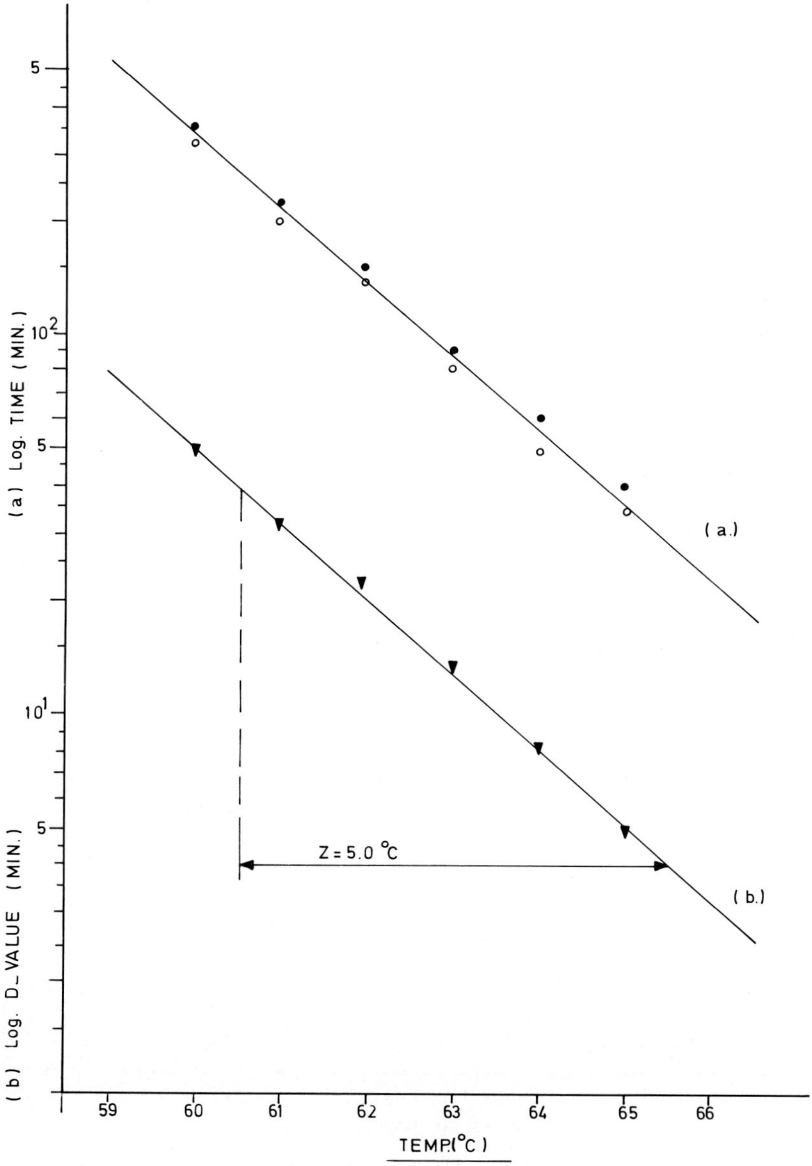

FIG. 2b *Thermal death time curves of* K. bulgaricus *120 in 1/20 M phosphate-citrate buffer of pH 4.5. Initial number of ascospores 5 x 10^5 per ml.* o, *survival time;* ●, *destruction time;* ▼, *D-values (min). (Put and de Jong, 1980)*

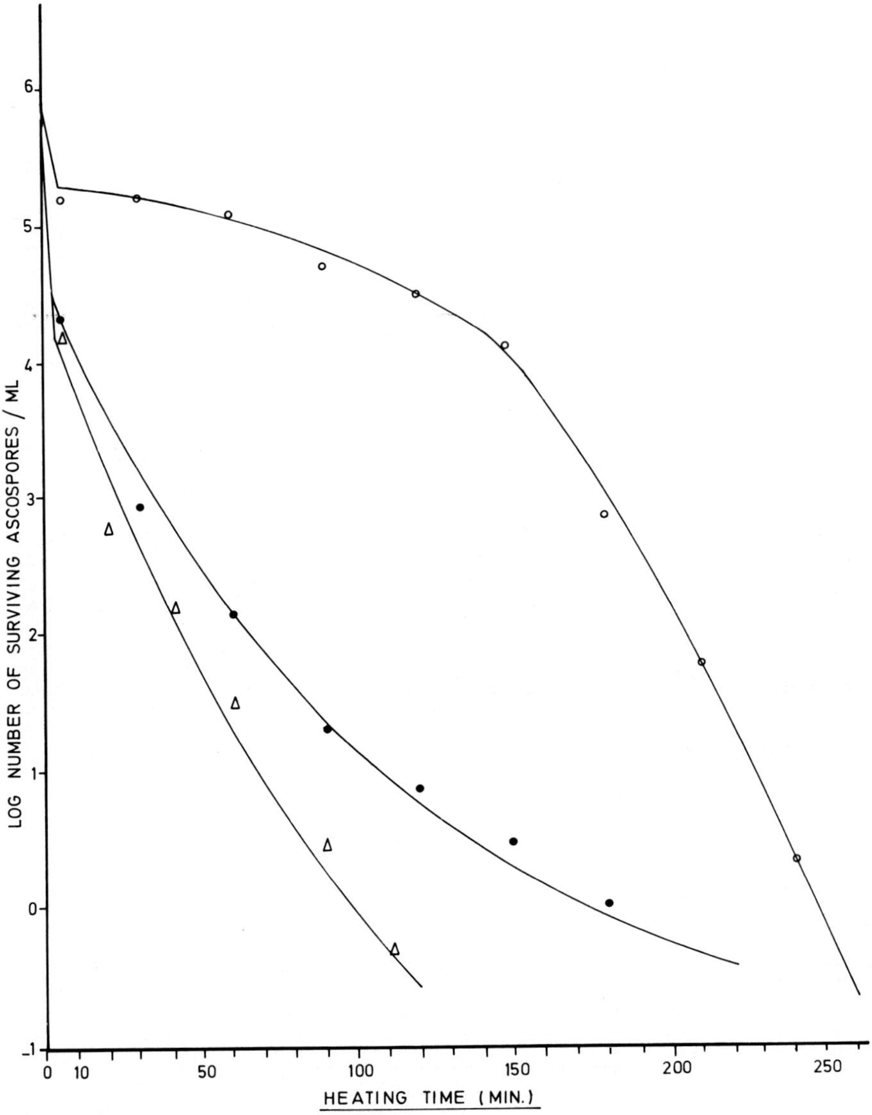

FIG. 3 *Thermal death rate curves of* K. cicerisporus *277 in 1/20 M phosphate-citrate buffer of pH 4.5. Initial number of ascospores 10 per ml. Heating temp. (°C):* o, *60°;* ●, *62.5°;* Δ, *65°. Ascospores: veg. cells = 1 : 3.*

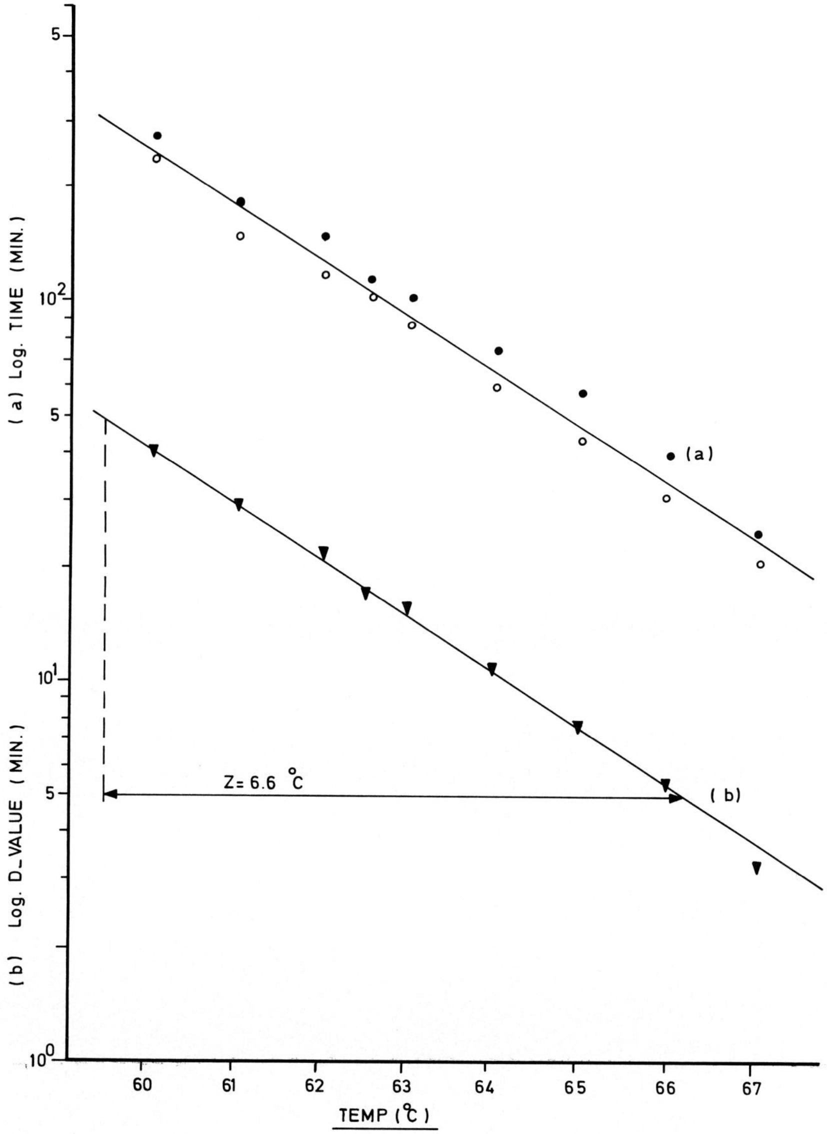

FIG. 4 *Thermal death time curves of* K. cicerisporus *277 in 1/20 M phosphate-citrate buffer of pH 4.5. Initial number of ascospores:* 10^6 *per ml.* o, *survival time;* ●, *destruction time,* ▼, *D-values (min).*

the death rate curves showing linear lines. The z-values were 4.5° and 4.8°C (Put and de Jong, 1982b).

$D_{60°C}$ Values of Decimal Dilutions of Ascospores Produced on PDA

Thermal death rate (TDR) studies at 60°C using decimal dilutions of the same stock suspensions of *K. bulgaricus* and *K. cicerisporus* showed that the shape and the slope of the survivor curves remain almost equal and parallel over the whole range, indicating that the heat resistance is homogeneously distributed among the ascospore population tested, although the individual spores are not equally heat resistant.

TABLE 5 *Heat resistance of* Kluyveromyces *ascospores (D- and z-values) in different solutes*+

Solute pH	% w/v	a_w	*Kluyveromyces* strain 120 D 60°C (min)	120 z (°C)	277 D 60°C ǂ (min)	277 z (°C)
Buffer 4.5	1/20 M		44 *8-9	5-6	24-30 1	6.5-7.0
NaCl pH 4.5	1	0.995	50-55 *9-10	5.0	24	7.0-7.5
	4	0.975	50-55 *9-10	5.0	24-30	7.5-8.0
	8	0.950	65-70 *10-11	6.0	36	8.0-8.5
Sucrose pH 4.5	10	0.995	NT *8	NT	30-35	6.5-7.0
	30	0.975	NT *10	NT	36-40	4.0-5.0
	45	0.950	NT *16	NT	50.54	7.0-7.2
Strawberry syrup	30	0.975	NT	NT	70	4.0-5.0

* $D_{64°}$; NT: not tested; + end point method of Schmidt (1957).
ǂ shape of survivor curves: Fig. 5.

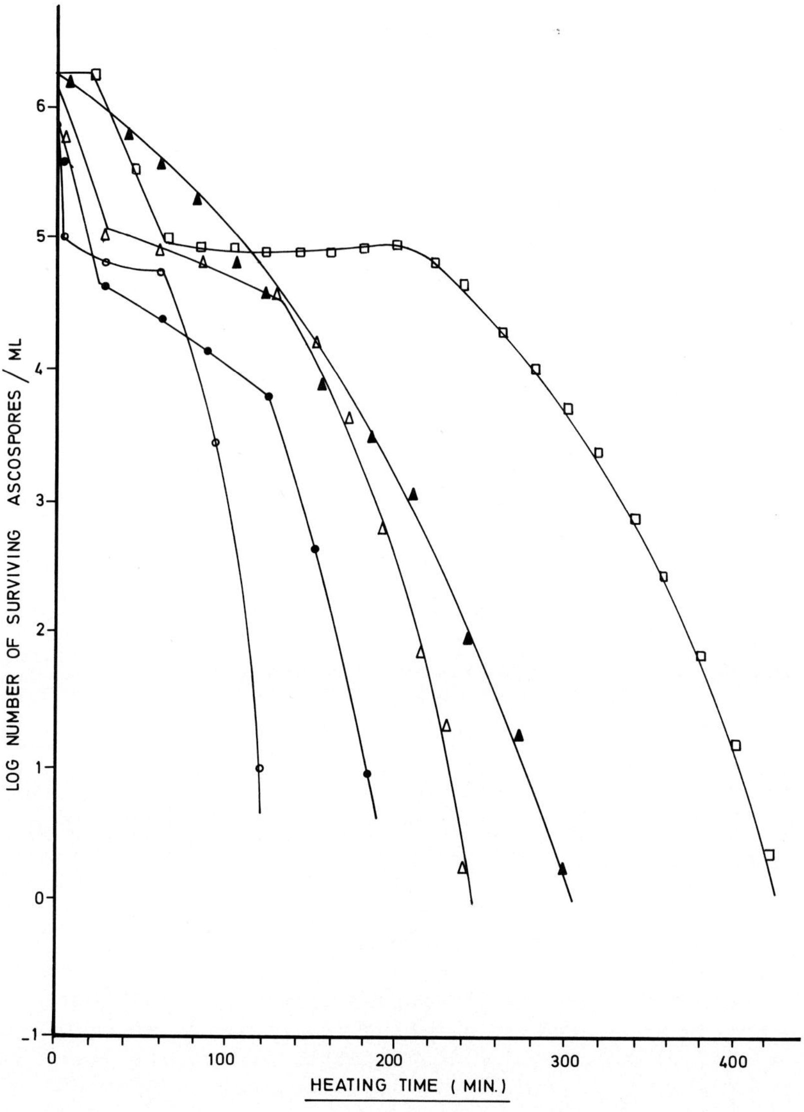

FIG. 5 *Thermal death rate curves of* K. cicerisporus *277 at 60°C. Initial number of ascospores c* 10^6 *per ml. Heating menstruum:* o, *1/20 M, phosphate-citrate buffer of pH 4.5;* ●, *sucrose 10%,* $a_w = 0.995$, *pH 4.5;* Δ, *sucrose 30%,* $a_w = 0.975$, *pH = 4.5;* ▲, *sucrose 45%,* $a_w = 0.950$, *pH = 4.5;* □, *strawberry juice,* $a_w = 0.975$, *pH = 3.5.*

TABLE 6 *Differentiation of D-curves of ascospores of* K. cicerisporus *277 at 60°C.*

Suspension code	Heating medium pH 4.5	A_w	D (min) Values per log cycle 1st	shoulder †	2nd-5th	average §
IV	1/20 M buffer	0.998	5	60	* 15	24
IV	Sucrose 10% w/v	0.995	20	-	ǂ 40	36
IV	Sucrose 30% w/v	0.975	25	60	ǂ 36	40
IV	Sucrose 45% w/v	0.950	100	-	ǂ 40	60
IV	Strawberry juice	0.975	50	200	* 48	75

IV: TDR 60 C curves, Fig. 5;
† : shoulder, non or slight heat inactivation;
* : straight logarithmic part of TDR curve;
ǂ : concave upwards part of TDR curve;
- : no shoulder;
§ : END point method (Schmidt, 1957).

Moreover, it was observed that the death rate curves at 60 and 62.5°C are triphasic and non-linear, showing two break points (Figs. 2 and 3): firstly, a rapid heat inactivation of circa 90% of the ascospores during the first few minutes of the heating process ($D_{60°C}$ of circa 5 min); secondly, a relatively large shoulder over circa 30% of the total thermal destruction time needed, shifting the curve by about a half-log cycle, slowly inactivating another 50% of the ascospores present; thirdly, a linear curve, the slope of which is almost parallel with the death rate curve of the other decimal dilutions of the ascospores tested. This latter part of the curve shows $D_{60°C}$ values of circa 40 min (*K. bulgaricus* 120) and circa 30 min (*K. cicerisporus* 277) being the resistance to heat of the remaining 5% of ascospores initially present.

By the end point method of Schmidt (1957) calculated $D_{60°C}$ values of 5 decimal dilutions of the same stock suspension vary between 36-50 min (*K. bulgaricus* 120) and 30-45 min (*K. cicerisporus* 277).

The Influence of Sugar and Salt (a_w) on the Heat Resistance of Ascospores (D and z values)

$D_{60°C}$ values, calculated by the end point method of Schmidt (1967) are tabulated in Table 5. Some z-values are given in the same table. The thermal death rate curves are not similar in shape (Fig. 5). The shapes and slopes of the curves are influenced by the a_w of the heating substrate as well as by the type of a_w decreasing solute (Table 6).

Transmission Electron Microscope Studies of *K. bulgaricus* 120 Ascospore

The outer wall layer of the young ascospore is thin and smooth, the inner ascospore wall layer is broad and light (Plate 1a,b).

The mature ascospore, on the contrary, clearly shows a thicker and darker outer wall layer, and the plasmalemma is darker (Plate 2a,b).

Differences between the high electron density of the outer ascospore wall layer as compared to the lower density of the vegetative cell wall is clearly visible in Plate 3a,b.

Breaking of the ultrastructure of the ascospore wall during early germination is shown in Plate 4a. The end-phase of ascospore germination is seen in electron micrograph 4b.

In a heat inactivated ascospore the outer part of the spore wall seems to be intact, though a greyish layer around the protoplasm does probably indicate leakage of the plasmalemma as a result of multiple membrane injuries (Plate 5, Kreger-van Rij, 1979 and unpublished).

DISCUSSION

$D_{60°C}$ values of ascospores of nine *Kluyveromyces* strains tested vary between 0.5 - 40 min (Table 2). From the data obtained it can be concluded that heat resistance of *Kluyveromyces* ascospores is less dependent on the species than on the strain. Heat selection and heat adaptation may have occurred after industrial- or laboratory-applied sublethal heat treatments. It should thus be taken into account that industrially applied sublethal heat processes may select yeasts which have a high heat resistance, either innate or acquired by adaptation.

In standard thermal death time (TDT) studies the maximal D60°C value of *Kluyveromyces* ascospores observed, is higher ($D_{60°C}$ = 50 min) than of *Saccharomyces* ascospores ($D_{60°C}$ = 22 min; Put and de Jong, 1982a). In was also observed that $D_{60°C}$ values of *Kluyveromyces* and *Saccharomyces* ascospores are significantly higher (350- and 160-fold respectively) than D60 C values of the corresponding vegetative cells (Put and de Jong, 1982a,b).

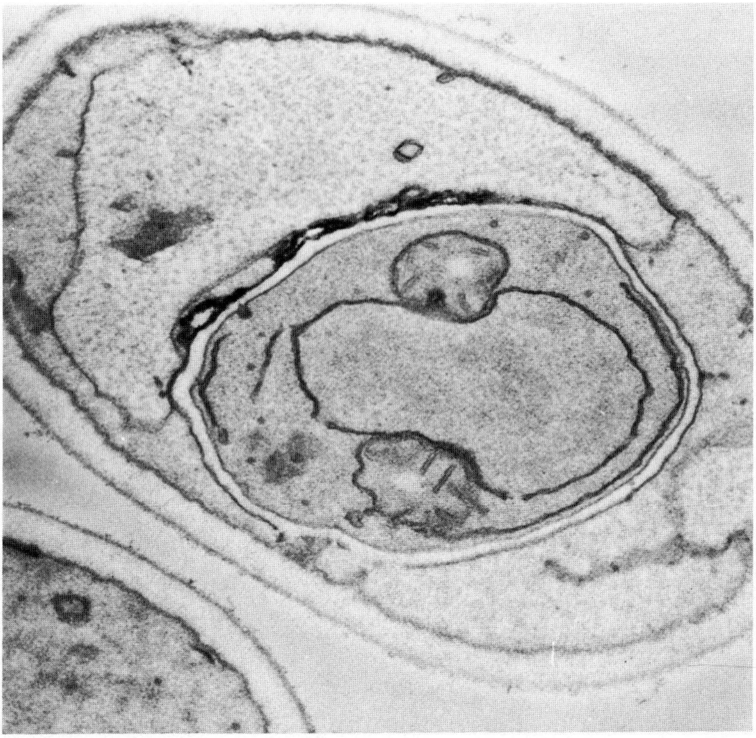

PLATES 1a and b: Kluyveromyces bulgaricus *strain 120.*
(a) very young ascospore x *145.600.*

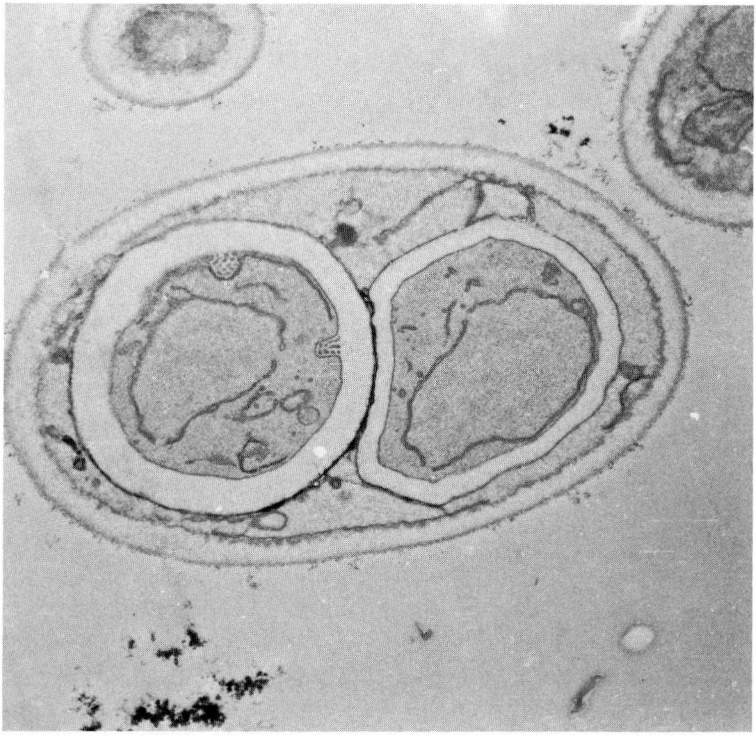

(b) young ascospores in ascus x 97.200 (Kreger-Van Rij unpublished).

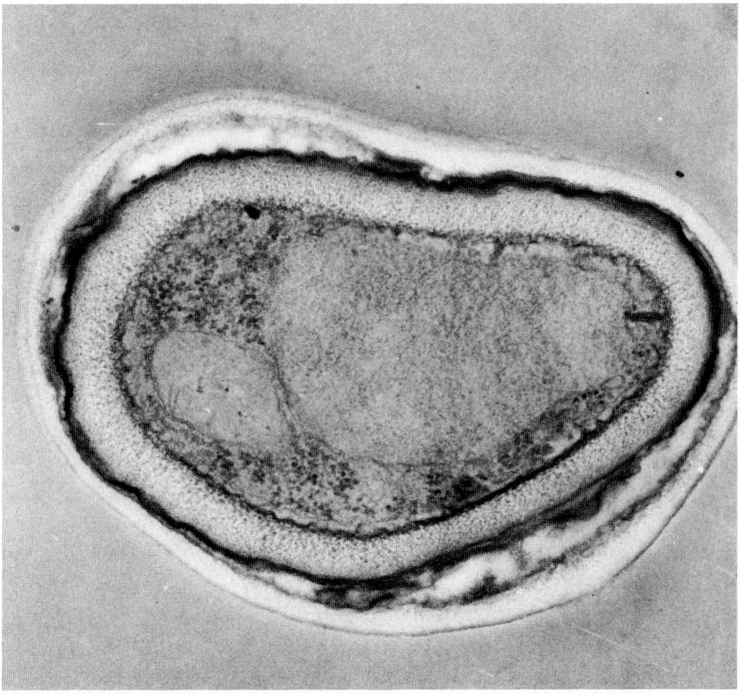

PLATES 2a and b: Kluyveromyces bulgaricus *strain 120.*
(a) mature ascospore in the ascus x *145.600.*

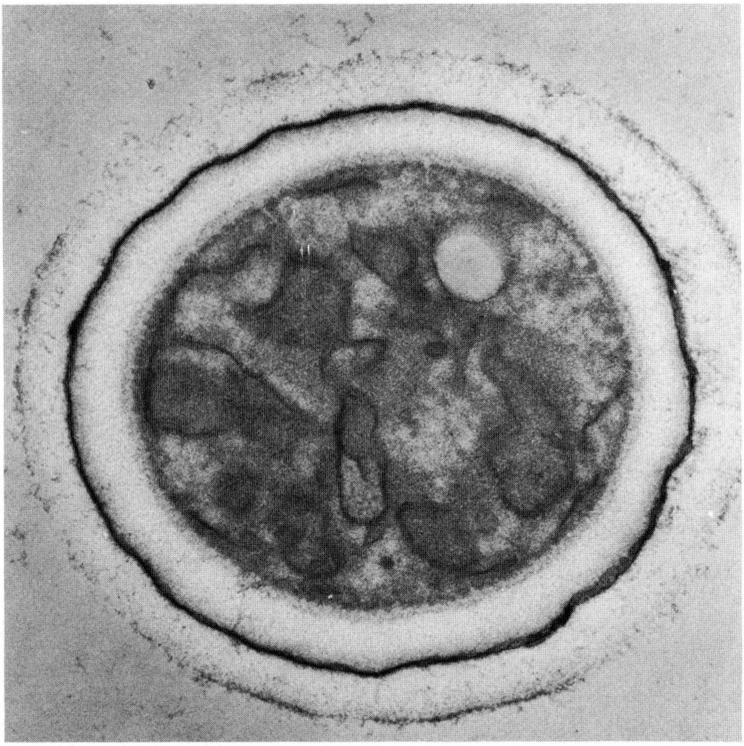

(b) mature ascospore in ascus remnants x *145.600 (Kreger-Van Rij unpublished).*

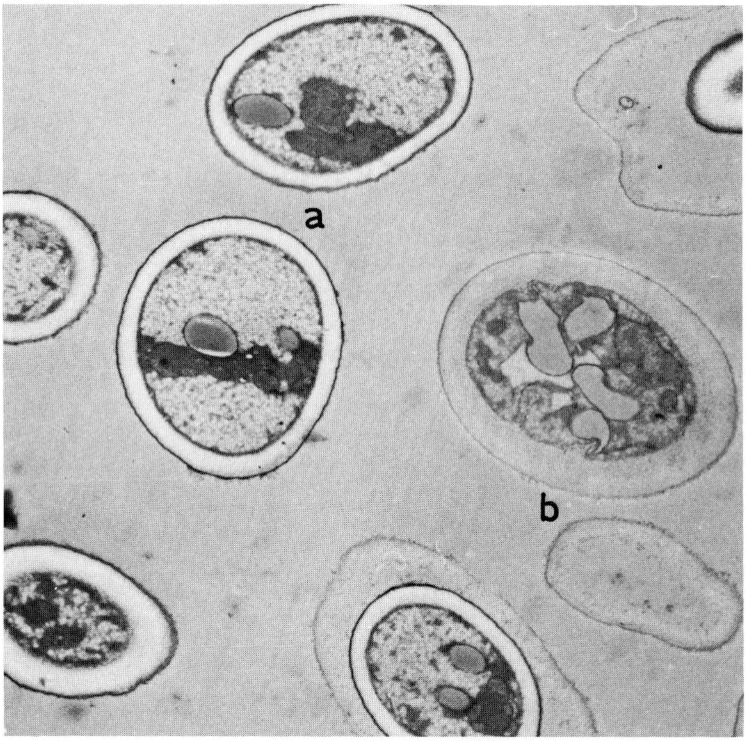

PLATE 3: Kluyveromyces bulgaricus *strain 120.*
(a) free ascospores.
(b) vegetative cell x *51.200 (Kreger-Van Rij unpublished).*

In heat resistance studies of vegetative yeast cells at pH 4.5 by Beuchat (1981b,c) $D_{50°C}$ values of 16 min (*K. fragilis*) was observed. The calculated z-value was 10.1°C. Tailing of survivor curves was occasionally observed, but D-values were calculated from straight-lined portions of the death rate curve.

In other recently published papers on heat resistance studies of vegetative yeast cells much lower D-values are reported. Cerny (1980) reported $D_{60°C}$ of 0.05 min for *Saccharomyces cerevisiae* at pH 4.5 and a z-value of 5°C. Tsang and Ingledew (1982) reported $D_{50°C}$ of 0.2 and 7 min (*Saccharomyces carlsbergensis* and *S. williams*) and z-values of 4.5 and 2.7°C. Besides it was noted by Tsang and Ingledew (1982) that z-values of beer spoilage yeasts are lower than used by Del Vecchio *et al.* (1951) for the classic beer pasteurization unit (PU) calculation which is related to z = 7°C.

In the reported studies the survivor curves of vegetative *Kluyveromyces* cells were straight-lined. The calculated $D_{60°C}$ values were 0.2 min (*K. bulgaricus* 120) and 0.1 min (*K. cicerisporus* 277). The corresponding z-values were 4.5 and 4.8°C.

Tailing of survivor curves of vegetative cells led us to believe that low numbers of ascospores might have been formed, even in typical vegetative growth medium, reducing the difference between D-values of vegetative yeast cells and their ascospores (Cerf, 1977). To obtain a pure vegetative yeast culture, sporulation can be suppressed adequately by erythromycin ethylsuccinate (1000 μg per ml growth medium, Puglisi and Zennaro, 1971), Put and de Jong, 1982b). However, it should be taken into account that after prolonged incubation (>48 h) of *K. bulgaricus* in the erythromycin-containing vegetative growth medium, some sporulation may occur and hence some tailing of the survivor curve. It may therefore be concluded that, without the addition of erythromycin, the incidence of spontaneous yeast sporulation, although in low numbers, is relatively high, even in a medium with poor environmental sporulation conditions. It is not surprising therefore that in previous publications a pronounced difference between the heat resistance of vegetative yeast cells and of the corresponding ascospores has only rarely been reported (Put and de Jong, 1982b).

The typical observation that the maximum ascospore heat resistance of *Kluyveromyces* is not yet reached at the point of morphological maturity but after some two days of further incubation at 25°C (Table 3) may probably be related to the time which is needed for the formation of a typical heat resistant biochemical structure of the ascospore. In this context it may be postulated that there may exist some

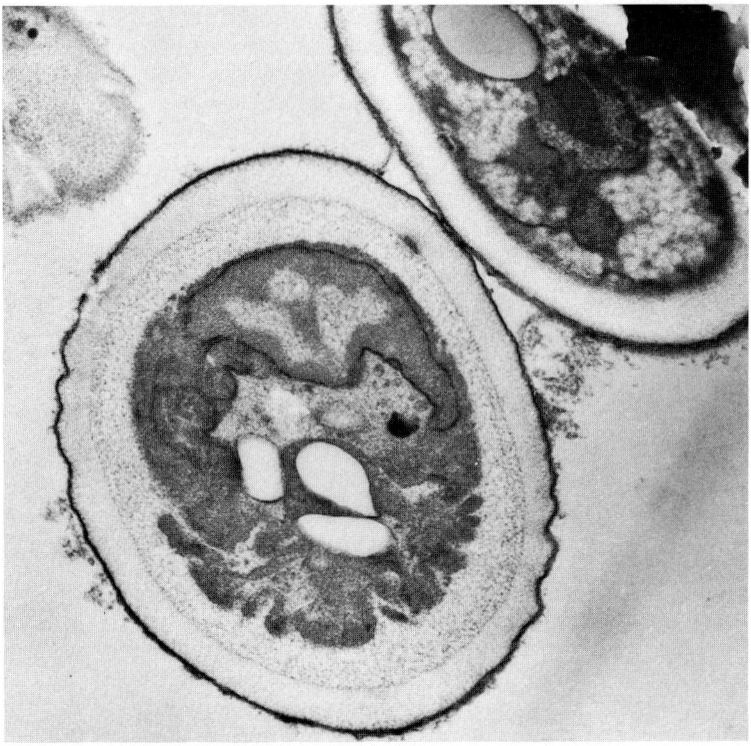

PLATES 4a and b: Kluyveromyces bulgaricus *strain 120.*
(a) beginning of spore germination x *97.200*

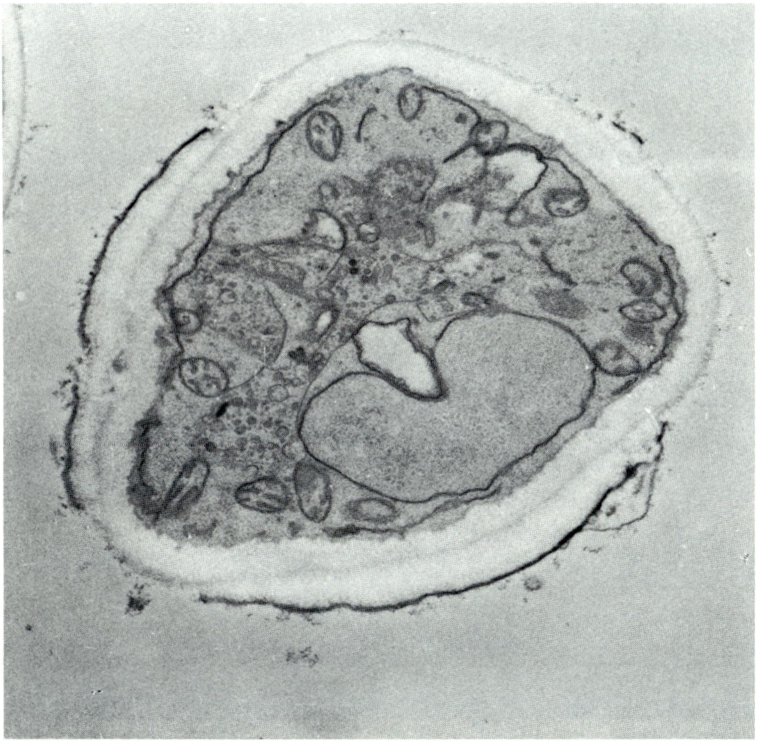

(b) germinating ascospore end-phase x 97.200 (Kreger-Van Rij unpublished)

below $a_w = 1.0$. Gibson (1973), Corry (1976, 1978) and Beuchat (1981a) observed an increased tolerance of vegetative yeast cells to heat when suspended in solutions at $a_w = 0.95$ of sugars and polyols, while heat resistance was maximum in solutions of sucrose.

In the experiments presented in this paper *Kluyveromyces* ascospores were protected against heat inactivation by sucrose and to a lesser extent by sodium chlrode at a_w of 0.95 and a pH of 4.5 (Fig. 5). Heat protection in a strawberry sugar syrup solution of pH 3.5, however, was maredly better than observed in a buffered sucrose solution at the same a_w (0.975) and pH 4.5 (Tables 5,6).

Although it may be possible that heat treated cells and spores may undergo an osmotic shock as a result of transfer from the heating medium of $a_w = 0.95$ to a dilution and recovery medium of $a_w = 0.998$, it has been pointed out by recovery experiments at $a_w = 0.95$ that it is unlikely that an increased death rate (concave downwards curve, Fig. 5) may have been the consequence of an osmotic shock phenomenon. Sucrose addition to the recovery medium up to $a_w = 0.95$ was inhibitory to ascospore recovery of *Kluyveromyces* or did not affect colony counts (Stevenson and Graumlich, 1978).

For vegetative yeast cells equilibrated in solutes of sucrose a marked reduction in the size of the cells was observed. The short axis of the yeast cells was reduced by up to 50%, and the long axis somewhat less (Corry, 1976). Electron microscopy of freeze-etched cells indicated a space between cell wall and cytoplasm, the result of plasmolysis. The degree of heat protection afforded by the solutes correlated with the degree of plasmolysis and cell shrinkage, which indicated that heat resistance is associated with dehydration of the cell more than with replacement of cell water by solutes. Corry (1978) concluded that molecules that are unable to penetrate the cell membrane, or evidently show a low penetration ability, promote increased heat resistance to a greater extent than those that are able to permeate the cytoplasm.

Studies on osmoregulation in heat resistance of bacterial spores support a hypothesis that the extent of spore heat resistance is based in whole or part on: (1) the extent of diminution and dehydration of the inner protoplast, without implications about a physiological mechanism; (2) constraint of molecular motion, preventing denaturation, associated with high concentrations of protoplast constituents; (3) non covalent interactions (Gould and Dring, 1977; Murrell, 1981a).

In our experiments with *K. bulgaricus* 120 ascospores equilibrated in solutes of sucrose and NaCl of $a_w = 0.95$, an increased length of the long axis of the spore was observed.

This observation may be related to spore water loss, though not identical to that observed in vegetative cells. The number of observations, however, are too small to associate the changes in spore shape at $a_W = 0.95$ with loss of cell-water thus protecting the ascospore against heat in a mannar analogous to that proposed for the mechanism of bacterial spore heat resistance (Murrell, 1981b).

The most important practical consequences of the heat resistance of yeast ascospores as presented in this paper are: (1) calculations of F-values for industrially applied heat processing of acid food products and drinks by the PU method of Del Vecchio *et al.* (1951) as well as by the method of Shapton *et al.* (1971) are not correct, for the z-values used are 7 and 10°C respectively (F = time in min to destroy a given number of microorganisms at T°C); (2) for F-value calculations statistical data should be made available; (3) for determinations of D- and z-values in food substrates, laboratory models are needed (Reichert, 1973); (4) during heat processing of food and drinks, physical and chemical factors (pH, a_W, viscosity) may change and subsequently also D and z values of the yeast microflora present, which should be taken into account for F-value calculations (Pflug and Christensen, 1980); (5) heat injured ascospores which show damage of the membrane that becomes the vegetative cell wall membrane, may take a relatively long recovery or thermorestoration time during incubation of the heated food substrate, or may need specific nutrients for germination and outgrowth in laboratory recovery media (Graumlich, 1981), see Plate 4; (6) ascospores equilibrated for a long period at ultra low a_W (dry materials, soil, fruit epiderm, syrups, fruit concentrates) may maintain their initial low water content and hence the corresponding high heat protection rate, even after reconstitution or dilution at higher a_W levels, shortly before the heat treatment occurs in many high-speed industrial preservation processes.

ACKNOWLEDGEMENT

Thanks are expressed to Dr N.J.W. Kreger-van Rij (Laboratorium voor Medische Microbiologie, Rijks Universiteit Groningen) for the beautiful micrographs made of strain *K. bulgaricus* 120.

REFERENCES

Baird-Parker, A.C. and Kooiman, W.J. (1980) Soft drinks, fruit juices, concentrates, and preserves. In *Microbial Ecology of Foods*, vol. II, Food Commodities pp.643-668.

The international commission on microbiological specifications for foods. Academic Press, London.
Ball, C.O. (1923) *Industrial and Engineering Chemistry* **35**, 71-85.
Ball, C.O. (1928) Mathematical solution of problems on thermal processing of canned food. University of California Publications in Public Health **1**, 15-245.
Banner, M.J., Mattick, L.R. and Splittstoesser, D.F. (1979) *J. Fd. Sci.* **44**, 545-548.
Bestic, P.B. and Arnold, W.N. (1976) *Appl. Environ. Microbiol.* **32**, 640-641.
Beuchat, L.R. (1981a) *Appl. Environ. Microbiol.* **41**, 472-477.
Beuchat, L.R. (1981b) *J. Fd. Protection* **44**, 765-769.
Beuchat, L.R. (1981c) *J. Fd. Sci.* **46**, 771-777.
Bostian, M. and Gilliland, S.E. (1981) *J. Fd. Sci.* **46**, 964-965.
Cerf, O. (1977) *J. Appl. Bact.* **42**, 1-19.
Cerny, G. (1980) *Zeitschrift für Lebensmittel-Untersuchung und Forschung* **170**, 173-179.
Corry, J.E.L. (1976) *J. Appl. Bact.* **40**, 269-276.
Corry, J.E.L. (1978) in *Food and Beverage Mycology* L.R., ed.), pp.45-82, AVI publ: Westport.
Davenport, R.R. (1976a) in *Microbiology of Aerial Plant Surfaces* (Dickinson, C.H. and Preece, T.F., eds.), pp.199-215, Academic Press, London.
Davenport, R.R. (1976b) in *Microbiology of Aerial Plant Surfaces* (Dickinson, C.H. and Preece, T.F., eds.), pp.325-351, Academic Press, London.
Davenport, R.R. (1980) in *Biology and Activities of Yeasts.* SAB symposium papers no.9 (Skinner, F.A., Passmoore, S.M. and Davenport, R.R., eds.), pp.215-230, Academic Press, London.
Del Vecchio, H.W., Dayharsh, C.A. and Baselt, F.C. (1951) *Proc. of American Society of Brewing Chemistry 1951*, 45-50.
Fowell, R.R. (1969) in *The Yeasts* vol.1 (Rose, A.H. and Harrison, J.S., eds.), pp.303-383, Academic Press, London.
Fowell, R.R. (1975) in *Spores VI* (Gerhardt, P., Costilow, R.N. and Sadoff, H.L., eds.), pp.124-131, American Society for Microbiology, Washington D.C.
Gal, S. (1975) in *Water Relations of Foods* (Duckworth, R.B., ed.), pp.139-154, Academic Press, London.
Gibson, B. (1973) *J. Appl. Bact.* **36**, 365-376.
Gould, G.W. (1977) *J. Appl. Bact.* **42**, 297-309.
Gould, G.W. (1978) in *Spores VII* (Chambliss, G. and Vary, J.C., eds.), pp.21-26, American Society for Microbiology, Washington D.C.
Gould, G.W. and Dring, G.J. (1977) in *Spore Research 1976* (Barker, A.N., Wolf, J., Ellar, D.J., Dring, G. and Gould, G.W., eds.), pp.421-429, Academic Press, London.

Graumlich, T.R. (1981) *J. Fd. Sci.* **46**, 1410-1411.
Herrmann, J., Al-Khayat, M. and Schleusener, H. (1978) *Die Nahrung* **22**, 89-99.
Janssens, J.H., Burris, N., Woodward, A. and Bailey, R. (1983) *Appl. Environ. Microbiol.* **45**, 598-602.
Johannsen, E. (1980) *Antonie van Leeuwenhoek* **46**, 177-189.
Juven, B.J., Kanner, J. and Weisslowicz, H. (1978) *J. Fd. Sci.* **43**, 1074-1080.
Kreger-Van Rij, N.J.W. (1969) in *The Yeasts* vol.i, (Rose, A.M. and Harrison, J.S., eds.), pp.5-78, Academic Press, London.
Kreger-Van Rij, N.J.W. (1979) *Archives of Microbiol.* **57**, 91-96.
Kreger-Van Rij, N.J.W. (1980) in *Biology and Activities of Yests.* SAB symposium papers no.9 (Skinner, F.A., Passmore, S.M. and Davenport, R.R., eds.), pp.53-62, Academic Press, London.
Kreger-Van Rij, N.J.W. and Veenhuis, M. (1976) *Antonie van Leeuwenhoek* **42**, 445-455.
Leistner, L. and Rödel, W. (1975) in *Water Relations of Foods* (Duckworth, R.B., ed.), pp.309-324, Academic Press, London.
Lin, J. (1979) *American Society of Brewing Chemists* **37**, 66-69.
Lodder, J. (ed) (1970) *The Yeasts - A Taxonomic Study*, 1st edn, North-Holland publ., Amsterdam.
Miller, J.J. and Kingsley, V.V. (1961) *Stain Technology* **36**, 1-4.
Murrell, W.G. (1981a) in *Sporulation and Germination* (Levinson, H.S., Sonenshein, A.L. and Tipper, D.J., eds.), pp.64-77, American Society for Microbiology, Washington D.C.
Murrell, W.G. (1981b) *Spore Newsletter* **7**(5), 75-81.
Pflug, I.J. and Christensen, R. (1980) *J. Fd. Sci.* **45**, 35-40.
Phaff, H.J., Miller, M.W. and Mrak, E.W. (1978) *The Life of Yeasts: their Nature, Activity, Ecology and Relation to Mankind*, 2nd ed. Harvard University Press, Cambridge, Mass.
Phaff, H.J. and Starmer, W.T. (1980) in *Biology and Activities of Yeasts*, SAB Symposium no.9, (Skinner, R.A., Passmore, S. and Davenport, R.R., eds.), pp.79-102, Academic Press, London.
Puglisi, P.P. and Zennaro, E. (1971) *Experimentia* **27**, 963-964.
Put, H.M.C. and Wybinga, S.J. (1963) *J. Appl. Bacteriol.* **26**, 428-434.
Put, H.M.C., de Jong, J., Sand, F.E.M.J. and Van Grinsven, A.M. (1976) *J. Appl. Bacteriol.* **40**, 135-152.
Put, H.M.C., de Jong, J. and Sand, F.E.M.J. (1977) in *Spore Research 1976 II* (Barker, A.N., Wolf, J., Ellar, D.J., Dring, G.J. and Gould, G.W., eds.), pp.545-563, Academic Press, London.

Put, H.M.C. and de Jong, J. (1980) in *Biology and Activities of Yeasts*, SAB symposium no.9 (Skinner, F.A., Passmore, S.M. and Davenport, R.R., eds.), pp.181-241, Academic Press, London.
Put, H.M.C. and de Jong, J. (1982a) *J. Appl. Bacteriol.* **52**, 235-243.
Put, H.M.C. and de Jong, J. (1982b) *J. Appl. Bacteriol.* **53**, 73-79.
Reichert, J.E. (1973) *Konserventechnische Informationen* **24**, 182-194.
Roberts, T.A. and Hitchins, A.D. (1969) in *The Bacterial Spore* (Gould, G.W. and Hurst, A., eds.), pp.611-670, Academic Press, London.
Röcken, W., Finken, E., Schulte, S. and Emeis, C.C. (1981) *Zeitschrift für Lebensmittel-Untersuchung und Forschung* **173**, 26-31.
Sand, F.E.M.J. (1976) *Brauwelt* **116**, 220-230.
Savarese, J.J. (1974) *Can. J. Microbiol.* **20**, 1517-1521.
Schmidt, C.F. (1950) *J. Bacteriol.* **59**, 433-437.
Schmidt, C.F. (1957) in *Antiseptics, Disinfectants, Fungicides, Chemical and Phtsical Sterilization* (Reddish, C.F., ed.), pp.831-900, Lea and Febiger, Philsdelphia.
Shapton, D.A., Lovelock, D.W. and Laurita-Longo, R. (1971) *J. Appl. Bacteriol.* **34**, 491-500.
Smith, J.E. and Berry, D.R. (1974) *An Introduction to Biochemistry of Fungal Development*. Academic Press, London.
Sognefest, P. and Benjamin, H.A. (1944) *Fd. Res.* **9**, 234-243.
Stevenson, K.E. and Graumlich, T.R. (1978) *Adv. Appl. Microbiol.* **23**, 203-215.
Stumbo, C.R. (1948) *Fd. Technol.* **2**, 115-132.
Stumbo, C.R. (1965) *Thermobacteriology in Food Processing*. Academic Press, London.
Suriyaracheh, V.R. and Fleet, G.H. (1981) *Appl. Environ. Microbiol.* **42**, 574-579.
Tingle, M., Singklar, A.J., Henry, S.A. and Halvorson, H.O. (1973) in *23rd Symposium of the Society for General Microbiology (London)* (Ashworth, J.M. and Smith, J.E., eds.), pp.209-243, Cambridge University Press, London.
Toda, K. (1970) *J. fermentation Technol.* **48**, 811-818.
Tsang, E.W.T. and Ingledew, W.M. (1982) *American Society of Brewing Chemists* **40**, 1-8.
Van der Walt, J.P. (1980) in *Biology and Activities of Yeats*. SAB Symposium papers no.9 (Skinner, F.A., Passmore, S. and Davenport, R.R., eds.), pp.63-78, Academic Press, London.

Van der Walt, J.P. and Johannsen, E. (1978) *Abstracts of the XII International Congress of Microbiology*, München, p.38,
Van der Walt, J.P. and Johannsen, E. (1979) *Antonie van Leeuwenhoek* **45**, 281-291.
Von Barwald, G. (1972) *Brauwissenschaft* **25**, 192-195.

QUALITATIVE AND QUANTITATIVE ANALYSIS OF AEROBIC SPORE-FORMING BACTERIA IN HUNGARIAN PAPRIKA

I. FABRI*, V. NAGEL*, V. TABAJDI-PINTER*, Zs. ZALAVARI*, J. SZABAD, and T. DEAK*****

**Centre for Food Control for the Ministry of Agriculture, H-1355 Budapest Pf.8,*
***Paprika Manufacturing Company, Szeged,*
and
****University of Horticulture, Budapest, Hungary*

SUMMARY

The bacterial flora of paprika consisted mainly of spore-forming bacteria. There was no significant difference between the counts of mesophilic aerobes and spore-formers surviving heat treatment at 80°C for 10 minutes. Dominant *Bacillus* species in paprika were *B. licheniformis*, *B. subtilis* and *B. pumilis*. In samples of different lots the microbial count showed a log normal distribution with a generalized mean value of $\bar{\bar{x}} = 5.02$. The generalized standard deviation, $s_1 = 0.6$ and the 95% value, $\phi = 6.02$ were calculated. The ϕ value is recommended as the reference value of sampling inspection.

INTRODUCTION

Spices are of interest to microbiologists because they may contain large numbers of microorganisms that occasionally cause spoilage or, more rarely, disease when introduced into foods (1). Spices may contain spores from mesophilic aerobes, mesophilic anaerobes and flat sour thermophiles. Excessive numbers of such spores introduced with spice into foods that are subsequently canned have caused spoilage (2). Thus, a major concern of spice processors is that the microbial load does not contribute to spoilage of foods in which the spice is used.

In general the bacterial count of spices varies from 10^2 to 10^7 per g and in the case of paprika from 10^4 - 10^7 (1), However, extremely large variations occur in the microbial content of different lots of the same spice (3).

Considering the high degree of contamination and the large variation of microbial count in samples of paprika, the elaboration of an appropriate quality control system is rather difficult. For the purpose of international trade, ICMSF (4) recommended sampling inspection by attributes because it is independent from the standard deviation of the microbial counts in different lots.

The aim of our work was to analyse quantitatively and qualitatively the aerobic mesophilic spore-formers in paprika produced in Hungary in order to determine the assumptions for developing an appropriate sampling inspection.

First the aerobic mesophilic counts (AMC) and the counts of spore-formers surviving different heat treatments were compared, as ICMSF recommends to determine AMC (4) while the Hungarian Food Law specifies the spore count (5) for checking microbial quality of spices. However, the reference method for the determination of spore count has not yet been standardised and heat treatments at 70°C and up to 100°C are applied in different laboratories.

As a further step, strains of aerobic spore-formers isolated from paprika were identified to see whether the spore association of paprika contains species which may risk public health (e.g. *B. cereus*) or the shelf-life of heat processed food (e.g. thermophilic flat sour bacteria). Finally, in order to get information about the distribution of microbial counts and the degree of their variation in samples of different lots, monitoring data from two paprika manufacturing companies were analysed statistically.

MATERIALS AND METHODS

Counts of aerobic mesophilic and spore-forming bacteria were determined by plating on tryptone-glucose-yeast extract agar. The heat resistances of spores were compared in terms of the numbers of survivors after treatments at 80, 90 and 100°C for 10 minutes. The incubation temperatures were 30°C for mesophiles and 55°C for thermophiles.

Strains of aerobic spore-formers surviving different heat treatments were isolated from paprika samples and were identified by a simplified key using the following characteristics: microscopic appearance, anaerobic growth, fermentation of glucose, hydrolysis of starch and reaction in Voges-Proskauer's test.

The aerobic mesophilic microbial count (AMC) and spore count of different lots produced in two factories were investigated as follows:

Series 1. AMC of 3 sampling units from each of 22 lots produced in 1980 in factory No.1.

Series 2. AMC of 3 sampling units from each of 33 lots produced in 1981 in factory No.1.

Series 3. Spore count after heat treatment at 80°C for 10 minutes in one sampling unit from each of 187 lots produced in 1981 in factory No.2.

Statistical analyses of data were performed (6) after logarthmic transformation. First, the mean values and standard deviations of AMC of Series 1 and 2 were calculated. The homogeneity of variances were checked by Bartlett's test. The data of all three series were evaluated by one-way analysis of variance. The nature of the distribution of microbial counts was studied by the Chi-Squared test.

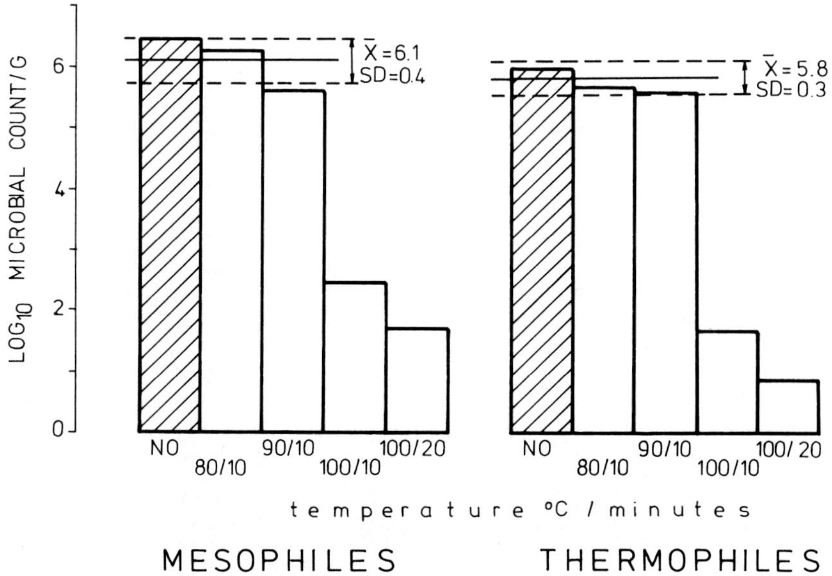

FIG. 1 *Comparison of counts of aerobic mesophilic and spore-forming bacteria in paprika applying different heat treatments and incubation temperatures. Seven samples of paprika were investigated in 3 replicates using TGY medium and standard plate count method. Striped column aerobic mesophilic count, empty column, spore count. SD = significant difference,* $\overline{X}$ *= mean value.*

RESULTS AND DISCUSSION

Comparison of Counts of Aerobic Mesophilic Bacteria and Spore-formers

As Fig. 1 shows there was no significant difference between AMC and spore count after heat treatment at 80°C, both reached 10^5 and 10^6 per g. Heat treatment at 90°C reduced the spore

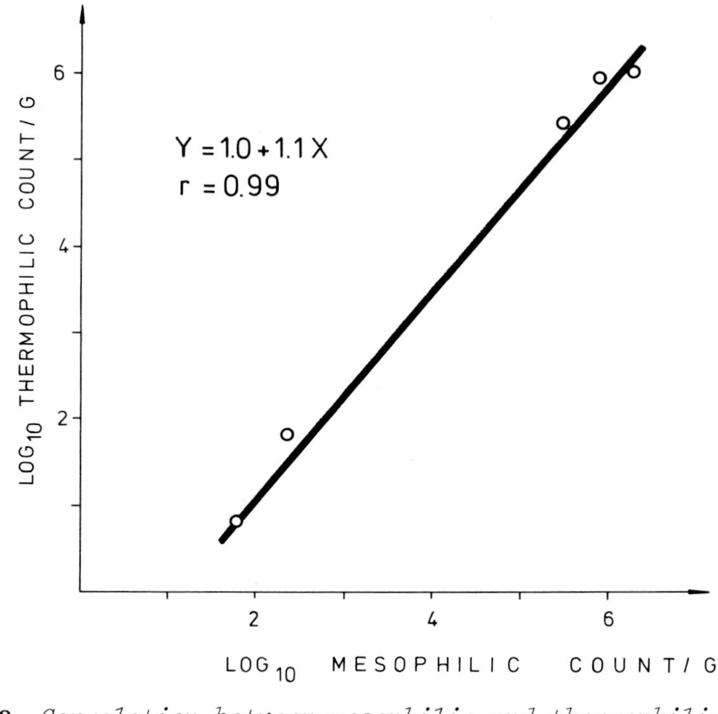

FIG. 2 *Correlation between mesophilic and thermophilic spore counts in paprika. Experimental details as in Fig. 1.*

TABLE 1 *Frequency of* Bacillus *species isolated from paprika*

Species	No. of strains	Frequency (%)
B. licheniformis	44	46.3
B. subtilis	22	23.2
B. pumilus	19	20.0
B. brevis	7	7.4
Not identified	3	3.1

count slightly but significantly while treatment at 100°C reduced it drastically. Incubation at 30 or 55°C caused no significant difference in figures of AMC and spores. A strict relationship between the counts of mesophiles and thermophiles is indicated by the correlation coefficient, r = 0.99 (Fig. 2).

Identification of *Bacillus* species

All of the 95 strains, except 3 isolated after heat treatment were identified as *Bacillus* species. Their frequency is summarized in Table 1. The dominant species was *B. licheniformis*. Only this species survived the heat treatment at 100°C for 10 minutes. All strains were able to grow both at 30 and 55°C. Neither *B. cereus* nor obligate thermophilic spore-formers were found.

TABLE 2 *Percentage distribution of microbial counts in paprika samples of 3 monitoring series*

Series	No. of samples	Percentage distribution of counts			
		10^4	10^4- 10^5	10^5-10^6	10^6
1	22	10	28	56	6
2	33	1	16	71	12
3	187	3	51	42	4
Total	242	3	45	47	5

TABLE 3 *Analysis of variance of microbial counts in paprika samples of 3 monitoring series*

Source of variance	SQ	DF	s^2	F	F_{95}
Total	81.17	30			
Between series	1.98	2	0.99	2.86	3.04
Between samples	79.21	228	0.35		

Distribution and Variability of Microbial Count in Paprika

The distribution of microbial count in samples taken in three years of monitoring at two paprika factories is shown in Table 2. The data were evaluated by one-way analysis of variance (Table 3). Comparing the calculated and tabulated F values the difference between the means of microbial counts of three series are not significant. Consequently it can be concluded that there was no significant difference between AMC and count spores surviving heat treatment at 80°C. This supports the finding of our comparative investigation (Fig. 2) and is in agreement with data of the literature (1). Therefore the data of all three monitoring series can be taken, and analysed as the aerobic mesophilic microbial counts of the same population. Table 2 shows that the microbial count of the 242 paprika samples varies from 10^4 to 10^7, and it was only in the 3% of the samples less than 10^4, the *m* value of ICMSF specification, and in 5% of the samples higher than 10^6, the *M* value recommended by ICMSF (4). Accordingly, the ICMSF specification for spices seems to be unreasonably harsh (n = 5, c = 2, m = 10 , M = 10) and its modification can be suggested.

A necessary assumption for the elaboration of a modified microbiological sampling plan is to prove a normal distribution of microbial counts in the samples, and to determine their generalized standard deviation. The nature of the distribution of AMC in different samples when analysed by

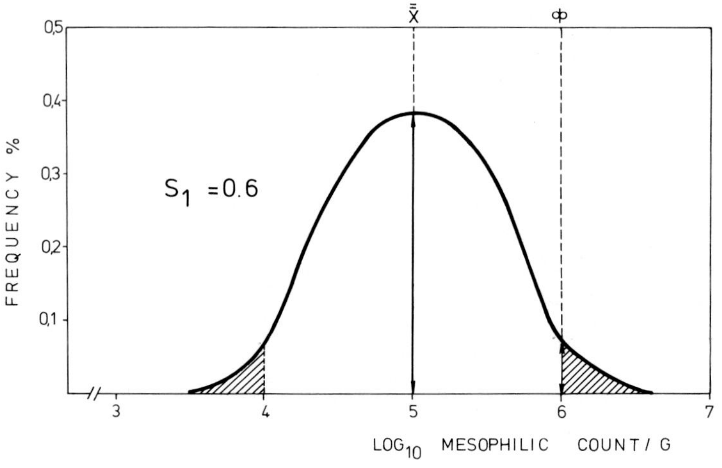

FIG. 3 *Distribution curve of aerobic mesophilic microbial count in paprika samples (n = 242).* $\overline{X}$ *= mean value,* s_1 *= standard deviation,* ϕ *= 95% value.*

Chi-Squared test showed indeed a normal distribution as the calculated χ value was less than the tabulated one, $\chi = 11.3 < \chi = 12.7$ (Fig. 3). The generalized mean value of AMC in different samples was $\bar{\bar{x}} = 5.02$, and their generalized standard deviation was $s_1 = 0.6$. This fairly small value proves the homogeneity of AMC in paprika samples. The low value of standard deviation can be ascribed to good manufacturing practica and similar technologies applied in the two firms.

From the standard normal distribution of data the so called 95th percentile or ϕ value (7) can be calculated:

$$\phi = \bar{X} + u_{95} \times s_1$$

$$\phi = 5.02 + 1.645 \times 0.6 = 6.02$$

The ϕ value is defined as the count not exceeded by 95% of the samples examined (7).

The value (6.02) calculated from the standard normal distribution agrees with the range of count not exceeded by 95% of the samples according to the empirical distribution of AMC given in Table 2. This value is also considered traditionally as the quality control limit for paprika. Therefore the ϕ value can be used as the reference value of AMC in paprika.

By the help of the ϕ value and the generalized standard deviation of AMC not only the adequate attributive sampling plan but also the variables sampling plan can be determined. The latter is more appropriate for the purpose of in-plant control and, at the same time, it needs less examination. This will be the subject of a separate paper.

ACKNOWLEDGEMENTS

The authors express their gratitude to Dr J. Farkas for stimulating and helpful discussions. Acknowledgement is made to J. Kara, M. Szabo and L. Ligeti for skilled technical assistance.

REFERENCES

1. Pivnick, H. (1980) in *Microbial Ecology of Foods*, ICMSF 3. vol. 2, pp.731-751, Academic Press, New York.
2. Jensen, L.B., Wood, I.H. and Jensen, C.E. (1934) *Ind. Eng. Chem.* **26**, 1118-1120.
3. Warmbroad, F. and Fry, L. (1966) *J. Assoc. Off. Anal. Chem.* **49**, 678-680.
4. ICMSF 2 (1974) *Microorganisms in Foods*, pp.110-117. Univ. of Toronto Press.
5. Hungarian Food Law (1978) EüM 6/1978,VII,14.

6. Schifler, W.C. (1979) *Statistic for the Biological Sciences*, Addison-Wesley Publ. Co., London.
7. Mossel, D.A.A. (1977) *Microbiology of Foods and Dairy Products*, p.82. Univ. of Utrecht.

THE CLOSTRIDIA-TUMOUR PHENOMENON: FUNDAMENTALS IN ONCOLYTIC TUMOUR RESEARCH

H. BRANTNER

Department of Microbiology, Hygiene Institute, University of Graz, Austria.

SUMMARY

The theoretical concept of the Clostridia-Tumour Phenomenon (CTP) points out that the examination of clostridial energy metabolism, especially the influence of oxygen and redox potential during activation, germination, out growth and replication, can give more information on the energy balance of mitotic cells, on cell kinetics and pathogenicity of growing tumours.

INTRODUCTION

The use of microorganisms and their metabolites as possible therapeutic agents in the control of cancer had its beginning in the early days of bacteriology. Many investigations were prompted by the clinical observation that a concurrent bacterial infection frequently retarded the development of malignant processes in man (18).

The first record of an attempt to treat human tumours with an induced bacterial infection is that of Busch in 1868 (4), whilst the first attempt to use clostridia in cancer treatment was made in 1935 (5).

In 1955, Malmgren and Flanigan (13) reported observations on the germination of *Cl. tetani* spores after intravenous injection in different animal tumours.

In 1959, Möse and Möse (14) were the first to recognize the use of this phenomenon in tumour therapy and demonstrated oncolysis by non-pathogenic *Cl. butyricum* M 55 which was renamed *Cl. oncolyticum* at the suggestion of the late Professor

Fredette (7). This strain satisfied following demands: non-pathogenicity; germination and outgrowth only in the tumour area; no formation of metabolites injurious to man; injectability of large spore amounts; pyrogen-free spore suspensions; slow elimination by the immune system.

The method is based on the intravenous injection of *Cl. oncolyticum* spores, their distribution by circulation, and specific germination to form vegetative rods within the tumour, followed by lysis of the tumour by metabolic activities of the clostridia. Warburg (24) recognized the oncolysis as confirmation of his theory of anaerobic glycometabolism in tumour cells. At first Kayser (11) suggested a theory of oncolysis based on his results on the influence of oxygen partial pressure leading to germination of *Cl. butyricum* ATCC 859 (10) and subsequent formation of hydrogen peroxide in the well oxygenated peripheral zones of the tumour. However, Kayser's opinion that hydrogen peroxide destroys both the tumour and clostridial rods was disproved by histological data (17).

Extended biological and enzymological examinations carried out in our laboratory have shown the involvement of enzymatic mechanisms in oncolysis (Table 1; 1,3,9). A theoretical model was developed which was based on the microbiological and biochemical results showing correlations between tumour and clostridial cells (Fig 1; 2).

MECHANISMS OF ONCOLYSIS

The existence of malignant cells in the host tissue implies the transmission of an unknown stimulus to the host, resulting in angiopathic and fibroblastic reactions. Though it is not characteristic only for tumour growth, the basic condition is the development of new capillary endothelium from the host vessels which allows oxygen and nutrient supply, and the elimination of catabolites. Therefore, future application of CTP can only be considered in vascularized tumours.

After intravenous injection the spores are spread by circulation and transported into the tumour capillaries and further to the interstice. There the spores begin to germinate depending on the oxygen partial pressure and a suitable redox potential existing in the micro-niches between the tumour cells. From the *in vitro* results reported by Kayser (10) can be deduced an oxygen partial pressure of 0.004 bar and a redox potential of -440 mV.

Examinations show the influence of tumour metabolism on germination, replication and sporulation of *Cl. oncolyticum*. Tumour membrane-permeable precursors and intermediates of anaerobic energy metabolism, as well as protein and nucleic acid synthesis, stimulate the growth and the replication of the

TABLE 1 *Characteristic enzyme pattern and parameters of oncolytic clostridia*

Clostridium	Lactose fermentation	H_2S formation	Lipolysis	Proteolysis	Gelatinolysis	Collagenolysis	Bradykinin degradation	Haemolysis	DNA-ase	E_h mV	Oncolysis
Cl. oncolyticum M 55 ATCC 13,732	−	+	+	+	+	+	+	+	+	− 400	+
Cl. butyricum M 82	+	+	+	+	+	+	+	+	+	− 360	+
Cl. butyricum Jena H 8	−	+	+	+	+	+	+	+	+	?	+
Cl. acetobutylicum ATCC 10,132	+	−	+	−	+	−	+	±	+	− 400	+
Cl. sporogenes ATCC 3,584	−	+	+	+	+	+	+	+	+	− 245 − 285	+

disturbance of the steady-state of the tumour kinin cascade, results in an increased capillary permeability, vasodilatation, and deficiency of oxygen and nutrient supply (Fig. 2). The kinin cascade is also activated by immunoreactions. The formation of antigen antibody complexes has an activating effect on plasma prekallikrein resulting in an increased kinin liberation (22).

After injection of clostridial spores the cell-mediated immune response is activated leading to spore reduction by phagocytosis. The injection of large spore numbers represents an immunogenic stimulus which is intensified by germination and replication of the spores. The replication in contact with the malignant cells results in a local focus of inflammation and migration of macrophages (Fig. 3).

Besides the cellular immune response the humoral immune response is also stimulated. The linkage of excessive amounts of antigen with soluble antibodies results in the formation of tissue-damaging immune complexes also giving rise to inflammation. The complement cascade on the one hand, and the kinin cascade on the other, are involved in those mechanisms and intensify the cell-mediated immune response.

The increased vascular permeability allows polymorphonuclear

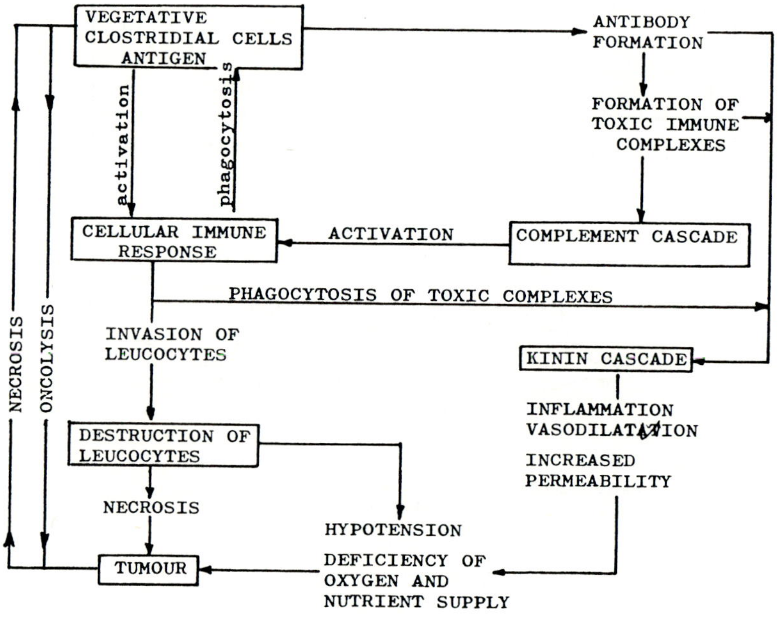

FIG. 3 *Correlations between tumour, clostridial cells, and cellular and humoral immune response.*

leucocytes to migrate into the tumour area and to phagocytize the immune complexes formed (12). The humoral immune response initially mobilized against the migration of clostridial spores allows, with other factors, increased replication of the vegetative rods and influences the tumour vessels. The dilatation of the tumour capillaries, followed by decreased oxygen and nutrient supply, results in an increased necrosis which, in its turn, forms the basis of excessive bacterial metabolism.

The spore reduction after intravenous injection is attributable to phagocytosis which partly eliminates the clostridial spores before they reach a tumour area suitable for germination. It seems that the residual and reduced spore numbers

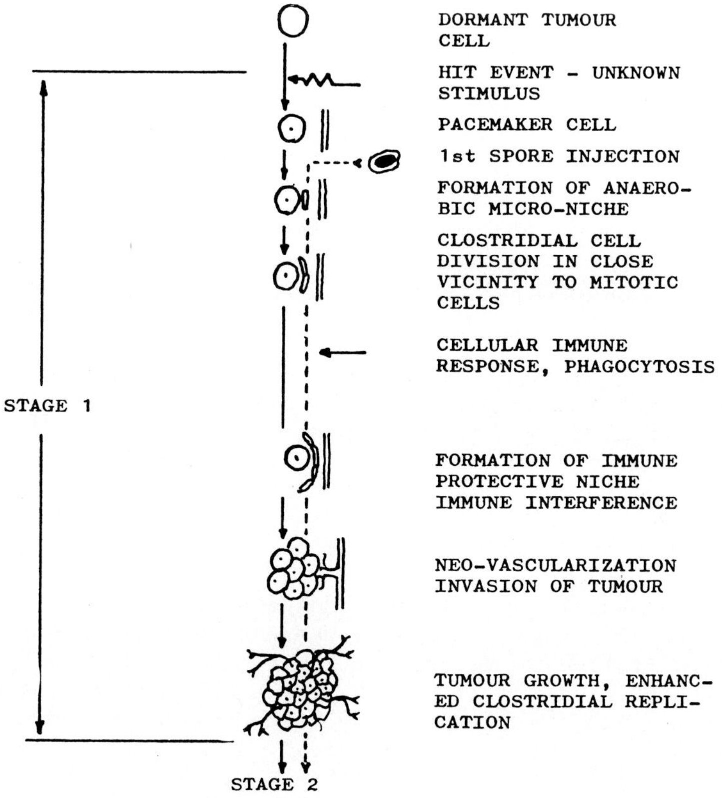

FIG. 4 *Clostridia-Tumour Phenomenon. Primary stage. A dormant cell is activated to a "pacemaker cell" by unknown stimuli which, in its turn, allows oncolytic clostridia to germinate and replicate in dependence on mitosis. Based on Schneeweiss (1980) and Brantner and Schwager (1980).*

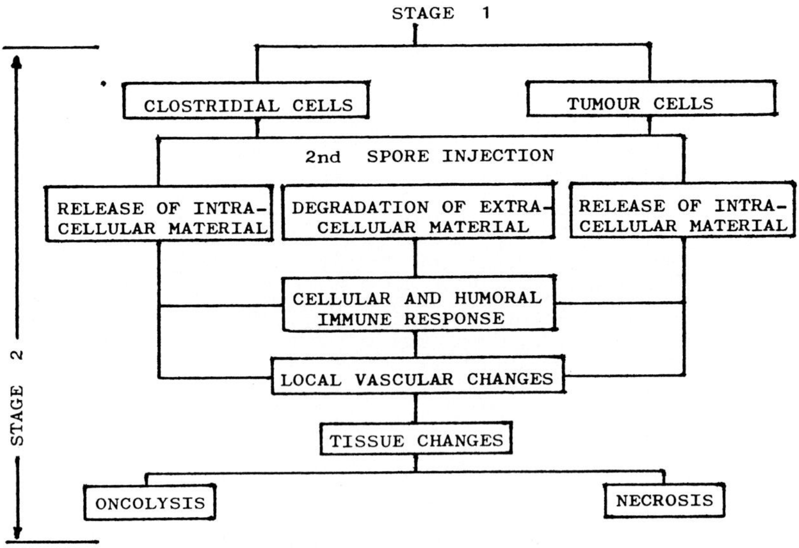

FIG. 5 *Clostridia-Tumour Phenomenon. Secondary stage. A second intravenous injection of clostridial spores triggered an excessive immunogenic impulse after enhanced proliferation of tumour and clostridial cells. Based on Brantner and Schwager (1980).*

can start the oncolysis at the end of activation phase and the beginning of germination and replication.

In their examinations on the use of clostridial spores in tumour diagnosis Schneeweiss *et al.* (20) have developed a tumour-tetanus model which has shown replication of clostridia dependent on mitosis (Fig. 4). This phenomenon is highly specific and is influenced by the different cloning behaviour of tumour and normal cells.

Thus, it has to be supposed that the CTP can be divided in two stages (Fig. 5): a primary phase of germination and outgrowth controlled by mitosis of the tumour cells; and a secondary stage of clostridial growth within the tumour area independent of mitosis and triggered by an excessive immunogenic impulse.

The transition from the primary to secondary stage is connected with the basic conditions of the oncolysis model we postulated (2). In contrast to dormant eukaryotic cells dividing ones show, at least temporarily, an enhanced oxygen consumption. Therefore, it has to be supposed that the microenvironment of the dividing cells is characterized by oxygen depletion. The steep oxygen gradient cannot be compensated for immediately because of limited oxygen diffusion from the supplying capillaries. The fluctuation of the oxygen gradient, regulated bioenergetically, represents an oxygen-shift-up-shift-down phenomenon for clostridial energy metabolism. The strong electro-positive influence of oxygen prevents the formation of a negative redox potential sufficient for clostridial growth. Thus, *in vivo* the clostridial metabolism is stimulated only for a very short period sufficient for germination, outgrowth, and a single replication step in accordance with the mitosis of the tumour cell. Only the increasing oxygen deficiency of the cell environment provides the anaerobic niche for oncolytic and heterotrophic clostridia in connection with the temperature, the pH value, and the carbon and nitrogen sources for nutrient supply.

During intervals without cell division the normalization of the oxygen concentration within the interstice induces total retardation of the clostridial metabolism. In this period the clostridial rods are exposed to toxic effects of oxygen radicals which are formed as short-lived, highly reactive intermediates arising from univalent reduction steps. It has to be supposed that tissue peroxidases, catalases, and superoxide dismutases, respectively, have protective functions towards the clostridial rods. Possibly, the vegetative rods produce one of these enzymes themselves (21).

CONCLUSIONS

In our opinion the knowledge of the *in vivo* conditions within the tumour can be extended by examination of clostridial metabolism. The high specificity of the CTP, i.e. one cell division exactly corresponding to one division of the clostridial rods (19), seems most suitable for cell kinetic and metabolic examinations.

Investigation of the mechanisms and parameters of activation, germination, outgrowth, and replication will also allow conclusions on the conditions within growing tumours. Presumably, the transition from the first stage of the CTP to the secondary phase has its beginning in the formation of an immunoprotective niche. The survival of clostridia could be provided by enzymatic reactions which parallel the cellular immune response corresponding with the immune interference phenomenon described by Voisin (23).

The inclusion of immunological mechanisms in oncolysis was confirmed by animal experiments. For instance, the effect of spores applied intravenously was compared with that of a single injection of living and heat-inactivated vegetative rods intratumourally (8). Tumour-lysis started at the same time in both groups, i.e. 40 hours after inoculation of the antigen. The

TABLE 2 *Experimental data resulted after intertumoral injection of vegetative and heat-inactivated rods. NMRI mice, Ehrlich ascites tumour. Mean values n = 3.*

Time after injection h	Vegetative clostridial rods		Heat-inactivated clostridial rods	
	No lysis	Lysis	No lysis	Lysis
0	22	0	22	0
40	3	19	13	9
72	6	16	13	9
168	11	11	16	6

results showed that a single injection of inactivated rods led to oncolysis, but the degree of lysis was greater in those tumours in which living clostridial rods were applied.

It can also be concluded that the massive introduction of immunogenic material stimulates the different interacting cycles leading to oncolysis.

At this time reactions are initiated, in both the tumour and clostridial cells, which lead to necrosis and oncolysis, respectively.

The obvious adaptation of clostridial growth to the mitotic cycle of the proliferating tumour cell during the first stage of the CTP, and the similarity of mechanisms in both systems during the secondary stage, indicate analogous basic reactions. Further *in vitro* and *in vivo* investigations in progress on the oxygen demand, redox potential, and metabolic pathways of *Cl. oncolyticum* and *Cl. sporogenes* dependent on tumour metabolism will show the validity of the model we postulate.

ACKNOWLEDGEMENTS

We gratefully acknowledge financial support by Austrian Research Funds, Nr. 4467, and cooperation and discussion with Professor Dr. Schneeweiss, Academy of Sciences, German Democratic Republic.

REFERENCES

1. Brantner, H. and Fischer, G. (1973) *Path. Microbiol.* **39**, 99-106.
2. Brantner, H. and Schwager, J. (1980) *Arch. Geschwelstforsch.* **50**, 601-612.
3. Brantner, H. and Wenzl, W. (1976) in *Spore Research 1986* (Barker, A.N., Wolf, J., Ellar, D.J., Dring, G.J. and

Gould, G.W., eds.), Vol. I, pp.391-405, Academic Press, London.
4. Busch, N. (1868) *Berlin. klin. Wschr.* **5**, 137-138.
5. Connell, H.C. (1935) *Canad. Med. Ass. J.* **33**, 364-370.
6. Fischer, G., Brantner, H. and Platzer, P. (1975) *Z. Krebsforsch.* **84**, 203-206.
7. Fredette, V. and Planté, C. (1970) *Proceed. Xth Intern. Congress Microbiol.*, Mexico, p.193.
8. Haller, E.-M. (1979) Thesis Univ. of Graz.
9. Haller, E.-M. and Brantner, H. (1979) *Zbl. Bakt. Hyg., I. Abt. Orig.* **A 243**, 522-527.
10. Kayser, D. (1962) *Z. Naturforsch.* **17 b**, 658-660.
11. Kayser, D. (1967) in *The Anaerobic Bacteria*, Intern. Workshop, Inst. Microbiol. Hyg., Univ. of Montreal, Laval-des-Rapides.
12. Keller, R. (1977) *Immunologie und Immunpathologie*, G. Thieme Verlang, Stuttgart.
13. Malmgren, R.A. and Flanigan, C.C. (1955) *Cancer Res.* **15**, 473-378.
14. Möse, J.R. and Möse, G. (1959) *Z. Krebsforsch.* **63**, 63-74.
15. Möse, J.R. and Fischer, G. (1974) *Z. Krebsforsch.* **82**, 143-152.
16. Möse, J.R., Fischer, G. and Briefs, C. (1972) *Zbl. Bakt. Hyd., I. Abt. Orig.* **A 221**, 474-491.
17. Propst, A. and Möse, J.R. (1966) *Z. Krebsforsch.* **68**, 337-251.
18. Reilly, H. Ch. (1953) *Cancer Res.* **13**, 821-834.
19. Schmidt, W., Schneeweiss, U. and Fabricius, E.-M. (1980) *J. theor. Biol.* **86**, 783-852.
20. Schneeweiss, U., Fabricius, E.-M. and Schmidt, W. (1980) *Tumorforschung am biologischen Modell.* VEB Fischer Verlag, Jena.
21. Shanklin, D.R. (1969) *Persp. Biol. Med.* **13**, 80-100.
22. Vogel, R. and Werle, E. (1970) Kallikren Inhibitors. In *Handbook Exp. Pharmakol.* (Erdös, E.G., ed.), Vol. XXV, pp. 213-249, Springer Verlag, Berlin.
23. Voisin, G.A. (1971) *Progr. Allergy* **15**, 328-485.
24. Warburg, O. (1962) *Arb. Paul-Ehrlich-Inst.*, Frankfurth/M., H.59, 19-26.

INSECTICIDAL METABOLITES OF SPORE-FORMING BACILLI

P. LÜTHY, H.-R. EBERSOLD, J.L. CORDIER and H.M. FISCHER

Institute of Microbiology, Swiss Federal Institute of Technology, 8092-Zürich, Switzerland

SUMMARY

Insecticidal metabolites produced by *Bacillus thuringiensis* and *Bacillus sphaericus* are of increasing importance as control agents of agricultural insect pests and vectors of human infectious diseases. The principal metabolites are toxic proteins synthesized by the sporulating cells. Safety aspects, high specificity and the lack of observed resistance to date are the main advantages of these insecticides. A great number of strains with different bioactivities represents a considerable potential for the development of new and more defined products with higher efficacy using modern microbial technology. A complete understanding of the biosynthesis and mode of action of the insecticidal metabolites is however necessary. This paper includes a short review of the milky disease organisms. There is some evidence that the pathogenesis of this group of spore formers is also toxin mediated.

INTRODUCTION

The number of spore formers associated with insects and producing insecticidal metabolites is limited essentially to the two species *Bacillus thuringiensis* and *Bacillus sphaericus* as well as to the rather ill-defined group of the milky disease organisms. *B. thuringiensis* which comprised at the end of 1982 29 varieties, classified by serotyping and where necessary by biochemical properties (1), is by far the most important species from an applied and economic point of view. The host spectrum of *B. thuringiensis* is limited to insect species within the two orders *Lepidoptera* and *Diptera*. So far, *B. thuringiensis* products have an excellent safety record. As a rule, only the target insects are reduced in their number, leaving the balance of the other populations including the beneficials untouched.

B. sphaericus, is highly active against the genus *Culex*, but is still in the experimental stage as an insecticide. Since the toxin is a component of the sporulated cell and not produced as a separate entity as in the case of *B. thuringiensis*, analysis and mode of action studies show rather slow progress.

The milky disease organisms, pathogenic to a few species within the family *Scarabaeidae* (order: *Coleoptera*), cannot be cultured successfully *in vitro*. *Bacillus popilliae*, the best known organism within this group, has been used extensively for control of Japanese beetle (*Popillia japonica*) larvae, a grassland pest of north-eastern USA. All the infectious spore material had to be produced in field-collected host larvae. The formation of a parasporal inclusion with biological activity in some of the milky disease bacteria, and the mode of infection suggest the presence of a toxin.

Exotoxins with a rather broad spectrum and enzymes with some biological activity are produced by strains of *B. thuringiensis*. They are however not linked to spore formation and are of no economic importance since commercial products are free of such compounds.

Recently published books and review articles (2,3,4) cover extensively the spore forming insect pathogens and their metabolites. We will present here only a general review and include some of the latest results and trends in this field.

BACILLUS THURINGIENSIS

The Delta-endotoxin

All the *B. thuringiensis* strains produce during the early sporulation phase a protein crystal which is deposited within the sporangium but outside the exosporium. The variety *finitimus* represents the only exception with the parasporal inclusion inside the exosporium. Here, spore and crystal are not separated following lysis of the sporangium as in the case of all the other varieties. The formation of a crystalline inclusion is the only criterion of differentiation between *B. thuringiensis* and *Bacillus cereus*.

The biosynthesis of the parasporal inclusion is closely linked to the sporulation process. It is initiated during phase II, concomitant with septum formation and concluded in the late phase IV or early phase V. Crystal synthesis must be controlled by a very efficient genetic system since about 30% of the cell's dry weight is turned over into the parasporal body. The crystal genes have been found on plasmids (5) as well as on the chromosome (6). Klier *et al.* (7) have isolated from a strain of the variety *thuringiensis* a sporulation specific mRNA with a long half life which is responsible for crystal synthesis.

The latest results on crystal analysis give an increasingly complex picture leading away from a highly pure metabolite exclusively produced in the parasporal compartment. Besides the crystal which is a protein and responsible for the insecticidal activity, we find for example proteases within or at the surface of the parasporal bodies. These enzymes interfere with analytical investigations or even modify the polypeptides prior to crystallization (8). Furthermore, carbohydrates are found in association with the crystal protein (9). The reported amounts range from traces up to 12%. There is however general agreement that carbohydrates are not involved in the pathogenesis, even if they should represent an integral constituent. The quantity and kind of impurities might depend on the strain and be influenced by the culture conditions.

Strains of *B. thuringiensis* producing more than one inclusion have been described. Some strains of the variety *kurstaki*, in addition to the regular crystal, toxic for lepidopterous insects, synthesise a so-called ovoid inclusion body (10) containing a protein active against mosquito larvae (11). Nickerson (12) identified poly-β-hydroxybutyric acid granules as a component closely associated with crystals of *kurstaki* strains. The variety *israelensis* gives rise to inclusions surrounded by an envelope. In can clearly be demonstrated by electron microscopy that at least two different metabolites are present in this variety since they vary in their electron densities. Thus, only one might be the crystal toxin responsible for the mosquito and black fly activity.

There is convincing evidence that the delta-endotoxin is also a constituent of the spore. This has been proved by immunology and by bioassays. We found in our laboratory that spores of the variety *israelensis* were toxic in the bioassay with *Aedes aegypti* larvae (LC_{50} for L_2 larvae within 24 h: 2.10^4 spores/ml). The question arises whether the delta-endotoxin was originally a spore component and the parasporal body the result of overproduction.

The shape of the parasporal inclusion is not uniform but differs from one variety to the other. For example, the bipyramid-shaped crystals of the variety *thuringiensis* possess pointed ends whereas those of *sotto* are rounded. The crystalline inclusions of the variety *israelensis* are without a

like misformed antennae, mouthparts or legs. Beta-exotoxin is toxic for mammals upon injection but inactive if applied per os. The compound is readily inactivated by dephosphorylation by alkaline phosphatase in the gut. Updated and comprehensive reviews on the beta-exotoxin have been written by Sebesta *et al.* (19) and by Lecadet and de Barjac (20).

Reported mutagenesis and the unspecific mode of action will prevent a commercial use of the beta-exotoxin for insect control. Already for many years *B. thuringiensis* products marketed in the US and Western Europe are free of beta-exotoxin.

Other Metabolites with Insecticidal Activity

Some few strains of *B. thuringiensis* produce during the vegetative growth phase a protein which is toxic upon injection to insects and mice (21). It has been designated as alpha-exotoxin. It is of no significance and does not interfere with *B. thuringiensis* products since this compound is absent from sporulated cultures.

Enzymes such as chitinases, phospholipases and proteases may support the pathogenicity of vegetative cells during proliferation within the host. But vegetative cells are usually only present in a late stage of the poisoning process since the gut juice is an adverse environment for microbial growth.

BACILLUS SPHAERICUS

Several strains of *Bacillus sphaericus* are highly toxic against some species of mosquito larvae. The most potent strains have been isolated from the genera *Culex* and *Anopheles* (1).

The pathogenicity of *B. sphaericus* is also based on a toxin whose formation is controlled by the sporulation process. In contrast to *B. thuringiensis*, the bulk of the sphaericus toxin is an integral part of the spore. Recently inclusion bodies with bioactivity have been found in some *B. sphaericus* strains.

The pathogenesis can be described as follows. The toxin is released upon ingestion of the spores by host larvae. Where inclusion bodies are present within the sporangium, they are readily dissolved in the gut. The primary target of the toxin is the gut epithelium whose cells show swelling and separation from each other at their bases. Death of larvae may start within 12 h depending on the bacterial strain, the dose, the target insect and the environmental conditions. Determination of the viable *B. sphaericus* cells during the poisoning process shows an initial rapid decrease due to digestion and defecation. In a later stage the cell number increases along with other microorganisms, indicating that the intestine changes into a more suitable environment for microbial growth. Good growth and sporulation within the host is the reason why *B. sphaericus* shows considerable persistence. Under favourable circumstances enough toxin-bearing spores are produced to ensure prolonged control.

The toxin was recently extracted from spores and subjected to analysis (22,23). It was found to be a protein, probably a polypeptide with a MW of 55,000. It is not yet known if proteolytic modification is necessary to obtain an active moiety.

Larvae of mosquito species belonging to the genus *Culex* are the most susceptible to *B. sphaericus*. The numerous strains which have been isolated from mosquito larvae are classified by phage-typing and serology (24). The most potent strains against *Culex* are designated as SSII-1, 1593, 2297 and MR-4. The LC_{50} is as low as 10^1-10^2 spores/ml. Species among the genera *Anopheles* and *Aedes* show a much weaker response.

Activity, safety and a specific host spectrum represent a considerable potential for *B. sphaericus* strains to be produced on a large scale. The recycling capacity of this organism is of special interest. Until now, only a limited number of field studies with experimental powders have been conducted (25). The results are encouraging and justify the continuation of the *B. sphaericus* work.

THE MILKY DISEASE ORGANISMS

The milky disease organisms are a group of spore formers which infect larvae within the family of the *Scarabaeidae* (order: *Coleoptera*). They have a world wide distribution but are so specific that one species or variety is able to infect only a single host species. The best known species is *Bacillus popilliae*, a pathogen of the Japenese beetle (*Popillia japonica*) larva, a grassland pest of the north-eastern USA. The successful use of *B. popilliae* against *P. japonica* is one of the classical examples of microbial insect control.

Despite great efforts, all attempts to produce infectious *B. popilliae* spores *in vitro* have failed, and all the material for Japanese beetle control had to be grown in field-collected host larvae. The general failure of *in vitro* cultivation with the exception of a few selected strains where vegetative proliferation and some rare sporulation were achieved, made a classification of the milky disease organisms difficult. Host spectrum, cellular morphology and serological comparisons had to serve as main criteria for taxonomic classification. Reviews on this group of spore formers, which have lost much interest during the last few years, have been published by Splittstoesser and Kawanishi (26), by Milner (27) and by Klein (28).

Milky disease bacteria are generally considered as pathogens not producing a toxin. But a review of the pertaining literature and a better understanding of the mode of action of other insect pathogens indicate that a toxin might be present. Some milky disease varieties produce a parasporal body which exhibits a fine structure. The shape of the inclusion is characteristic for a given variety (27). It was shown by Ebersold (29) that purified and solubilized parasporal material irreversibly damaged primary hemocyte cultures of *Melolontha melolontha* at a concentration of 10 μg/ml.

A high number of spores (10^5-10^6) is necessary to initiate the infection. In the midgut of the larvae the parasporal bodies are dissolved and the spores germinate. In the next stage histopathological changes appear in the midgut epithelium which must be caused by the action of a toxin that is very likely present in the parasporal inclusion, or in the spore where a parasporal body is not present, as for example in the case of *B. popilliae* var. *lentimorbus* or *Bacillus euloomarahae*. There are sites where epithelial cells are peeled off, leading to an increased mitotic activity of the nidi which replace the lost cells. The regions of the nidi become colonized by cells of the pathogen which goes through a first cycle of multiplication. Vegetative cells appear in the hemolymph 2-3 days after infection where they continue to proliferate at a rather slow pace. The first spores are formed before the vegetative growth is concluded. The parasporal bodies are formed during the sporulation process but remain within the sporangia since the walls of the former vegetative cells do not undergo lysis. The steadily increasing number of spores changes the transparent hemolymph into a milky-white liquid. The insect is finally nutritionally exhausted by the pathogen and dies within 2-3 weeks after the onset of the infection.

THE ROLE OF THE TOXINS

A review of the toxins produced in close association with spore formation shows some striking parallels. The main task of the metabolites of *B. thuringiensis*, *B. sphaericus* and probably the milky disease organisms is the destruction or damaging of the gut epithelium. Healthy insects are well protected from bacterial invasion by the high alkalinity and digestive enzymes of the gut juice and by the barrier of the intestinal epithelium. The toxin mediated damage of the gut epithelium gives not only direct access to the body cavity which represents a rich nutritional source for microbial growth, but leads also to an exchange between hemolymph and gut juice improving the conditions for bacterial development in the intestine.

ACKNOWLEDGEMENT

Our research on bacterial insecticides is supported by the Swiss National Science Foundation, Project Nr. 3.098-0.81.

REFERENCES

1. de Barjac, H. (1982) *Proc. 3rd Int. Coll. Invertebr. Pathol.* University of Sussex, Brighton, UK. pp.451-453.
2. Burges, H.D. (1981) *Microbial Control of Pests and Plant Diseases*, p.949, Academic Press, New York.
3. Davidson, E.W. (1981) *Pathogenesis of Invertebrate Microbial Diseases*, p.562, Allanheld, Osmun, Totowa, N.J.
4. Kurstak, E. (1982) *Microbial and Viral Pesticides*, Marcel Dekker, New York.
5. Schnepf, H.E. and Whiteley, H.R. (1981) *Proc. Natl. Acad. Sci. USA* **78**, 2893-2897.
6. Klier, A. and Rapoport, G. (1982) *Abstracts 3rd Int. Coll. Invertebr. Pathol.* University of Sussex, Brighton, UK, 86.
7. Klier, A., Lecadet, M.M. and Rapoport, G. (1978) in *Spores VII*, (Chambliss, G. and Vary, J.C., eds.), pp.205-212, American Society for Microbiology.
8. Chestukhina, G.G., Zalunin, I.A., Kostina, I., Kotova, S., Kattrukha, S.P. and Stepanov, M. (1980) *Biochem. J.* **187**, 457-465.
9. Bulla, L.A., Kramer, K.J. and Davidson, L.I. (1977) *J. Bacteriol.* **130**, 375-383.
10. Bechtel, D.B. and Bulla, L.A. (1976) *J. Bacteriol.* **127**, 1472-1481.
11. Yamamoto, T. and McLaughlin, R.E. (1981) *Biochem. Biophys. Res. Commun.* **103**, 414-421.
12. Nickerson, K.W., Zarnick, W.J. and Kramer, V.C. (1981) *FEMS Microbiol. Lett.* **12**, 327-331.
13. Huber-Lukac, H.E. (1982) *Diss. Nr. 7050*, Swiss Federal Institute of Technology, Zurich, Switzerland.
14. Krywienczyk, J. (1977) *Report IP-X-16*, Insect Pathology Research Institute, Canadian Forestry Service, Sault Ste. Marie, Ont., Canada.
15. Harvey, W.R. and Wolfersberger, M.G. (1979) *J. exp. Biol.* **83**, 293-304.
16. Dulmage, H.T. (1981) in *Microbial Control of Pests and Plant Diseases* (Burges, H.D., ed.), pp.192-222, Academic Press, New York.
17. Lereclus, D., Lecadet, M.M., Ribier, J. and Dedonder, R. (1982) *Mol. Gen. Genet.* **186**, 391-398.
18. Dean, H.D., Clark, B.D., Lohr, J.R. and Chu, C.Y. (1982) *Proc. 3rd Int. Coll. Invertebr. Pathol.*, University of Sussex, Brighton, UK, pp.11-13.

19. Sebesta, K., Farkas, J., Horská, K. and Vanková, J. (1981) in *Microbial Control of Pests and Plant Diseases* (Burges, H.D., ed.), pp.249-281.
20. Lecadet, M.M. and de Barjac, H. (1981) in *Pathogenesis of Invertebrate Microbial Diseases* (Davidson, E.W., ed.), pp. 293-321.
21. Krieg, A. (1971) *J. Invertebr. Pathol.* **17**, 133-135.
22. Davidson, E.W. (1982) *J. Invertebr. Pathol.* **39**, 6-9.
23. Tinelli, R. and Bourguoin, C. (1982) *FEBS Lett.* **142**, 155-158.
24. Yousten, A.A., de Barjac, H., Hedrick, J., Cosmao-Dumanoir, V. and Myers, P. (1980) *Ann. Microbiol. Inst. Pasteur* **131B**, 297-308.
25. Lacey, L.A. (1982) *Proc. 3rd Int. Coll. Invertebr. Pathol.* University of Sussex, Brighton, UK, pp.490-493.
26. Splittstoesser, C.M. and Kawanishi, C.Y. (1981) in *Pathogenesis of Invertebrate Microbial Diseases* (Davidson, E.W., ed), pp.189-208.
27. Milner, R.J. (1981) in *Microbial Control of Pests and Plant Diseases* (Burges, H.D., ed.), pp.45-59.
28. Klein, M.G. (1981) in *Microbial Control of Pests and Plant Diseases* (Burges, H.D., ed.), pp.183-192.
29. Ebersold, H.R. (1976) *Diss. Nr. 5699*, Swiss Federal Institute of Technology, Zurich, Switzerland.

SOLVENT FERMENTATION PRECEDES ACID FERMENTATION IN ELONGATED CELLS OF *CLOSTRIDIUM THERMOSACCHAROLYTICUM*

S.L. LANDUYT* and E.J. HSU**

**Department of Biology, University of Missouri-Kansas City, Kansas City, Missouri, U.S.A.*
***Institute of Botany, Academia Sinica, Taiwan*

SUMMARY

Elongated cells of *Clostridium thermosaccharolyticum* strain SD 105 were prepared by shifting a batch culture that had grown for one generation or less, to a continuous dilution culture. The elongated cells showed a step-wise growth pattern with concomitant solvent fermentation, not preceded by acid fermentation. The amount of ethanol increased 13,846-fold after the shift and the elongated cells continued to produce large amounts of ethanol, proportional to the increase in cell mass, for more than one generation. However, with the appearance of multi-septations in the majority of the cells, and the reversal to vegetative multiplication that followed, there was an abrupt decrease in solvent production and a shift to the acid fermentation that is characteristic of this organism.

INTRODUCTION

Although the use of thermophilic microorganisms for the industrial production of solvents from renewable resources dates to the 1920s (1), very little modern research has been performed until the petroleum shortages of the 1970s. The cost and availability of petroleum, however, has initiated renewed interest in both basic and applied studies of solvent production from cellulosic derivatives (i.e., α-cellulose, hemicellulose, and other saccharide components). Production by thermophilic bacteria offers several advantages including: high growth and metabolic rates; low cellular growth yield; high physico-chemical stability of enzymes and organisms; and facilitated reactant activity and product recovery (2,3). However, previous experience suggests that certain criteria must be satisfied prior to the application of a fermentation process to complex, carbohydrate sources such as cellulosic biomass. These include: (i) the process must be relatively simple; (ii) the yield of solvents must be high; (ii) the carbohydrate source used as a substrate must have few competing markets; and (iv) significant quantities of the carbohydrate must be available at a central location, because substrate cost is critical (4). A possible alternative to direct biological conversion would be the production of solvents from hemicellulose containing industrial waste residues. Hemicellulose-derived pentoses, hexoses, and uronic acids could be made available from these residues by dilute hydrolysis at moderate temperatures and atmospheric pressure (4). Also, due to the fact that this source of carbohydrates would be available using a relatively simple process, they could become economical fermentable substrates available for conversion to solvents.

However, regulation of end-product formation in pure cultures of anaerobic bacteria is poorly understood and appears to involve multiple factors interacting to control the rates and yields of specific products. These include parameters influencing the flow of carbon and electrons in a given metabolic path as well as species tolerance to the fermentation products formed (5). It is well established that acetate is the first product of carbohydrate fermentation which, upon accumulation, can be converted to organic acids such as butyrate, propionate, succinate, etc, or be reduced to organic solvents such as acetone, butanol, ethanol, etc. This is true even in the so-called "solvent producing strains" or the sporulating cultures of *C. thermosaccharolyticum* (6).

The present study was prompted by the need to elucidate whether it was possible to obtain solvent fermentation without acid fermentation in strains of *C. thermosaccharolyticum*.

MATERIALS AND METHODS

Organism and Media

C. thermosaccharolyticum, National Canners Association strain 3814, was used in this investigation. *C. thermosaccharolyticum* strain SD 105 was an oligosporogenic mutant derived from the above parental strain, which was isolated and purified in this laboratory and has been previously described (M. Lu, M.S. Thesis, University of Missouri, Kansas City, Missouri, U.S.A., 1982).

Stock cultures were prepared by the methods of Hsu and Ordal (6). For each experiment, a stock culture was removed from 3°C storage and was incubated at 56°C overnight. Following this activation, 1 ml of the culture was transferred to 10 ml of fresh medium as a batch culture.

All of the studies were performed in the modified medium of Hsu and Ordal (7) to which 0.2% of various carbohydrates were added. The pH was adjusted prior to sterilization in all cases, unless the pH was below 5.0. At pH 5.0 and lower, the adjustment was made following sterilization. All media were maintained at 56°C and used within 1 day.

Cultivation Methods

Batch cultures were grown in large test tubes (16 x 150 mm) fitted with rubber septa and placed in a heating block at 56°C. Strictly anaerobic conditions were obtained by the procedure of Hungate (8). Samples were obtained aseptically with sterile syringes through the rubber septa.

Continuous dilution cultures were started as batch cultures. The incubation time was carefully monitored and after one generation of growth 30 ml of the actively growing cultures were diluted into 1500 ml of fresh medium containing 0.00522% carbohydrate. The diluted culture was maintained in a bench top chemostat (New Brunswick Model C-32, New Brunswick Scientific; Edison, New Jersey) with agitation of 200 RPM, temperature of 56°C, and continuous pH monitoring within a range of set limits by a pH-stat. Strictly anaerobic conditions were obtained with the procedure of Bauchop and Elsden (9) except that the amount of alkaline pyrogallol was doubled. Samples were obtained aseptically with a sterile syringe connected to the sampling port. Continuous dilution was begun after one generation of growth at a rate of 70 ml/h and was continued over a period of several days. Contamination was not detected during growth.

Measurement of Growth

Growth was measured by following the changes in optical density (OD) of cultures at 600 nm by use of a Bausch and Lomb Spectronic-20 colorimeter. Uninoculated medium was used as the reference. Dilutions of cultures with the medium were made when necessary.

Spore and Vegetative Cell Counts

Direct microscopic counts, made with a Petroff-Hauser counting chamber were used to determine the total number of spores and vegetative cells using an Olympus phase contrast microscope with a magnification of 1,000x.

Determination of Fermentation End-products

The supernatant fluid of cultures was analyzed by gas-liquid chromatography (GLC) with a Hewlett-Packard GLC model 5700 equipped with a flame ionization detector. Separation of primary metabolites was accomplished on a stainless steel column (0.31 cm by 1.82 m) packed with 100/120 Chromosorb W AW coated with 10% SP-1000/1% H_3PO_4 (Supelco, Inc.; Bellefonte, PA) and was recorded on an accompanying strip-chart recorder. The column was initially conditioned for 48 h at 200°C. For analysis the injection port temperature was 200°C and the detector temperature was 250°C. The carrier gas was prepurified nitrogen with a flow rate of 40 ml/min; the flow rates of hydrogen and compressed air were 30 and 300 ml/min, respectively, as measured by a soap bubble flowmeter. The electrometer range was set at 10^{-11} A/mv with an attenuation of 16 and a range of 10. The oven temperature was held at 80°C for 2 min after injection of the 10 1 sample, then programmed to 170°C at a range of 80°C/min, followed by a 2-min hold period. This temperature programming achieved greater separation of acetone, methanol, ethanol, butanol, lactic acid, acetic acid, propionic acid, iso-butyric acid, and butyric acid.

To better estimate the actual amount produced by each individual cell, the concentration of alcochols and organic acids was determined by dividing the percent of product (calculated from the area of the GLC peak and forming a ratio to the area of the standard reference peak by the OD of the

TABLE 1 *Concentration of ethanol produced and sporulation response to various carbon sources by strain 3814*

Carbon source	nmole ethanol/OD[a]	Percent sporulation[b]
Arabinose	0.09	45
Glucose	0.13	0
α-Methyl glucoside	0.24	60
Soluble starch (Difco)	0.21	0
Xylan	0.15	0
Xylose	0.13	55

[a]Concentration computed from GLC analysis data.
[b]Sporulation determined after 24 h of incubation at 56°C.

sample, and converting this percentage to the mathematical equivalent of moles/OD.

RESULTS

Effect of Carbon Source on Ethanol Production and Sporulation in Batch Cultures

The spore and ethanol yield of *C. thermosaccahrolyticum* 3814 varied with the kind of energy source fermented. Carbon sources which contained a glucosidic linkage resulted in the highest ethanol concentrations, although ethanol yield did not correlate directly with percent sporulation as was expected (Table 1). The highest ethanol concentration was 0.24 mM ethanol/OD with a 60% sporulation response, obtained from cultures grown in 0.2% α-methyl glucoside; the lowest ethanol concentration was 0.09 mM ethanol/OD with a 45% sporulation response, obtained from cultures grown in 0.2% arabinose. In all cases the pH prior to sterilization of the media was 7.0.

In contrast, cells of *C. thermosaccharolyticum* strain SD 105 grown in batch cultures produced a different response to the highest fermentable carbohydrates (Table 2). Although the highest ethanol concentration of 0.23 mM ethanol/OD was indeed obtained from cultures grown in 0.2% starch (with a glucosidic linkage) the second highest concentration of 0.13 mM ethanol/OD was found in cultures grown in 0.2% xylan or glucose (with and without a glucosidic linkage, respectively). Furthermore, all of the cultures which formed sporangia within 24 h of fermentation produced insignificant amounts of ethanol. It should

TABLE 2 *Concentration of ethanol produced and sporangia formation response to various carbon sources by strain SD 105*

Carbon source	nmole ethanol/OD[a]	Percent sporangia[b]
Arabinose	0.04	40
Glucose	0.13	0
α-Methyl glucoside	0.08	55
Soluble starch (Difco)	-.23	<10
Xylan	0.13	0
Xylose	0.07	45

[a]Concentration computed from GLC analysis data.
[b]Sporangia formation determined after 24 h of incubation at 56°C.

be noted that all cultures of strain SD 105 grew much slower than the parental strain with a generation time of 2.5 h in xylan as compared to a generation time of 2.0 h for the wild-type cells.

However, all these findings were based on batch culture experiments that are extremely complex in terms of substrate concentration, growth rate, morphological change, pH change, and the sequence by which different fermentation end-products are produced. What was most interesting to note, was the apparent interruption in cell division for an extensive period of time, which lead to a substantial increment in cell length, that may be the cause or the consequence of solvent production.

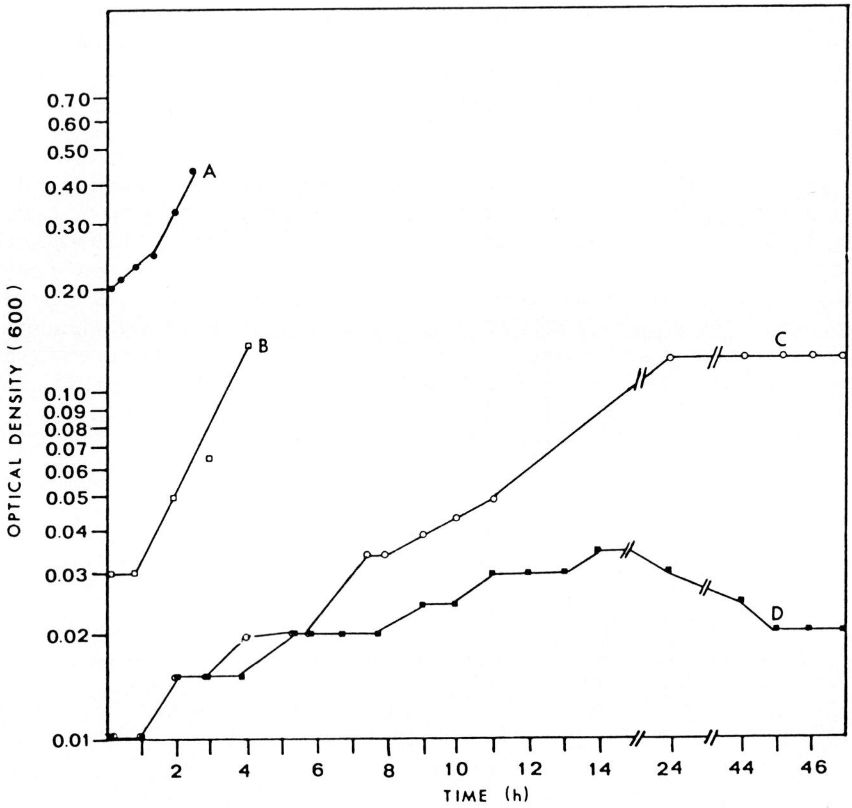

FIG. 1 *Growth curves of* C. thermosaccharolyticum. *Curve A (●) and Curve C (o) represent strain 3814 in batch culture and continuous dilution culture, respectively. Curve B (□) and Curve D (■) represent strain SD 105 in batch culture and continuous dilution culture, respectively.*

Effect of pH on Cultures of *C. thermosaccharolyticum* Strains 3814 and SD 105

Initially, batch culture experiments were performed to determine the effect of pH on growth and ethanol production in these organisms. A series of cultures were grown in triplicate with the pH adjusted prior to inoculation with a range of 3.9 - 7.0 in 0.1 increments. With each strain of *Clostridium* studied, the findings indicated that no growth occurred at a pH below 4.2 and that the concentration of ethanol produced in batch cultures was affected only minimally by any change in pH.

However, in the continuous dilution cultures pH did, in fact, play a very significant role. If the pH of the medium rose above 5.0, there was a drastic decrease in ethanol concentration and an increase in butyric acid concentration, resembling the amount produced in batch cultures. It was further evident that between pH 4.5 to 5.0 there was a lesser, yet significant decrease in the concentration of ethanol produced (data not shown). This correlated very well with the work of Davies and Stephenson (10).

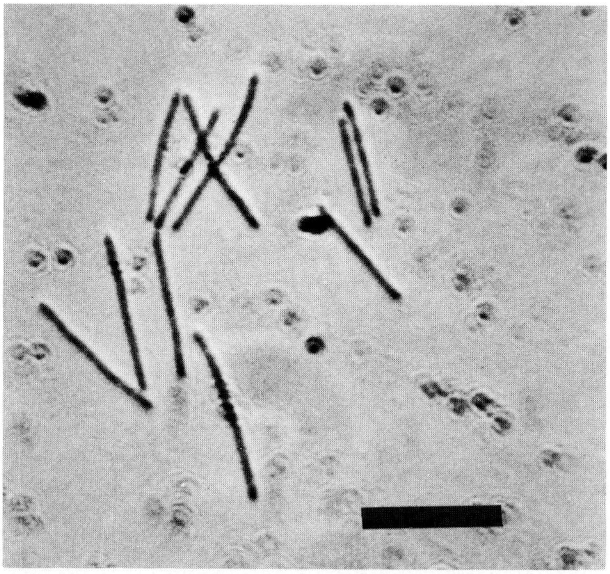

FIG. 2 *Elongated cells of* C. thermosaccharolyticum *strain 3814 observed during the first step (Fig. 1, Curve C) in the continuous dilution culture. Bar represents 10 μm.*

Growth and Morphology of *C. thermosaccharolyticum* Strains 3814 and SD 105 in Continuous Dilution Cultures

Growth curves A and B (Fig. 1) are from batch cultures of *C. thermosaccharolyticum* strains 3814 and SD 105, respectively, grown in 0.2% xylan. Curve C shows the growth of *C. thermosaccharolyticum* strain 3814 in a continuous dilution culture containing 0.00522% xylan. The generation time increased from 2.0 h in the batch culture to 4.0 h in the continuous dilution culture, furthermore, there was a step-wise increase (3 steps) in cell mass for 1.5 generations. It should be noted that at each step in the curve 80-90% of the cell population appeared elongated (3-4x) (Fig. 2). Although dilution with fresh medium was begun after one generation of growth, the step-wise pattern of increase in cell mass did not continue and after 7.8 h the culture deteriorated back to a vegetative pattern of growth that was similar to the previous batch culture experiments.

Curve D (Fig. 1) indicates a similar increase in generation time for *C. thermosaccharolyticum* strain SD 105 from 2.5 h in the batch culture to 5.0 h in the continuous dilution culture. Again there was a step-wise increase in cell mass which continued for nearly 2 generations (greater than 10 h). Furthermore, there was also a concomitant elongation at each step that exceeded the elongation of the parental strain by at least 10 fold (Fig. 3A). At each step the subsequent elongation was 40-50 fold in 90% of the cell population. The elongated cells of strain SD 105 showed increasing multiple septations (Fig. 3B) as the step-wise growth deteriorated.

Cultures in which the organisms were grown under identical conditions, except that 0.00522% soluble starch was the fermentable substrate, showed no significant differences between the two strains (data not shown). The generation time in batch culture was 1.1 h as compared to 4.0 h in the continuous dilution culture. A step-wise increase was observed in the cell mass that was precisely identical to the increase in strain 3814 when grown in xylan (continued for one generation) and was followed by a vegetative growth pattern. There was detectable elongation of cells, however, less than 10% of the population elongated to 5x.

The Fermentation Pattern in the Continuous Dilution Culture of *C. thermosaccharolyticum* Strain 3814

As soon as the cells in the batch culture (30 ml) were transferred to the continuous dilution culture (1500 ml) large amounts of ethanol, butyrate, acetate, and iso-butyrate were formed simultaneously (Fig. 4). What is most interesting to note, was that ethanol concentration exceeded acetate

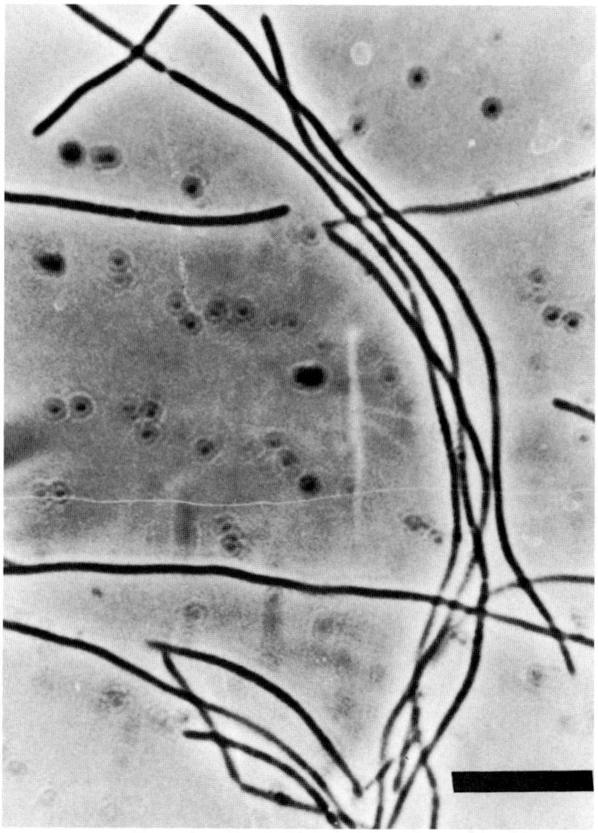

FIG. 3-A *Highly elongated cells of* C. thermosaccharolyticum *strain SD 105 observed during the first two steps (Fig. 1, Curve D) in the continuous dilution culture. Bar represents 10 μm.*

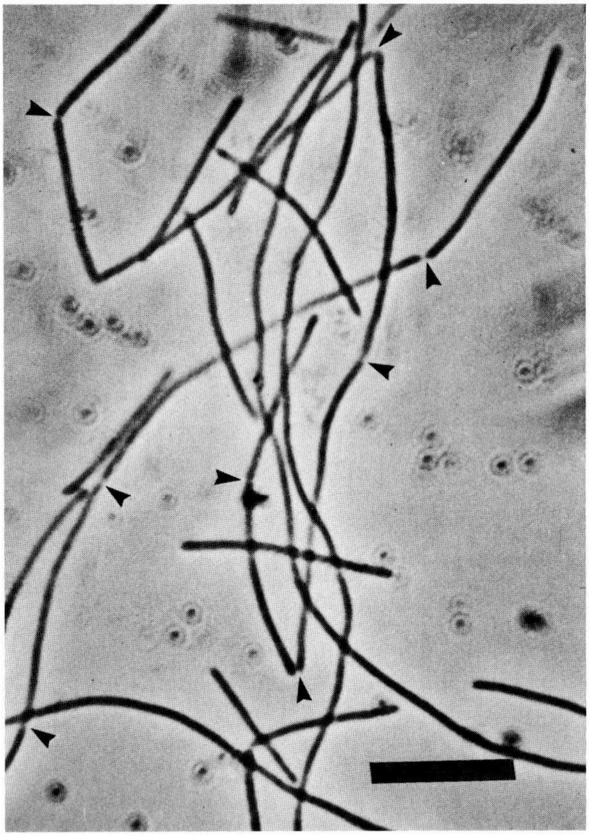

FIG. 3-B *Elongated cells of* C. thermosaccharolyticum *strain SD 105 in continuous dilution culture. Note symmetric multiple septations (indicated by arrows) along the filaments which appeared as the stepwise growth deteriorated after 6 h (Fig. 1, Curve D). Bar represents 10 μm.*

concentration by 2.4x and yet was not preceded by acetate production. Considering the extent of dilution (50x) at the time the shift was made, rather than a proportional reduction of metabolites, ethanol production increased from 0.15 mM/OD to 1100 mM/OD, equivalent to a 7,333 fold increase in terms of ethanol production by each individual cell. There were also significant amounts of acetone, methanol and butanol produced, and the concentrations changed only slightly during the entire fermentation. Propionic acid was never detected.

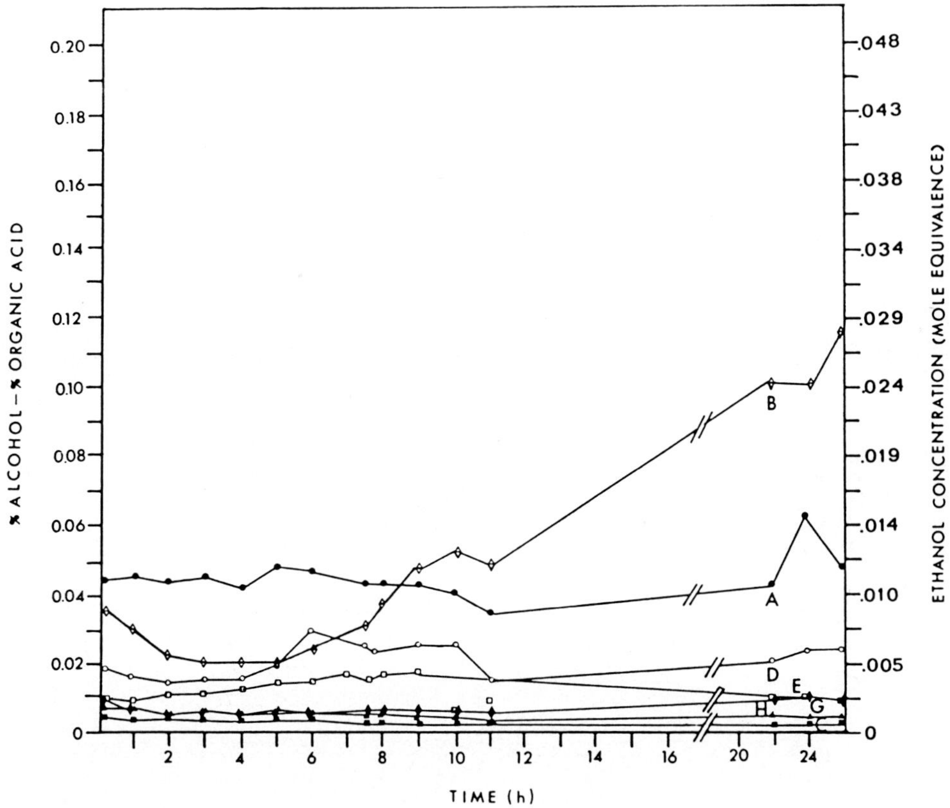

FIG. 4 *Concentrations of alcohols and organic acids produced during continuous dilution fermentation of 0.00522% xylan by* C. thermosaccharolyticum *strain 3814, as determined by GLC analysis. Curve A (●) represents ethanol, Curve B (⌽) represents butyric acid, Curve C (O) represents acetone, Curve D (o) represents acetic acid, Curve E (□) represents isobutyric acid, Curve G (♦) represents butanol, and Curve H (▲) represents methanol.*

Although ethanol appeared very early in the fermentation, there was only a slight variation in the concentration, which never rose above 0.015 moles during the entire fermentation period of several days. However, the butyric acid concentration continued to rise throughout the fermentation, reaching a peak at 26 h of 0.018 moles. Although the solvent fermentation preceded the acid fermentation, it was quickly surpassed by butyrate fermentation at 8 h, which correlates with the time during the fermentation that the cells entered a pattern of vegetative growth.

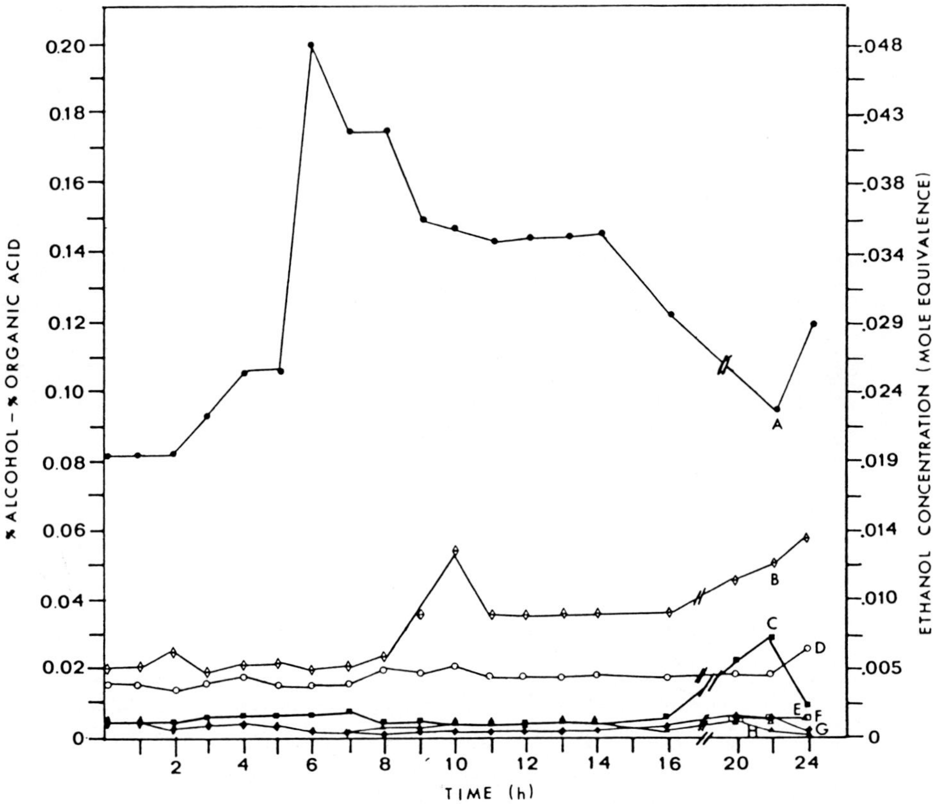

FIG. 5 *Concentration of alcohols and organic acids produced during continuous dilution fermentation of 0.00522% xylan by* C. thermosaccharolyticum *strain SD 105, as determined by GLC analysis. Curve A (●) represents ethanol, Curve B (⍚) represents butyric acid, Curve C (■) represents acetone, Curve D (○) represents acetic acid, Curve E (□) represents iso-butyric acid, Curve F (Δ) represents propionic acid, Curve G (◆) represents butanol, and Curve H (▲) represents methanol.*

The Fermentation Pattern in the Continuous Dilution Culture of *C. thermosaccharolyticum* Strain SD 105

For more than 4 h after the shift the fermentation pattern of the mutant strain SD 105 was similar to the previous experiment in which the parental strain 3814 was used. Initially, much more ethanol (1800 mM/OD or a 13,846 fold increase over the batch culture) was produced and dominated the entire fermentation period reaching the highest concentration of 0.048 moles at 6 h (1,930 mM/OD), (Fig. 5). However, much less acetate and butyrate as well as similar amounts of other solvents were all formed simultaneously. Propionate and isobutyrate were not detected until 20 h, followed by relatively steady concentrations.

It is important to note that the increase in ethanol concentration at 6 h was proportional to the step-wise increase in cell mass which was just opposite to the observation in the wild type cells where ethanol concentration decreased proportionally with the increased cell mass. Unlike the parental strain, the mutant strain continued to elongate in the continuous dilution culture to a point where cells appeared filamentous (Fig. 3A). The fact that this elongation coincided with the peak in ethanol production, strongly indicates that the morphological changes may be related to the solvent production without preceding acid production. Furthermore, it is likely that elongated cells continued to produce ethanol because of the lack of septum formation. The decline in ethanol production after 6 h was accompanied by multi-septations along the entire filament (Fig. 3B), indicating a reversal of metabolism and a return to normal vegetative growth.

The pattern of solvents and acids produced in the continuous dilution culture with soluble starch as the carbon source, by these two strains of clostridia, was nearly identical to the pattern produced by strain 3814 when the fermentable substrate was xylan.

DISCUSSION

The evidence presented by this investigation has confirmed the following: (i) the carbon source (its chemical structure, molecular weight, and the presence or absence of glucosidic linkage) that was fermented, as well as the concentration in the growth medium had a direct effect on the ethanol yield; (ii) utilization of carbon sources by the organisms which resulted in longer generation times increased ethanol production; (iii) morphological change was related to ethanol production (cells induced to elongate 4-5-fold, resembling stage I of sporulation in these clostridial strains, supported higher ethanol concentrations); (iv) by continuously diluting the cultures with fresh medium, thereby maintaining a very low level of readily utilizable carbon source as well as slower growth rate, in addition to the initial dilution due to the increased culture volume (50-fold) in the continuous dilution cultures, step-wise increases in cell mass and constant levels of ethanol production could be maintained; and (v) pH was a significant factor in the shift between acid fermentation and solvent fermentation and decreased the lag time generally preceding solvent fermentation in *Clostridium* species.

Previous investigations have shown that sporulation in *C. thermosaccharolyticum* was directly related to the carbon source(s) by the fact that carbon sources which reduced the growth rate supported higher percent sporulation (11). It was also shown previously that readily fermentable substrates, when available in limited amounts supported a high degree of sporulation (7). Finally it had been shown that sporulation was related to cell size; elongation to a critical length was required before sporulation was induced (12). In this study we have shown that it was possible to induce the organism to elongate to the critical length and to be maintained at that length over a period of time without forming mature spores or reverting to cell division. This was achieved through a combination of the substrate that was utilized and restriction of the availability of that particular fermentable substrate. Furthermore, by transferring the cells after one generation of growth into an environment with reduced substrate concentration and increased culture volume, it was possible to maintain a precise physiological state whereby the cells elongated extensively. These elongated cells appeared to be blocked between commitment to sporulation and vegetative growth in which there was no repression of the spore genome. If the batch culture "starters" were allowed to grow for more than one generation, this physiological state could not be reached even by excessive dilution or multiple transfers of the cells. It was also shown that a step-wise increase of cell mass was never obtained in these older batch cultures when they were inoculated into the continuous dilution cultures.

In the experimentation involving cells of the parental strain (3814) that reverted to vegetative growth or continued to form mature spores in the diluted cultures, there was a shift to acid fermentation and reduced ethanol concentrations. However, the mutant strain SD 105, which appeared blocked at stage I of sporulation (elongation), continued to produce maximum amounts of ethanol over a period of several days and did not revert to vegetative growth or the acid fermentation characteristic of the wild type cells. Similar results have been obtained by Ljungdahl *et al.* (1) with *Thermoanaerobacter ethanolicum* strain JW 200 which shifted its fermentation pattern to acetate and lactate formation when the substrate concentrations were greater than 10%. Ben-Bassat *et al.* (5) found that cultures of *Thermoanaerobium brockii* produced ethanol as the major end-product of starch fermentation (doubling time of 5 h), while on the contrary, lactate was the major end product of glucose fermentation (doubling time of approximately 1.7 h).

It has been previously shown that sporulating cultures of *C. thermosaccharolyticum* formed simple patterns of metabolite production, especially if the culture was partially synchronized and that ethanol was a major metabolite present in that pattern (6). In contrast, the vegetative cells produced complex patterns of metabolite production with reduced ethanol concentrations (6,13). Jones *et al.* (14) found in their work with *Clostridium acetobutylicum* that solvent production occurred with a concomitant change in cellular morphology and the appearance of clostridial forms (stage III of sporulation). Formation of mature spores, however, was not required. The investigation presented here correlates well with the findings of Jones *et al.* (14) in that strain SD 105 formed elongated sporangia, but never mature spores and produced the highest ethanol concentrations in the continuous dilution cultures. It is, therefore, postulated that by continuously diluting the cultures, restricting the amount of available substrate, and partially synchronizing the cells, the metabolic machinery required for sporulation is derepressed and the events necessary for the maintenance of a particular growth rate leading to stage I of sporulation proceed uninhibited. It should be noted that the stepwise increase of cell mass is indicative of synchronous growth cycles, involving elongation, multiseptation, and rapid bursts in cell division.

In the study of acetone-butanol fermentations, it has been shown that the pH of the fermentation must fall to a critical level before solvent production occurs (10,14). By mechanically controlling the pH at this low level from the beginning of the fermentation, we have consequently initiated solvent fermentation prior to acid fermentation contrary to the behaviour of *C. thermosaccharolyticum* strains 3814 and SD 105 in batch cultures with excess of carbon source. However, it must be noted that the simple lowering of pH without the concomitant morphological and physiological changes of the organisms resulted in no significant shift to solvent fermentation. We, therefore, conclude that although carbon source, restricted growth pattern, and pH are significant, no one is effective alone. Rather a complex combination of environmental conditions and physiological state must be maintained before solvent fermentation will occur. It is further postulated that lack of cell division during prolonged fermentation could result in extended periods of solvent fermentation with a drastically reduced level of acid fermentation. The following model is, therefore, suggested:

VEGETATIVE CELLS → ELONGATION without septation →

- BINARY FISSION with no solvents (ACID FERMENTATION)
- MATURE SPORES with low solvent levels (ACID FERMENTATION SOLVENT FERMENTATION)
- CONTINUED ELONGATION with high solvent levels (SOLVENT FERMENTATION ACID FERMENTATION)

Finally, it has been suggested that rapid growth rates in *C. thermosaccharolyticum* exerted a catabolic repression of the NADPH generating system, and that this repression is sufficient to prevent sporulation (6,13). However, if derepression occurred, sporulation was initiated resulting in an excess of NADPH which could be used for the reduction reactions leading to production of ethanol. Ljungdahl *et al.* (1) in their study of the alcohol dehydrogenase of *T. ethanolicus* strain JW 200 have shown a specificity for $NADP^+$ in the reversible reaction

$$CH_3CH_2OH + NADP^+ \rightleftharpoons CH_3CHO + NADPH + H^+$$

In fact, at high concentrations $NADH^+$ acted as an inhibitor. In the study presented here, it was shown that fermentation of xylan (without complete sporulation) resulted in significantly increased ethanol concentrations whereas fermentation of starch (with formation of mature spores) resulted in insignificant ethanol concentrations. It is suggested, therefore, that utilization by the organism of substrate(s) degraded via the hexose monophosphate pathway resulted in increased levels of NADPH which was in turn available for the initiation of sporulation as well as for the reduction reactions which produced ethanol. If, indeed, the alcohol dehydrogenase of *C. thermosaccharolyticum* is similar to the alcohol dehydrogenase of *T. ethanolicus*, with respect to its specificity for $NADP^+$, studies into the function of the enzyme would not only become beneficial, but necessary.

The results presented by this investigation may further provide: (i) a definite direction for the search for mutants that are useful in industrial solvent production; (ii) studies directed towards cells that are retained in a physiological state that are neither dividing nor sporulating; and (iii) practical applications by means of immobilizing solvent-producing cells or immobilizing enzymes derived from these cells.

ACKNOWLEDGEMENTS

The authors of this work wish to thank Edward K.C. Chang for his excellent photographic assistance.

REFERENCES

1. Ljungdahl, L.G., Bryant, F., Carriera, L., Saiki, T. and Wiegel, J. (1981) in *Trends in the Biology of Fermentation for Fuel and Chemicals* (Hollaender, A., Rabson, R., Rogers, P., San Pietro, A., Valentine, R. and Wolfe, R., eds.), pp.397-419, Plenum Press, New York.
2. Ng, T.K., Ben.Bassat, A. and Aeikus, J.G. (1981) *Appl. Environ. Microbiol.* **41**, 1337-1343.
3. Zeikus, J.G., Ben-Bassat, A., Ng, T.K. and Lamed, R.J. (1981) in *Trends in the Biology of Fermantation for Fuel and Chemicals* (Hollaender, A., Rabson, R., Rogers, P., San Pietro, A., Valentine, R. and Wolfe, R., eds.), pp. 441-461, Plenum Press, New York.
4. Flickinger, M.C. (1980) *Biotechnol. Bioeng.* **22**, 27-48.
5. Ben-Bassat, A., Lamed, R. and Zeikus, J.G. (1981) *J. Bacteriol.* **146**, 192-199.
6. Hsu, E.J. and Ordal, A.J. (1970) *J. Bacteriol.* **102**, 369-376.
7. Hsu, E.J. and Ordal, Z.J. (1969) *J. Bacteriol.* **97**, 1511-1512.
8. Hungate, R.E. (1969) in *Methods in Microbiology* (Norris, J.R. and Ribbons, D.W. eds.), Vol. 3B, pp.117-132, Academic Press Inc., New York.
9. Bauchop, R. and Elsden, S.R. (1960) *J. Gen. Microbiol.* **23**, 457-469.
10. Davies, R. and Stephenson, M. (1941) *J. Bacteriol.* **35**, 1320-1331.
11. Hsu, E.J. and Ordal, Z.J. (1969) *Appl. Environ. Microbiol.* **18**, 958-960.
12. Hsu, E.J. (1977) in *Spore Research* (Gould, B.W. and Wolf, J., eds.), pp.223-242, Academic Press Inc., London.
13. Hoffmann, J.W., Chang, E.K. and Hsu, E.J. (1978) in *Spores VII* (Chambliss, G. and Vary, J.C., eds.), pp.312-318. American Society of Microbiology, Washington D.C.
14. Jones, D.T., Westhuizen, A. van der, Lond, S., Allcock, E. R., Reid, S.J. and Woods, D.R. (1982) *Appl. Environ. Microbiol.* **43**, 1434-1439.

Index

F

G

H

I

M

O